GREENHOUSE TECHNOLOGY

GREENHOUSE TECHNOLOGY

Editor

Benedikt Hasek

scitus
academics

Greenhouse Technology

Edited by **Benedikt Hasek**

This Book is Printed on Acid-Free Paper. ♾

ISBN: 9781681179124
Library of Congress Control Number: 2017948940

SCITUS Academics LLC,
616, Corporate Way, Suite 2, 4766,
Valley Cottage, NY 10989

www.scitusacademics.com

Notice

Printed in United States of America.

Preface

It is clear that for survival of living plants on the earth, there should be a favorable environment in terrestrial region controlled by short wave length radiation transmitted by the atmosphere. However, using the concept for optimum growth of living plants, a micro climate can also be created for maximum medicinal crops production in a controlled environment. Growing of vegetable and floriculture crops under protected cultivation is receiving utmost attention and gaining popularity among farming community across the world. Greenhouse technology implies production of plants for economic use in a covered structure that allows rapid harvesting of solar radiation and modification of agroclimatic conditions conducive for plant growth and development. Trends in greenhouse growing continue to evolve in new ways to meet the demands of the planet's growing populations. From low-tech tunnel and hoop houses to vertical farming and hydroponic greenhouses, food growers are combining best growing practices with the latest advancements to make their greenhouses more efficient. The purpose of growing crops under greenhouse conditions is to extend their cropping season and to protect them from adverse environmental conditions, such as extreme temperatures and precipitation, and from diseases and pests. Greenhouse production normally requires a high level of technology to obtain adequate economic returns on investments. Quality is a high priority for greenhouse crops, requiring much care in pest and disease management, not only to secure yields but also to obtain a high cosmetic standard. Although technological changes are ultimately intended to reduce production costs and maximize profits, precise environmental and nutritional control push plants to new limits of growth and productivity. This can generate chronic stress conditions, which are difficult to measure but apparently conducive to some pests and diseases.

This volume 'Greenhouse Technology' deals with the scope of greenhouse technology in agricultural engineering (cultivation, drying and space heating). The greenhouse can provide control environment for high value crops like medicinal plants, flowers etc. crops grown inside a greenhouse are healthy and give better results. It takes a look on several climatic factors that need care in the management of a greenhouse i.e. the control of temperature, light, air composition, and nature of the root medium.

Table of Contents

CHAPTER 1

Greenhouses for Food Production and the Environment

Abdeen Mustafa Omer

Energy Research Institute (ERI), Forest Road West, Nottingham NG7 4EU, UK

ABSTRACT

A greenhouse is essentially an enclosed structure, which traps the short wavelength solar radiation and stores the long wavelength thermal radiation to create a favorable microclimate for higher productivity. The sun's radiation incident on the greenhouse has two parts: direct radiation and an associated diffuse sky radiation. The diffuse part is not focused by the lenses and goes right through Frensel lenses onto the surface of the absorbers. This energy is absorbed and transformed into heat, which is then transported via the liquid medium in copper pipes to the water (heat) storage tanks or, if used, open fish tanks. In this way, an optimal temperature for both plant cultivation and fish production can be maintained. Stable plant growth conditions are light, temperature and air humidity. Light for the photosynthesis of plants comes from the diffuse radiation, which is without substantial fluctuations and variation throughout most of the day. The air temperature inside the greenhouse is one of the factors that have an influence on the precocity of production. The selective collector acts in a more perceptible way on extreme air temperatures inside the greenhouse. Hence, the system makes it possible to avoid the excessive deviation of the temperature inside the greenhouse and provides a favorable microclimate for the precocity of the culture. Sediment and some associated water from the sediment traps are used as organic fertilizer for the plant cultivation. The present trend in greenhouse cultivation is to extend the crop production season in order to maximize use of the equipment and increase annual productivity and profitability.

However, in many Mediterranean greenhouses, such practices are limited because the improper cooling methods (mainly natural or forced ventilation) used do not provide the desired micro-climatic condition during the summer of a composite climate. Also, some of these greenhouses have been built where the meteorological conditions require some heating during the winter, particularly at night. The worst scenario is during the winter months when relatively large difference in temperature between day and night occurs. However, overheating of the greenhouse during the day is common, even in winter, requiring ventilation of the structure. Hence, several techniques have been proposed for the storage of the solar energy received by the greenhouse during the day and its use to heat the structure at night. Reviews of such techniques are presented in this chapter. Air or water can be used for heat transport. The circulating water is heated during the day via two processes. The water absorbs part of the infrared radiation of the solar spectrum. Since the water is transparent in the visible region, they do not compete with the plants that need it. Alternatively, the water exchanges heat with the greenhouse air through the walls. At night, if the greenhouse temperature goes down below a specified value, the water begins to circulate acting as heat transfer surfaces heating the air in the greenhouse. This chapter describes various designs of low energy greenhouses. It also, outlines the effect of dense urban building nature on energy consumption, and its contribution to climate change. Measures, which would help to save energy in greenhouses, are also presented. It also enabled the minimization of temperature variation and, hence avoided the hazard of any sudden climatic change inside the greenhouse.

Keywords: Greenhouse environment; Energy efficient comfort; Ventilation; Humidity: Sustainable environmental impact

INTRODUCTION

Globally, buildings are responsible for approximately 40% of the total world annual energy consumption [1]. Most of this energy is for the provision of lighting, heating, cooling, and air conditioning. Increasing awareness of the environmental impact of $CO_2$and NOx emissions and CFCs triggered a renewed interest in environmentally friendly cooling, and heating technologies. Under the 1997 Montreal Protocol, governments agreed to phase out chemicals used as refrigerants that have the potential to destroy stratospheric ozone. It was therefore considered desirable to reduce energy consumption and decrease the rate of depletion of world energy reserves and pollution of the environment.

One way of reducing building energy consumption is to design building, which is more economical in their use of energy for heating, lighting, cooling, **VENTILATION** and hot water supply. Passive measures, particularly natural or hybrid ventilation rather than air-conditioning, can dramatically reduce primary energy consumption [2]. However, exploitation of renewable energy in buildings and agricultural greenhouses can, also, significantly contribute towards reducing dependency on fossil fuels. Therefore, promoting innovative renewable applications and reinforcing the renewable energy market will contribute to preservation of the ecosystem by reducing emissions at local and global levels. This will also contribute to the amelioration of environmental conditions by replacing conventional fuels with renewable energies that produce no air pollution or greenhouse gases.

The provision of good indoor environmental quality while achieving energy and cost efficient operation of the heating, ventilating and air-conditioning (HVAC) plants in buildings represents a multi variant problem. The comfort of building occupants is dependent on many environmental parameters including air speed, temperature, relative humidity and quality in addition to lighting and noise. The overall objective is to provide a high level of building performance (BP), which can be defined as indoor environmental quality (IEQ), energy efficiency (EE) and cost efficiency (CE).

- Indoor environmental quality is the perceived condition of comfort that building occupants experience due to the physical and psychological conditions to which they are exposed by their surroundings. The main physical parameters affecting IEQ are air speed, temperature, relative humidity and quality.

- Energy efficiency is related to the provision of the desired environmental conditions while consuming the minimal quantity of energy.

- Cost efficiency is the financial expenditure on energy relative to the level of environmental comfort and productivity that the building occupants attained. The overall cost efficiency can be improved by improving the indoor environmental quality and the energy efficiency of a building.

An approach is needed to integrate renewable energies in a way to meet high building performance. However, because renewable energy sources are stochastic and geographically diffuse, their ability to match demand is determined by adoption of one of the following two approaches [2]: the utilization of a capture area greater than that occupied by the community to be supplied, or the reduction of the community's energy demands to a level commensurate with the locally available renewable resources.

For a northern European climate, which is characterized by an average annual solar irradiance of 150 Wm-2, the mean power production from a photovoltaic component of 13% conversion efficiency is approximately 20 Wm-2. For an average wind speed of 5 ms-1, the power produced by a micro wind turbine will be of a similar order of magnitude, though with a different profile shape. In the UK, for example, a typical office building will have a demand in the order of 300 kWhm-2yr-1. This translates into approximately 50 Wm-2 of façade, which is twice as much as the available renewable energies [3]. Thus, the aim is to utilize energy efficiency measures in order to reduce the overall energy consumption and adjust the demand profiles to be met by renewable energies. For instance, this approach can be applied to greenhouses, which use solar energy to provide indoor environmental quality. The greenhouse effect is one result of the differing properties of heat radiation when it is generated at different temperatures. Objects inside the greenhouse, or any other building, such as plants, re-radiate the heat or absorb it. Because the objects inside the greenhouse are at a lower temperature than the sun, the re-radiated heat is of longer wavelengths, and cannot penetrate the glass. This re-radiated heat is trapped and causes the temperature inside the greenhouse to rise. Note that the atmosphere surrounding the earth, also, behaves as a large greenhouse around the world. Changes to the gases in the atmosphere, such as increased carbon dioxide content from the burning of fossil fuels, can act like a layer of glass and reduce the quantity of heat that the planet earth would otherwise radiate back into space. This particular greenhouse effect, therefore, contributes to global warming. The application of greenhouses for plants growth can be considered one of the measures in the success of solving this problem. Maximizing the efficiency gained from a greenhouse can be achieved using various approaches, employing different techniques that could be applied at the design, construction and operational stages. The development of greenhouses could be a solution to farming industry and food security.

The move towards a de-carbonized world, driven partly by climate science and partly by the business opportunities it offers, will need the promotion of environmentally friendly alternatives, if an acceptable stabilization level of atmospheric carbon dioxide is to be achieved. This requires the harnessing and use of natural resources that produce no air pollution or greenhouse gases and provides comfortable coexistence of human, livestock, and plants. This study reviews the energy-using technologies based on natural resources, which are available to and applicable in the farming industry. Integral concept for buildings with both excellent indoor environment control and **SUSTAINABLE ENVIRONMENTAL IMPACT** are reported in the present communication. Techniques considered are hybrid (controlled natural and mechanical) ventilation including night ventilation, thermo-active building mass systems with free cooling in a cooling tower, and air intake via ground heat exchangers. Special emphasis is put on ventilation concepts utilising ambient energy from air ground and other renewable energy sources, and on the interaction with heating and cooling. It has been observed that for both residential and office buildings, the electricity demand of ventilation systems is related to the overall demand of the building and the potential of photovoltaic systems and advanced co-generation units. The focus of the world's attention on environmental issues in recent years has stimulated response in many countries, which have led to a closer examination of energy conservation strategies for conventional fossil fuels. One way of reducing building energy consumption is to design buildings, which are more economical in their use of energy for heating, lighting, cooling, ventilation and hot water supply. Passive measures, particularly natural or hybrid ventilation rather than air-conditioning, can dramatically reduce primary energy consumption. However, exploitation of renewable energy in buildings and agricultural greenhouses can, also, significantly contribute towards reducing dependency on fossil fuels.

The main advantages of solar greenhouse are summarized as follows:

- In the climatic conditions of Europe, the collector system equipped with linear raster lenses is able to absorb, on average, 12% of the total incoming global solar energy on the collector and convert this energy into heat at a temperature of between 30 to 50°C. The system, therefore, consumes approximately 50% less energy for heating purposes than would a traditional normal greenhouse.

- The system provides suitable, perhaps ideal, conditions for the cultivation of high quality vegetables, and even during periods of maximum solar energy absorption on the collectors, there still remains sufficient light for good vegetable growth under the area of the collectors.

- Due to the almost continuous high **HUMIDITY** levels and to the applied nutrient solution being rich in organic matter and microorganisms, organic matter is hardly mineralizing in the soil, hence, does not degrade in patches. On the contrary, organic matter content in the soil increased during cultivation.

- In comparison with a traditional greenhouse, the system does not overheat inside. Therefore, less ventilation is necessary, which brings the benefits of smaller losses of water. Furthermore, the system saves energy, allows the efficient recycling of water and nutrients, and provides suitable growth conditions with a smaller range of extreme humidity, temperature and light allowing the cultivated plants to face less stress and have a higher quality.

- Due to the relatively low temperature in the greenhouse, additional heating might be required. Therefore, vegetables will adapt to low radiation levels, and low temperatures and, consequently, quality is preserved even during failure of control system.

This study describes various designs of low energy buildings. It also, outline the effect of dense urban building nature on energy consumption, the problems related to inadequate ventilation in buildings, and its contribution to climate change. Measures, which would help to save energy in buildings, are also presented.

INDOORS ENVIRONMENT

The heating or cooling of a space to maintain thermal comfort is a highly energy intensive process accounting for as much as 60-70% of total energy use in non-industrial buildings. Of this, approximately 30-50% is lost through ventilation and air infiltration. However, estimation of energy impact of ventilation relies on detailed knowledge about air change rate and the difference in enthalpy between the incoming and outgoing air streams. In practice, this is a difficult exercise to undertake since there is much uncertainty about the value of these parameters [4]. As a result, a suitable datum from which strategic planning for improving the energy efficiency of ventilation can be developed has proved difficult to establish [4]. Efforts to overcome these difficulties are progressing in the following two ways:

- Identifying ventilation rates in a representative cross section of buildings
- The energy impact of air change in both commercial and domestic buildings

In addition to conditioning energy, the fan energy needed to provide mechanical ventilation can make a significant further contribution to energy demand. Much depends on the efficiency of design, both in relation to the performance of fans themselves and to the resistance to flow arising from the associated ductwork.

The building sector is an important part of the energy picture. Note that the major function of buildings is to provide an acceptable indoor environment, which allows occupants to carry out various activities. Hence, the purpose behind this energy consumption is to provide a variety of building services, which include weather protection, storage, communications, thermal comfort, facilities of daily living, aesthetics, work environment, etc. However, the three main energy-related building services are space conditioning (for thermal comfort), lighting (for visual comfort), and ventilation (for indoor air quality). Pollution-free environments are a practical impossibility. Therefore, it is often useful to differentiate between unavoidable pollutants over which little source control is possible, and avoidable pollutants for which control is possible. Unavoidable pollutants are primarily those emitted by metabolism and those arising from the essential activities of occupants. 'Whole building' ventilation usually provides an effective measure to deal with the unavoidable emissions, whereas 'source control' is the preferred and sometimes only practical, method to address avoidable pollutant sources [5]. Hence, achieving optimum indoor air quality relies on an integrated approach to the removal and control of pollutants using engineering judgment based on source control, filtration, and ventilation. Regardless of the kind of building involved, good indoor air quality requires attention to both source control and ventilation. While there are sources common too many kinds of buildings, buildings focusing on renewable energy may have some unique sources and, therefore, may require special attention [5]. In smaller (i.e., house size) buildings, renewable sources are already the primary mechanism for providing ventilation. Infiltration and natural ventilation are the predominant mechanisms for providing residential ventilation for these smaller buildings.

Ventilation is the building service most associated with controlling the indoor air quality to provide a healthy and comfortable environment. In large buildings ventilation is normally supplied through mechanical systems, but in smaller ones, such as single-family homes, it is principally supplied by leakage through the building envelope, i.e., infiltration, which is a renewable resource, albeit unintendedly so. Ventilation can be defined as the process by which clean air is provided to a space. It is needed to meet the metabolic requirements of occupants and to dilute and remove pollutants emitted within a space. Usually, ventilation air must be conditioned by heating or cooling in order to maintain thermal comfort and, hence, becomes an energy liability. Indeed, ventilation energy requirements can exceed 50% of the conditioning load in some spaces [5]. Thus, excessive or uncontrolled ventilation can be a major contributor to energy costs and global pollution. Therefore, in terms of cost, energy, and pollution, efficient ventilation is essential. On the other hand, inadequate ventilation can cause comfort or health problems for the occupants.

Good indoor air quality may be defined as air, which is free of pollutants that cause irritation, discomfort or ill health to occupants [6]. Since long time is spent inside buildings, considerable effort has focused on developing methods to achieve an optimum indoor environment. An almost limitless number of pollutants may be present in a space, of which many are at virtually immeasurably low concentrations and have largely unknown toxicological effects [6]. The task of identifying and assessing the risk of individual pollutants has become a major research activity [5-7]. In reality, a perfectly pollutants-free environment is unlikely to be attained. Some pollutants can be tolerated at low concentrations, while irritation and odour often provide an early warning of deteriorating conditions. Health related air quality standards are typically based on risk assessment and are either specified in terms of maximum-permitted concentrations or a maximum allowed dose. Higher concentrations of pollutants are normally permitted for short-term exposure than are allowed for long-term exposure [7].

Ventilation is essential for securing a good indoor air quality, but, as explained earlier, can have a dominating influence on energy consumption in buildings. Air quality problems are more likely to occur if air supply is restricted. Probably a ventilation rate averaging 7 l/s.p represents a minimum acceptable rate for normal odour and comfort requirements in office type buildings [8]. Diminishing returns are likely to be experienced at rates significantly above 10 l/s.p [8]. If air quality problems still persist, the cause is likely to be poor outdoor air quality (e.g., the entrainment of outdoor traffic fumes), poor air distribution or the excessive release of avoidable pollutants into space. However, the energy efficiency of ventilation can be improved by introducing exhaust air heat recovery, ground pre-heating, demand controlled ventilation, displacement ventilation and passive cooling [9]. In each case, a very careful analysis is necessary to ensure that the anticipated savings are actually achievable. Also, it is essential to differentiate between avoidable and unavoidable pollutant emissions. Achieving energy efficiency and optimum Indoor Air Quality (IAQ) depends on minimizing the emission of avoidable pollutants. Pollutants inside greenhouses are derived from both indoor and outdoor contaminant sources, as illustrated in **FIGURE 1**. Each of these tends to impose different requirements on the control strategies needed to secure good health and comfort conditions.

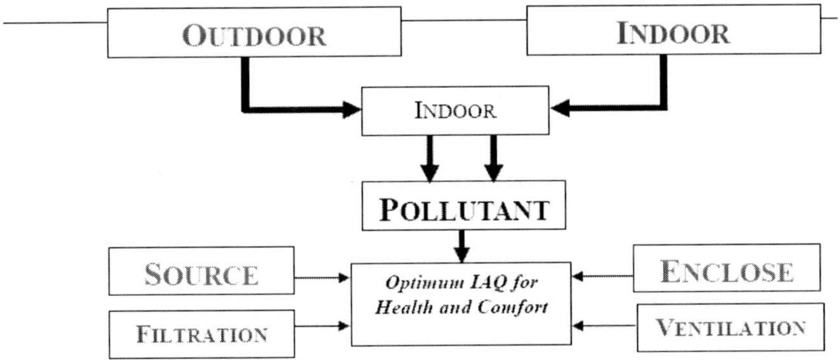

FIGURE 1: Strategies for controlling IAQ.

ACHIEVING ENERGY EFFICIENT VENTILATION

A recent review of the International Energy Agency [10] concludes that the thermal insulation characteristics of buildings improve; ventilation and air movement is expected to become the dominant heating and cooling loss mechanism in buildings of the next century. Poor air quality in buildings sometimes manifests itself; refer to a range of symptoms that an occupant can experience while present in the building. Typical symptoms include lethargy, headaches, lack of concentration, runny nose, dry throat and eye and skin irritation [10]. Other examples of sick building syndrome have been associated with the presence of specific pollutants, such as outdoor fumes entering through air intakes [11]. Improved ventilation is one way of tackling the problem. Many standards are being introduced to ensure the adequacy and efficiency of ventilation. However, to be effective, standards need to address the minimum requirements of comfort, operational performance, air tightness, provision for maintenance and durability [11]. Also, various methods have been introduced to improve the energy performance of ventilation. These include:

THERMAL RECOVERY

Recovery of thermal energy from the exhaust air stream is possible by means of air-to-air heat recovery systems or heat pumps. In theory, such methods can recover as much as 70% of the waste heat [12]. While these methods are exceedingly popular, their full potential is, often, not achieved. This is because buildings or ductwork are often excessively leaky and, hence, additional electrical energy load is needed. To be successful, the designer must integrate such ventilation design with that of the building itself and be thoroughly aware of all the energy paths [12].

GROUND PRE-CONDITIONING

Ventilation air can be pre-conditioned by passing the supply air ducts under-ground. This can provide a good measure of winter pre-heating and summer pre-cooling [13].

DEMAND CONTROL VENTILATION

Demand control ventilation, DCV, systems provide a means by which the rate of ventilation is modulated in response to varying air quality conditions. This is effective if a dominant pollutant is identifiable. In transiently occupied buildings, control by metabolic carbon dioxide concentration has become popular although must be introduced with caution. For success, it is essential to ensure that no other problem pollutants are present [14].

DISPLACEMENT VENTILATION

Displacement ventilation involves distributing clean air at low velocity and at a temperature of approximately 2°K below the ambient air temperature of the space [15]. This dense air mass moves at floor level until it reaches a thermal source such as an occupant or electric equipment, causing the air to warm and gently rise. Polluted air is then collected and extracted above the breathing zone. This process reduces the mixing effect of classical dilution ventilation, thus reducing the amount of ventilation needed to achieve the same air quality in the vicinity of occupants [15]. For success, very careful temperature control is needed and there is a limit on potential cooling capacity [15].

OUTDOORS AIR POLLUTION

Clean outdoor air is essential for good indoor air quality. Although air cleaning is possible, it is costly and not effective in the many offices and dwellings that are naturally ventilated, leaky or ventilated by mechanical extract systems. Some air quality problems are global and can only be controlled by international effort. Other pollutants are much more regional and may be associated with local industry and traffic. Nature, too, presents its own problems with large volumes of dust and gaseous emissions being associated with volcanic activity. Similarly, while naturally occurring, radon can penetrate buildings from the underlying geological strata. Even rural areas are not immune to pollution, where the presence of pollen, fungal spores and agricultural chemicals can result in poor health and cause allergic reactions. There are several pollution control strategies.

Some of them are discussed below:

FILTRATION

Filtration is applied primarily to remove particulates from the air. Filtration of outdoor air cannot be readily applied to the many buildings that are naturally ventilated or excessively leaky. To be effective, filtration systems must be capable of trapping the smallest of particles and of handling large volumes of airflow. Activated carbon and other absorbing filters are additionally able to remove gaseous pollutants [16]. Mechanical systems in residential buildings contribute little to the ventilation rate. The thermal comfort is an important aspect of human life.

POSITIONING OF AIR INTAKES

Air intakes must be located away from pollutant sources. Particular sources include street level and car parking locations [16]. Although urban air quality is normally much improved at elevations above street level, contamination from adjacent exhaust stacks and cooling towers must, also, be avoided [16]. Determining the optimum position for air intakes may require extensive wind tunnel or fluid dynamics analysis. Further information on the positioning of air intakes is reviewed by Limb 1995 [7].

AIR QUALITY CONTROLLED FRESH AIR DAMPERS

Traffic pollution in urban areas is often highly transient, with peaks occurring during the morning and evening commuting periods. At these times, it may be possible to improve indoor air quality by temporarily closing fresh air intakes and windows.

BUILDING AIR TIGHTNESS

None of the above control strategies will be effective unless the building is well sealed from the outdoor environment to prevent contaminant ingress through air infiltration. Underground parking garages must also be well sealed from occupied accommodation above. Evidence suggests that sealing is often inadequate [16].

INDOOR POLLUTANTS

Pollutants emitted inside buildings are derived from metabolism (odour, carbon dioxide, and bacteria), the activities of occupants (e.g., smoking, washing and cooking), emissions from materials used in construction and furnishing and emissions from machinery and processes. The preferred order of control is discussed below:

SOURCE CONTROL

Once a pollutant has entered a space, it can, at best, only be diluted [17]. Avoidable pollutants should, therefore, be eliminated. This means restricting or, preferably, eliminating potentially harmful pollutant emissions.

ENCLOSING AND VENTILATING AT SOURCE

Pollutants generated as part of the activity of occupants are usually highly localized. Wherever possible, source control should be applied, combined with the use of local extractors [17].

GENERAL VENTILATION

General ventilation of a space is needed to dilute and remove residual pollution primarily from unavoidable contaminant sources [17].

VENTILATION OF SPACES IN HUMID CLIMATE

The design of windows in modern buildings in a warm, humid climate can be influenced either by their use to provide physiological and psychological comfort via providing air and daylight to interior spaces or by using them to provide aesthetically appealing fenestration. Most spaces in modern buildings are not adequately ventilated and it is recommended that effort should be directed towards the use of windows to achieve physiological comfort. Evaluation of public housing has focused on four main aspects:

economics, social and physical factors, and resident's satisfaction. However, information about the physiological characteristics of spaces in a warm humid climate will aid the design of appropriate spaces with respect to the development and adequate choice of building materials and appropriate use of suitable passive energy. In this light there is a need to examine resident's satisfaction with respect to these physiological issues. Proper ventilation in a space is a primary factor in determining human health, comfort and wellbeing of the occupants. At present, getting a proper naturally ventilated space seems to be a difficult task. This is partly due to the specific environmental problems of high temperature, high humidity, low wind velocity, and variable wind direction- usually attributed to the warm humid climate, on the one hand, and the difficulty of articulating the design constraints of security, privacy and the desire of users for large spaces on the other hand. As pointed out by most researchers in the field of passive energy design, such as Givoni [18], Koenigsberger et al. [19] Boulet [20] and Szokolay [21], the types of spaces most suited to this climate are spaces, which are cross-ventilated. This implies that these spaces must have openings at least on opposite sides of a wall, but this condition is difficult to achieve in view of the design constraints mentioned above. So in most cases, the option left to the designer is to have openings on a wall or openings on adjacent walls. The effectiveness of the above arrangement for effective ventilation of a space still depends on other parameters. Therefore, in order to optimize comfort in spaces in warm humid climate, there is a need to re-examine the factors affecting proper ventilation with respect to these design issues. In order to be thermally comfortable in interior spaces, four environmental parameters, namely air temperature, relative humidity, mean radiant temperature and air velocity, need to be present in the space in adequate proportions [21]. In a warm, humid climate, the predominance of high humidity necessitates a corresponding steady, continuous breeze of medium air speed to increase the efficiency of sweat evaporation and to avoid discomfort caused by moisture on skin and clothes. Continuous ventilation is therefore the primary requirement for comfort [18]. From the above, it is apparent that the most important of these comfort parameters in a warm, humid climate is air velocity. It should, also, be noted that indoor air velocity depends on the velocity of the air outdoors [21]. The factors affecting indoor air movement are orientation of the building with respect to wind direction, effect of the external features of the openings, the position of openings in the wall, the size of the openings and control of the openings. Cross-ventilation is the most effective method of getting appreciable air movement in interior spaces in warm, humid climates. For comfort purposes, the indoor wind velocity should be set at between 0.15 and 1.5 m/s [22]. A mathematical model based on analysis of the experimental results [23], which established the relationship between the average indoor and outdoor air velocities with the windows placed perpendicular to each other, was adapted to suit a warm, humid climate and is, usually, used to evaluate the spaces. The formula states that:

$$V1=0.45\ (1\text{-}\exp^{-3.84x})\ Vo \tag{1}$$

Where:

V1=average indoor velocity

x=ratio of window area to wall area

Vo=outdoor wind speed

AIR MOVEMENT IN BUILDINGS

Natural ventilation is now considered to be one of the requirements for a low energy building designs. Until about three decades ago the majority of office buildings were naturally ventilated. With the availability of inexpensive fossil energy and the tendency to provide better indoor environmental control, there has been a vast increase in the use of air-conditioning in new and refurbished buildings. However, recent scientific evidence on the impact of refrigerants and air-conditioning systems on the environment has promoted the more conscious building designers to give serious considerations to natural ventilation in non-domestic buildings [24]. Two major difficulties that a designer has to resolve are the questions of airflow control and room air movement in the space.

Because of the problem of scaling and the difficulty of representing natural ventilation in laboratory, most of the methods used for predicting the air movement in mechanically ventilated buildings are not very suitable for naturally ventilated spaces [25]. However, computational fluid dynamics (CFD) is now becoming increasingly used for the design of both mechanical and natural ventilation systems. Since a CFD solution is based on the fundamental flow and energy equations, the technique is equally applicable to a naturally ventilated space as well as a mechanically ventilated one, providing that a realistic representation of the boundary conditions are made in the solution.

NATURAL VENTILATION

Generally, buildings should be designed with controllable natural ventilation. A very high range of natural ventilation rates is necessary so that the heat transfer rate between inside and outside can be selected to suit conditions [25]. The ventilation rates required to control summertime temperatures are very much higher than these required to control pollution or odour. Any natural ventilation system that can control summer temperatures can readily provide adequate ventilation to control levels of odour and carbon dioxide production in a building. Theoretically, it is not possible to achieve heat transfer without momentum transfer and loss of pressure. However, **FIGURES 2** and **3** show some ideas for achieving heat reclaim at low velocities. Such ideas work well for small buildings.

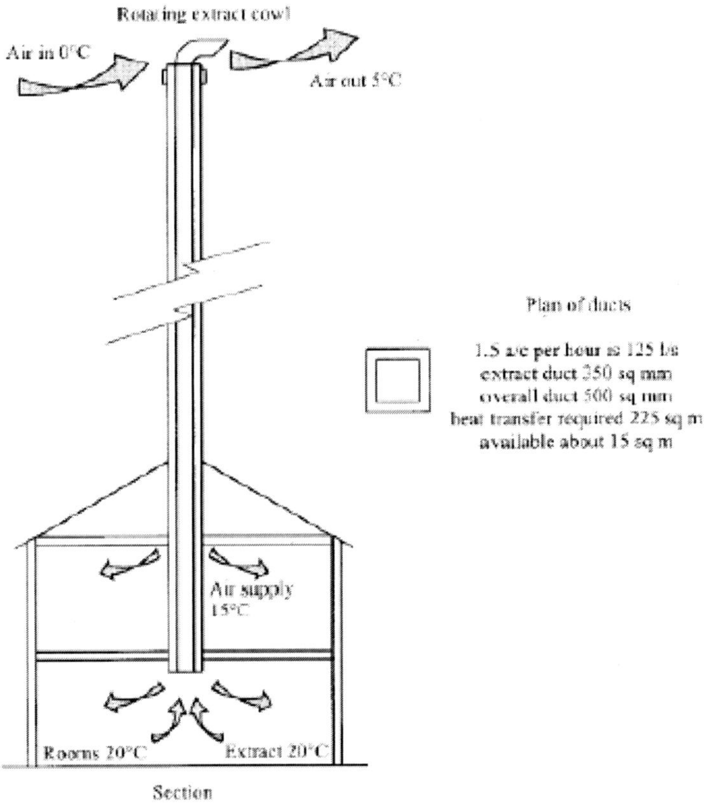

FIGURE 2: Small house natural ventilation with heat reclaims (a very tall chimney).

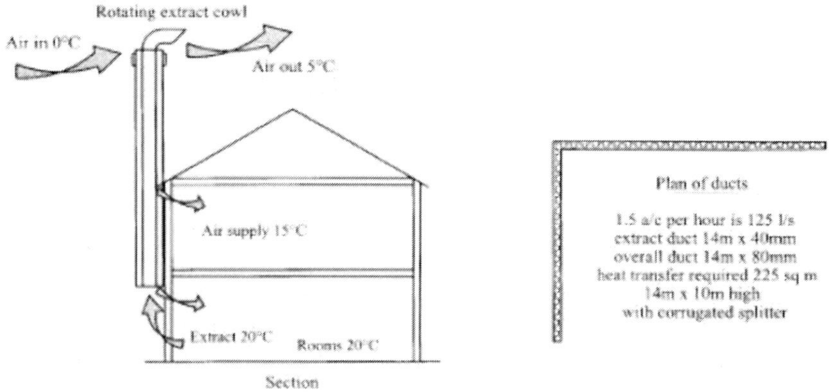

FIGURE 3: Ventilation duct supply and extract wraps around the building.

MECHANICAL VENTILATION

Most of the medium and large size buildings are ventilated by mechanical systems designed to bring in outside air, filter it, supply it to the occupants and then exhaust an approximately equal amount of stale air. Ideally, these systems should be based on criteria that can be established at the design stage. To return afterwards in attempts to mitigate problems may lead to considerable expense and energy waste, and may not be entirely successful [25]. The key factors that must be included in the design of ventilation systems are: code requirement and other regulations or standards (e.g. fire), ventilation strategy and systems sizing, climate and weather variations, air distribution, diffuser location and local ventilation, ease of operation and maintenance and impact of system on occupants (e.g. acoustically). These factors differ for various building types and occupancy patterns. For example, in office buildings, pollutants tend to come from sources such as occupancy, office equipment, and automobile fumes. Occupant pollutants typically include metabolic carbon dioxide emission, odors and sometimes smoking. When occupants (and not smoking) are the prime source. Carbon dioxide acts as a surrogate and can be used to cost-effectively modulate the ventilation, forming what is known as a demand controlled ventilation system. Generally, contaminant sources are varied but, often, well-defined and limiting values are often determined by occupational standards.

BIOCLIMATIC DESIGN

Bioclimatic design cannot continue to be a side issue of a technical nature to the main architectural design. In recent years started to alter course and to become much more holistic in its approach while trying to address itself to:

- The achievement of a sustainable development
- The depletion of non-renewable sources and materials
- The life cycle analysis of buildings
- The total polluting effects of buildings on the environment
- The reduction of energy consumption and
- Human health and comfort.

Hidden dimensions of architectural creation are vital to the notion of bioclimatic design. The most fundamental ones are:

TIME, which has been called the fourth dimension of architectural space, is of importance because every object cannot exist but in time. The notion of time gives life to an object and releases it to periodic (predictable) or unperiodic repetition.

Times relates to seasonal and diurnal patterns and thus to climate and the way that a building behaves or should be designed to couple with and not antagonize nature. It further releases to the dynamic nature of a building in contrast to the static image that we have created for it.

AIR, is a second invisible but important element. We create space and pretend that it is empty, oblivious of the fact that it is both surrounded by and filled with air. Air in its turn, due to air-movement, which is generated by either temperature or pressure differences, is very much there and alive. And related to the movement of air should be building shapes, sections, heights, orientations and the size and positioning of openings.

LIGHT, and in particular daylight, is a third important element. Architecture cannot exist but with light and from the time we have been able to substitute natural light with artificial lighting, many a building and a lot of architecture has become poorer so. It is not an exaggeration to say that the real form giver to architecture is not the architect himself but light and that the architect is but the forms moulder.

Vernacular architecture is beautiful to look at as well as significant to contemplate on. It is particularly interesting to realize the nature of traditional architecture where various devices to attain thermal comfort without resorting to fossil fuels can be seen.

Sun shading and cross ventilation are two major concerns in house design and a south-facing façade is mandatory to harness the sun in winter as much as possible.

Natural ventilation required higher ceilings to bring a cooling effect to occupants in buildings built fifty years ago, whereas modern high technology buildings have lower ceiling heights, thus making air conditioning mandatory.

Admitting the human right of enjoying modern lives with a certain level of comfort and convenience, it is necessary to consider how people can live and work in an ideal environment with the least amount of energy consumption in the age of global environment problems.

People in the modern age could not put up with the poor indoor environment that people in the old age used to live in.

In fact, in those days people had to live with the least amount of fuels readily available and to devise various means of constructing their houses so that they would be compatible with the local climate.

It is important; therefore, in designing passive and low energy architecture for the future to learn from their spirit to overcome difficulties by having their creative designs adapted to respective regional climatic conditions and to try to devise the ecotechniques in combination with a high grade of modern science.

Heat gain in the summer is the main problem as it overheats the indoor environment of residential buildings. This forces the residents to utilize mechanical air conditioning systems to satisfy their comfort. Under today's economic crisis, energy conversation programmes and acts for respect of environment are receiving more attention.

As a contribution to such efforts and in order to overcome the heat gain in houses, it is advisable to utilize passive systems, namely, producing ventilation by a solar chimney [26].

Room air is removed by ventilation produced by the metallic solar wall (MSW) as shown in **FIGURE 4**.

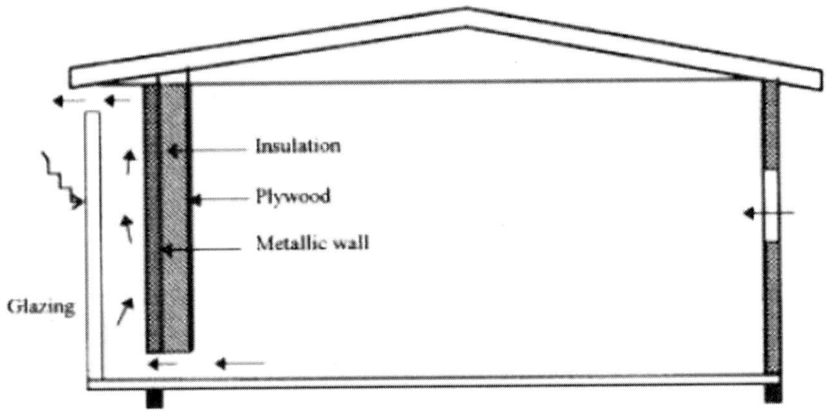

FIGURE 4: Schematic representations of the passive solar house and natural ventilation by metallic solar wall.

INFILTRATION

Infiltration is the process of air flowing in (or out) of leaks in the building envelope, thereby providing (renewable) ventilation in an uncontrolled manner [27]. All buildings are subject to infiltration, but it is more important in smaller buildings as many such buildings rely exclusively on infiltration when doors and windows are closed. In larger buildings there is less surface area to leak for a given amount of building volume, so the same leakage matters less. More importantly, the pressures in larger buildings are usually dominated by the mechanical system and the leaks in the building envelope have only a secondary impact on the ventilation rates [27]. Infiltration in larger buildings may, however, affect thermal comfort and control and systems balance. Typical minimum values of air exchange rates range from 0.5 to 1.0 h^{-1} in office buildings [27]. Buildings with higher occupant density will have higher minimum outside air exchange rates when ventilation is based on outdoor air supply per occupant, typically 7 to 10 ls-1 [27]. Thus, schools have minimum outdoor air ventilation rates 3 h^{-1}, while fully occupied theatres, auditoriums and meeting rooms may have minimum air exchange rates of 4 to 7 h^{-1} [27]. It is in low-rise residential buildings (most typically, single-family houses) that infiltration is a dominant force. Mechanical systems in these buildings contribute little to the ventilation rate.

PASSIVE VENTILATION SYSTEMS

Passive solar systems for space heating and cooling, as well as passive cooling techniques can significantly contribute to energy saving in the building sector when used in combination with conventional systems for heating, cooling, ventilation and lighting. The overall thermal behaviour of the building is dependent on the alternatives and interventions made on the building's shell. Passive ventilation systems share the use of renewable energy to provide ventilation with infiltration. But unlike air leakage and open windows, passive ventilation systems are designed to provide specific amounts of ventilation to minimise both energy liabilities due to excessive ventilation and periods of poor air quality due to under-ventilation [28]. However, the most common passive ventilation system is the passive stack, which is normally used to extract air from kitchen and bathrooms. In this method, prevailing wind and temperature differences are used to drive airflow through a vertical shaft. Various stack designs can be used to control or enhance the performance, based on local climate. However, careful design is required to avoid back draughting and to insure proper mean rates. Although there is significant experience with this approach in Europe, it has been rarely used in North America [28]. Well-designed passive ventilation systems can be used to provide whole-building ventilation as well as local exhaust. Some efforts are currently underway to develop passive ventilation systems that incorporate heat recovery to minimise the need for conditioning the ventilation air [28]. These approaches aim towards a fully renewable ventilation system in that it requires no non-renewable resources for either providing the ventilation air or conditioning it. A traditional, naturally ventilated building can readily provide a high ventilation rate. On the other hand, the mechanical ventilation systems are very expensive. However, a comprehensive ecological concept can be developed to achieve a reduction of electrical and heating energy consumption and optimize natural air condition and ventilation.

PASSIVE COOLING

In the office environment, high heat loads are commonly developed through lighting, computers and other electrical equipment. Further heat gains are developed through solar gains, occupants and high outdoor air temperature. Passive cooling methods attempt to reduce or eliminate the need for energy intensive refrigerative cooling by minimising heat gains. This involves taking advantage of thermal mass (night cooling) and introducing high levels of air change. Night ventilation techniques seem to be the most appropriate strategy for buildings. This arises as a consequence of the large diurnal temperature range during the cooling seasons and the relatively low peak air temperatures, which occur during the day [29]. Such a combination allows the thermal mass of the building to use the cool night air to discard the heat absorbed during the day. An initial examination of the weather conditions experienced during the summer months of June to September in the UK indicates that most peak conditions of external weather fall within the ventilation and thermal mass edge of the bioclimatic chart [29,30]. **FIGURE 5** shows that the summer (June to September) climatic envelope is within the heating, comfort, thermal mass and ventilation effectiveness areas of the chart. The key parameters influencing the effectiveness of night cooling are summarized into the following four categories [30]:

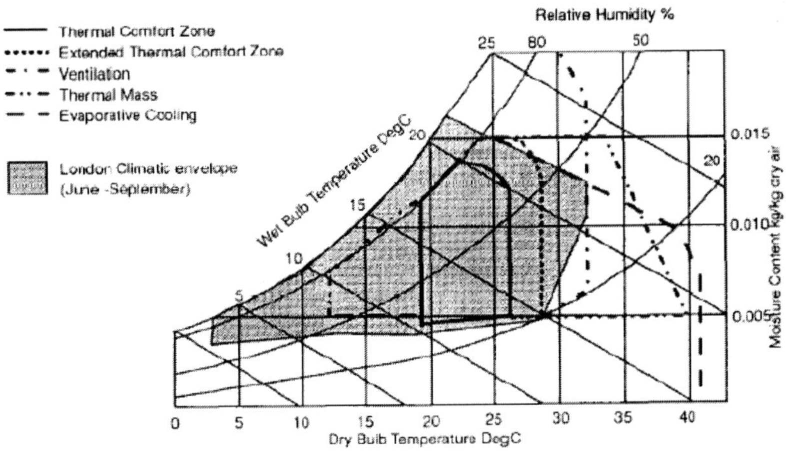

FIGURE 5: Bioclimatic chart with summer climatic envelope for London.

- Internal heat gains of 10, 25 and 40 W/m^2; representing occupancy only, occupancy plus lights, and occupancy together with lights and IT load respectively.
- Envelope gains.
- Thermal response.
- Ventilation gains/losses.

Due to the complexity resulting from all the interrelated parameters affecting the effectiveness of night ventilation, it is necessary for designers to have access to a simple user friendly and yet accurate model when assessing the viability of night ventilation during the initial design stage. The followings are the key output parameters [30]:

- Maximum dry resultant temperature during the occupied period.
- Dry resultant temperature at the start of the occupied period.
- Energy savings.

Maximizing the efficiency and benefit gained from a greenhouse can be achieved using various approaches, employing different techniques that could be applied at the design, construction or operational stages. Greenhouse cultivation is one of the most absorbing and rewarding form of gardening for anyone who enjoys growing plants. The enthusiastic gardener can adapt the greenhouse climate to suit a particular group of plants, or raise flowers, fruit and vegetables out of their natural season. The comfort in greenhouse depends on many environmental parameters. These include temperature, relative humidity, air quality and lighting. Although greenhouse and conservatory originally both meant a place to house or conserve greens (variegated hollies, cirrus, myrtles and oleanders), a greenhouse today implies a place in which plants are raised while conservatory usually describes a glazed room where plants may or may not play a significant role. Indeed, a greenhouse can be used for so many different purposes.

The input data required are the following:

- Thermal gains related data: solar protection is assumed good, thermal gains can be varied and the user specifies the occupancy period.
- Building fabric data: glazing ratio can be any value while thermal mass can be varied at three levels.
- Ventilation data: infiltration, day ventilation and night ventilation can be specified as necessary.
- Weather data: solar data are fixed but temperature is user specified for seven days although temperature profiles need not be the same for all days. The weather data are specified in the form of maximum and minimum temperature for each day and hourly values are calculated by sinusoidal fitting.

However, a primary strategy for cooling buildings without mechanical intervention in hot humid climates is to promote natural ventilation. To control the energy used for the cooling of buildings in hot-arid regions with ambient air temperatures during the hottest period between 42 to 47°C, passive cooling approaches should be implemented [30]. A solar chimney that employs convective currents to draw air out of the building could be used. By creating a hot zone with an exterior outlet, air can be drawn into the house, ventilating the structure as well as the occupants. Since solar energy in such a region is immense, the hot zone created with a black metal sheet on the glazing element can draw hotter air at a slightly higher speed [30]. Applications of solar chimneys in buildings were limited to external walls. Integrating a solar chimney with an evaporatively cooled cavity could result in a better cooling effect. However, this should be applied with care since water sources are limited [31]. **FIGURE 6** shows the combined wall-roof solar chimney incorporated into that building. Average room and ambient air temperatures are 23 and 27°C respectively. Air velocity required to achieve thermal comfort in the room should reach a maximum of 0.3 m/s [31]. **FIGURE 7** gives the cooling load versus air change per hour. This indicates that the inclined airflow by the combined wall-roof chimney is enough to overcome a high cooling load required to cool heavy residential buildings. This suggests that night ventilation could be improved, and incorporating a combined wall-roof solar chimney increases the cooling load. However, thermal mass and ventilation should be sufficient to cover cooling requirements in typical buildings. A high percentage of the cooling requirements can be met by night ventilation before another form of cooling is used. Finally, a simplified ventilation tool for assessing the applicability of night cooling in buildings, currently under development in terms of user inputs and typical outputs.

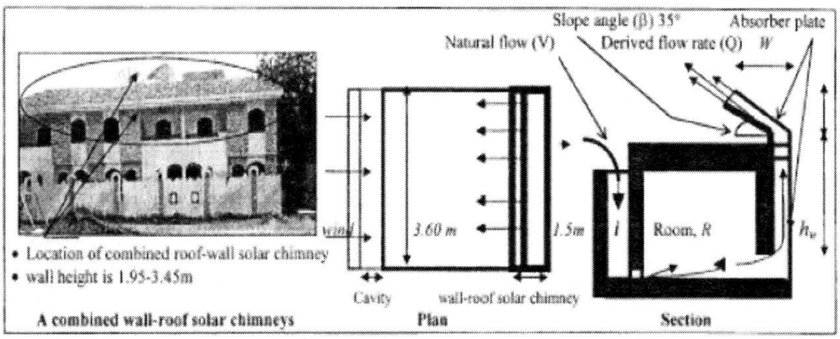

FIGURE 6: A combined wall-roof solar chimney incorporated into residential building night ventilation.

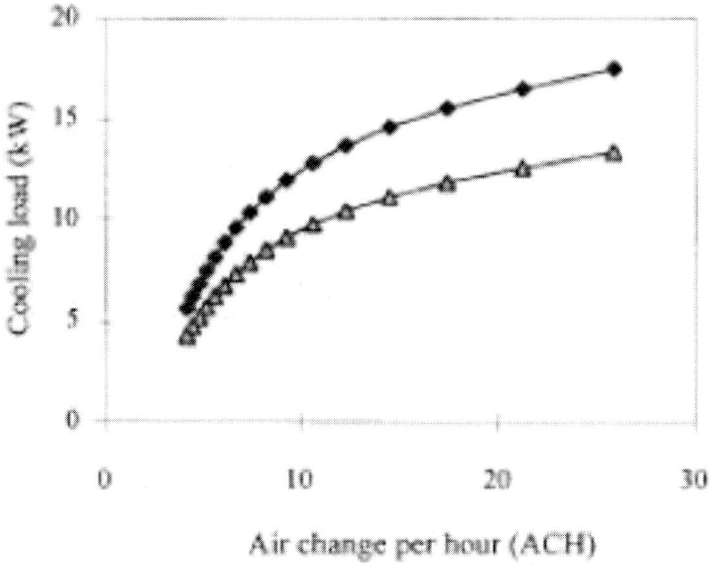

FIGURE 7: Cooling loads by night ventilation for desired room indoor air temperatures of 23 and 25°C.

Mean cooling load (kW)

AIR POLLUTANTS AND TRANSMUTATION

Controlling the pollution of the present civilization in an increasing concern. More importance is given to control global carbon dioxide, which is considered to be the main factor of greenhouse effect. Though the complete experimental result on the fact is yet to be debated, the immense heat, temperature and turbulence of nuclear explosion oxidizing the atmospheric nitrogen into nitric oxide, are considered to be similarly responsible for depletion of ozone layer [32]. At present, more importance is given for plantation to reduce the level of global carbon dioxide. The plantation over the whole earth surface may control only 50% of carbon dioxide disposed to atmosphere and its greenhouse effect. There are, also, explosions in the ozone layer time to time to add to the problem. Irrespective of the relative importance of each factor, the ozone layer protects us from harmful cosmic radiations and it is believed that the depletion of ozone layer increases the threat of outer radiations to human habitation if environmental pollution is not controlled or there is no possibility of self-sustainable stability in nature [32].

The presence of ionosphere in the outer-sphere is most probably for ionic dissociation of the gases of the outer-sphere in the presence of low pressure and cosmic radiation [32]. Moreover the ionosphere contains charged helium ions (alpha particle). Therefore, it may be concluded that the explosion in the ozone and transmission of radiations through it are the possible effects of transmutation of pollutants with exothermic reaction (emission of radiations) [32]. The existence of a black hole in the space, which is found in the photo camera of astrologist, is still unexplored. This black hole may be an effect of transmutation process with absorption of heat energy (endothermic reaction). The idea of transmutation of pollutants has been proposed for one or more of the following reasons:

- The experimental results support the transmutation of materials.
- To search the sinks of the remaining carbon dioxide not absorbed by plants or seawater.
- To find out the possible causes of explosion in the ozone layer other than the depletion of ozone layer.
- To investigate the possibilities of the self-sustaining stability of global environment.

To prove the portable of transmutation of pollutants, experimental investigations may be conducted to bombard C or CO_2 or CH_4or other air pollutants by accelerated alpha particles in a low-pressure vacuum tube in a similar condition of ionosphere. Heating them with gamma radiation can accelerate the alpha particles. The results of such experimental investigation may prove the probable transmutation of pollutants and self-sustaining equilibrium of the global environment.

GREENHOUSES

Population growth and less availability of food material have become global concerns. The world population increases exponentially whereas food production has increased only arithmetically, meaning that the availability of food per capita has decreased. This is more pronounced in the cases of oils, vegetables, fruits and milk, whereas it is marginal, rather than minimum, in cereals. The increase in population has also resulted in the use of more urban areas for habitation, less land available for cultivation and, hence, more food requirements. The resultant need is, therefore, to increase productivity and year round cultivation. To maximize production and meet the global demand on food, vegetables, flowers and horticultural crops, it is necessary to increase the effective production span of crops. The sun is the source of energy for plants and animals. This energy is converted into food (i.e., carbohydrates) by plants through a process called photosynthesis. This process is accomplished at suitable atmospheric conditions. These conditions are provided by nature in different seasons and artificially by a greenhouse. The primary objective of greenhouses is to produce agricultural products outside the cultivation season. They offer a suitable microclimate for plants and make possible growth and fruiting, where it is not possible in open fields. This is why a greenhouse is also known as a "controlled environment greenhouse". Through a controlled environment, greenhouse production is advanced and can be continued for longer duration, and finally, production is increased [33]. The off-season production of flowers and vegetables is the unique feature of the controlled environment greenhouse. Hence, greenhouse technology has evolved to create the favorable environment, or maintaining the climate, in order to cultivate the desirable crop the year round. The use of "maintaining the climate" concept may be extended for crop drying, distillation, biogas plant heating and space conditioning. The use of greenhouses is widespread. During the last 10 years, the amount of greenhouses has increased considerably to cover up to several hundred hectares at present. Most of the production is commercialized locally or exported. In India, about 300 ha of land are under greenhouse cultivation. On the higher side, however, it is 98600 ha in Netherlands, 48000 ha in China and 40000 ha in Japan [34]. This shows that there is a large scope to extend greenhouse technology for various climates.

However, the effective utilization of greenhouses has to deal with some specific climate problems like frost, during winter and overheating in summer days. These problems show the necessity of having a tool capable of predicting the thermal behaviour of a greenhouse under specific exterior conditions. Also, greenhouse industry has to deal with some problems related to a poor design of a great number of greenhouses. Such problems are mostly related to, on the one hand, its incapacity to deal with the problem of frost, which in the cold clear sky days of winter can destroy the whole work of a season, and, on the other hand, the question of overheating in the summer days.

EFFECTS OF URBAN DENSITY

Compact development patterns can reduce infrastructure demands and the need to travel by car. As population density increases, transportation options multiply and dependence areas, per capita fuel consumption is much lower in densely populated areas because people drive so much less.

Few roads and commercially viable public transport are the major merits. On the other hand, urban density is a major factor that determines the urban ventilation conditions, as well as the urban temperature. Under given circumstances, an urban area with a high density of buildings can experience poor ventilation and strong heat island effect.

In warm-humid regions these features would lead to a high level of thermal stress of the inhabitants and increased use of energy in air-conditioned buildings. With increasing urbanization in the world, cities are growing in number, population and complexity.

At present, 2% of the world's land surface is covered by cities, yet the people living in them consume 75% of the resources consumed by mankind. The reality of modern urbanization inevitably leads to higher densities than in traditional settlements and this trend is particularly notable in developing countries.

Today, the challenge before many cities to support large numbers of people while limiting their impact on the natural environment.

However, it is also possible that a high-density urban area, obtained by a mixture of high and low buildings, could have better ventilation conditions than an area with lower density but with buildings of the same height. Closely spaced or high-rise buildings are also affected by the use of natural lighting, natural ventilation and solar energy. If not properly planned, energy for electric lighting and mechanical cooling/ventilation may be increased and application of solar energy systems will be greatly limited. **TABLE 1** gives a summary of the positive and negative effects of urban density. All in all, denser city models require more careful design in order to maximize energy efficiency and satisfy other social and

Table 1: Effects of urban density on city's energy demand.

Positive effects	Negative effects
Transport: Promote public transport and reduce the need for, and length of, trips by private cars. Infrastructure: Reduce street length needed to accommodate a given number of inhabitants. Shorten the length of infrastructure facilities such as water supply and sewage lines, reducing the energy needed for pumping. Thermal performance: Multi-story, multiunit buildings could reduce the overall area of the building's envelope and heat loss from the buildings. Shading among buildings could reduce solar exposure of buildings during the summer period. Energy systems: District cooling and heating system, which is usually more energy efficiency, is more feasible as density is higher. Ventilation. A desirable flow pattern around buildings may be obtained by proper arrangement of high-rise building blocks.	Transport. Congestion in urban areas reduces fuel efficiency of vehicles. Vertical transportation: High-rise buildings involve lifts, thus increasing the need for electricity for the vertical transportation. Ventilation: A concentration of high-rise and large buildings may impede the urban ventilation conditions. Urban heat island: Heat released and trapped in the urban areas may increase the need for air conditioning. The potential for natural lighting is generally reduced in high-density areas, increasing the need for electric lighting and the load on air conditioning to remove the heat resulting from the electric lighting. Use of solar energy: Roof and exposed areas for collection of solar energy are limited.

development requirements. Low energy design should not be considered in isolation, and in fact, it is a measure, which should work in harmony with other environmental objectives. Hence, building energy study provides opportunities not only for identifying energy and cost savings, but also for examining the indoor and outdoor environment.

ENERGY EFFICIENCY AND ARCHITECTURAL EXPRESSION

The focus of the world's attention on environmental issues in recent years has stimulated response in many countries, which have led to a closer examination of energy conservation strategies for conventional fossil fuels. Buildings are important consumers of energy and thus important contributors to emissions of greenhouse gases into the global atmosphere. The development and adoption of suitable renewable energy technology in buildings has an important role to play. A review of options indicates benefits and some problems [27]. There are two key elements to the fulfilling of renewable energy technology potential within the field of building design; first the installation of appropriate skills and attitudes in building design professionals and second the provision of the opportunity for such people to demonstrate their skills. This second element may only be created when the population at large and clients commissioning building design in particular, become more aware of what can be achieved and what resources are required. Terms like passive cooling or passive solar use mean that the cooling of a building or the exploitation of the energy of the sun is achieved not by machines but by the building's particular morphological organization. Hence, the passive approach to themes of energy savings is essentially based on the morphological articulations of the constructions. Passive solar design, in particular, can realize significant energy and cost savings. For a design to be successful, it is crucial for the designer to have a good understanding of the use of the building. Few of the buildings had performed as expected by their designers. To be more precise, their performance had been compromised by a variety of influences related to their design, construction and operation. However, there is no doubt that the passive energy approach is certainly the one that, being supported by the material shape of the buildings has a direct influence on architectural language and most greatly influences architectural expressiveness [30]. Furthermore, form is a main tool in architectural expression. To give form to the material things that one produces is an ineluctable necessity. In architecture, form, in fact, summarizes and gives concreteness to its every value in terms of economy, aesthetics, functionality and, consequently, energy efficiency [31]. The target is to enrich the expressive message with forms producing an advantage energy-wise. Hence, form, in its geometric and material sense, conditions the energy efficiency of a building in its interaction with the environment. It is, then, very hard to extract and separate the parameters and the elements relative to this efficiency from the expressive unit to which they belong. By analyzing energy issues and strategies by means of the designs, of which they are an integral part, one will, more easily, focus the attention on the relationship between these themes, their specific context and their architectural expressiveness. Many concrete examples and a whole literature have recently grown up around these subjects and the wisdom of forms and expedients that belong to millennia-old traditions has been rediscovered. Such a revisiting, however, is only, or most especially, conceptual, since it must be filtered through today's technology and needs; both being almost irreconcilable with those of the past. Two among the historical concepts are of special importance. One is rooted in the effort to establish rational and friendly strategic relations with the physical environment, while the other recognizes the interactions between the psyche and physical perceptions in the creation of the feeling of comfort. The former, which may be defined as an alliance with the environment deals with the physical parameters involving a mixture of natural and artificial ingredients such as soil and vegetation, urban fabrics and pollution [32]. The most dominant outside parameter is, of course, the sun's irradiation, our planet's primary energy source. All these elements can be measured in physical terms and are therefore the subject of science. Within the second concept, however, one considers the emotional and intellectual energies, which are the prime inexhaustible source of renewable power [33]. In this case, cultural parameters, which are not exactly measurable, are involved. However, they represent the very essence of the architectural quality. Objective scientific measurement parameters tell us very little about the emotional way of perceiving, which influences the messages of human are physical sensorial organs. The perceptual reality arises from a multitude of sensorial components; visual, thermal, acoustic, olfactory and kinesthetic. It can, also, arise from the organizational quality of the space in which different parameters come together, like the sense of order or of serenity. Likewise, practical evaluations, such as usefulness, can be involved too. The evaluation is a wholly subjective matter, but can be shared by a set of experiencing persons [31]. Therefore, these cultural parameters could be different in different contexts in spite of the inexorable levelling on a planet- wide scale. However, the parameters change in the anthropological sense, not only with the cultural environment, but also in relation to function. The scientifically measurable parameters can, thus, have their meanings very profoundly altered by the non-measurable, but describable, cultural parameters.

However, the low energy target also means to eliminate any excess in the quantities of material and in the manufacturing process necessary for the construction of our built environment. This claims for a more sober, elegant and essential expression, which is not jeopardizing at all, but instead enhancing, the richness and preciousness of architecture, while contributing to a better environment from an aesthetic viewpoint [34]. Arguably, the most successful designs were in fact the simplest. Paying attention to orientation, plan and form can have far greater impact on energy performance than opting for elaborate solutions [34]. However, a design strategy can fail when those responsible for specifying materials for example, do not implement the passive solar strategy correctly. Similarly, cost-cutting exercises can seriously upset the effectiveness of a design strategy. Therefore, it is imperative that a designer fully informs key personnel, such as the quantity surveyor and client, about their design and be prepared to defend it. Therefore, the designer should have an adequate understanding of how the occupants or processes, such as ventilation, would function within the building. Thinking through such processes in isolation without reference to others can lead to conflicting strategies, which can have a detrimental impact upon performance. Likewise, if the design intent of the building is not communicated to its occupants, there is a risk that they will use it inappropriately, thus, compromising its performance. Hence, the designer should communicate in simple terms the actions expected of the occupant to control the building. For example, occupants should be well informed about how to guard against summer overheating. If the designer opted for a simple, seasonally adjusted control; say, insulated sliding doors were to be used between the mass wall and the internal space. The lesson here is that designers must be prepared to defend their design such that others appreciate the importance and interrelationship of each component. A strategy will only work if each individual component is considered as part of the bigger picture. Failure to implement a component or incorrect installation, for example, can lead to failure of the strategy and consequently, in some instances, the building may not like by its occupants due to its poor performance.

SUSTAINABLE PRACTICES

Within the last decade sustainable development and building practices have acquired great importance due to the negative impact of various development projects on the environment. In line with a sustainable development approach, it is critical for practitioners to create a healthy, sustainable built environment [31-33]. In Europe, 50% of material resources taken from nature are building-related, over 50% of national waste production comes from the building sector and 40% of energy consumption is building-related [30]. Therefore, more attention should be directed towards establishing sustainable guidelines for practitioners. Furthermore, the rapid growth in population has led to active construction that, in some instances, neglected the impact on the environment and human activities. At the same time, the impact on the traditional heritage, an often-neglected issue of sustainability, has not been taken into consideration, despite representing a rich resource for sustainable building practices.

Sustainability has been defined as the extent to which progress and development should meet the need of the present without compromising the ability of the future generations to meet their own needs [30]. This encompasses a variety of levels and scales ranging from economic development and agriculture, to the management of human settlements and building practices. This general definition was further developed to include sustainable building practices and management of human settlements. The following issues were addressed during the Rio Earth Summit in 1992 [30]:

- The use of local materials and indigenous building sources.
- Incentive to promote the continuation of traditional techniques, with regional resources and self-help strategies.
- Regulation of energy-efficient design principles.
- International information exchange on all aspects of construction related to the environment, among architects and contractors, particularly non-conventional resources.
- Exploration of methods to encourage and facilitate the recycling and reuse of building materials, especially those requiring intensive energy use during manufacturing, and the use of clean technologies.

The objectives of the sustainable building practices aim to:

- Develop a comprehensive definition of sustainability that includes socio-cultural, bio-climate, and technological aspects.
- Establish guidelines for future sustainable architecture.
- Predict the CO_2 emissions in buildings.
- The proper architectural measure for sustainability is efficient, energy use, waste control, population growth, carrying capacity, and resource efficiency.
- Establish methods of design that conserve energy and natural resources.

A building inevitably consumes materials and energy resources. The technology is available to use methods and materials that reduce the environmental impacts, increase operating efficiency, and increase durability of buildings (**TABLE 2**).

Table 2: Design, construction and environmental control description of traditional and new houses.

Design characteristics	Traditional houses	New houses
Form	Courtyard (height twice its width)-open to sky	Rectangle-closed
Construction	Brick walls 50 cm thick, brick roof with no insulation in either	Brick walls 25 cm thick, concrete roof-no insulation
Environmental control	Evaporative air coolers	Evaporative air coolers
Ease of climatic control	Difficult-rooms open into the climatically uncontrolled open courtyard	Moderate-rooms open into the enclosed internal corridor
Maintenance	Well conserved and maintained	New construction
Windows	Vertical and single glazing	Horizontal-single glazing
Urban morphology	Each house attached from 3 sides	Row houses attached from 2 sides
Orientation	Varies-irregular shapes and winding alleyways	North-south (row houses)
Orientation and solar gain	Solar gains less affected by orientation due to shading provided by the deep courtyard	Solar gain determined by orientation-no obstruction (shading)
Sharing solar gain	Significant - due to long-wave exchange or convective exchange between the 4 vertical walls surrounding the courtyard	Minimal- S-wall receives much more solar radiation than the N-wall due to the absence of any inter reflection and long-wave exchange
Occupant's social status	Low income families	Low and middle income families

Literature on green buildings reveals a number of principles that can be synthesized in the creation of the built environment that is sustainable. According to Lobo [14], these are: land development, building design and construction, occupant considerations, life cycle assessment, volunteer incentives and marketing programmes, facilitate reuse and remodeling, and final disposition of the structure. These parameters and many more are essential for analysis, making them an important element of the design decision-making process. Today, architects should prepare for this as well as dealing with existing buildings with many unfavorable urban environmental factors, such as many spaces have no choice of orientation, and, often, set in noisy streets with their windows opening into dusty and polluted air and surrounding buildings overshadowing them.

BUILDINGS AND CO_2 EMISSION

To achieve carbon dioxide, CO_2, emission targets, more fundamental changes to building designs have been suggested [35]. The actual performance of buildings must also be improved to meet the emission targets. To this end, it has been suggested that the performance assessment should be introduced to ensure that the quality of construction, installation and commissioning achieve the design intent. Air-tightness and the commissioning of plant and controls are the main two elements of assessing CO_2emission. Air-tightness is important as uncontrolled air leakage wastes energy. Uncertainties over infiltration rates are often the reason for excessive design margins that result in oversized and inefficient plants. On the other hand, commissioning to accept procedures would significantly improve energy efficiency. The slow turnover in the building stock means that improved performance of new buildings will only cut CO_2 emissions significantly in the long term. Consequently, the performance of existing buildings must be improved. For example, improving 3% of existing buildings would be more effective in cutting emissions than, say, improving the fabric standards for new non-domestic buildings and improving the efficiency of new air conditioning and ventilation systems [27]. A reduction in emissions arising from urban activities can, however, only be achieved by a combination of energy efficiency measures and a move away from fossil fuels.

ENERGY EFFICIENCY

Energy efficiency is the most cost-effective way of cutting carbon dioxide emissions and improvements to households and businesses. It can also have many other additional social, economic and health benefits, such as warmer and healthier homes, lower fuel bills and company running costs and, indirectly, jobs. Britain wastes 20 per cent of its fossil fuel and electricity use. This implies that it would be cost-effective to cut £10 billion a year off the collective fuel bill and reduce CO_2 emissions by some 120 million tones. Yet, due to lack of good information and advice on energy saving, along with the capital to finance energy efficiency improvements, this huge potential for reducing energy demand is not being realized. Traditionally, energy utilities have been essentially fuel providers and the industry has pursued profits from increased volume of sales. Institutional and market arrangements have favored energy consumption rather than conservation. However, energy is at the center of the sustainable development paradigm as few activities affect the environment as much as the continually increasing use of energy. Most of the used energy depends on finite resources, such as coal, oil, gas and uranium. In addition, more than three quarters of the world's consumption of these fuels is used, often inefficiently, by only one quarter of the world's population. Without even addressing these inequities or the precious, finite nature of these resources, the scale of environmental damage will force the reduction of the usage of these fuels long before they run out.

Throughout the energy generation process there are impacts on the environment on local, national and international levels, from opencast mining and oil exploration to emissions of the potent greenhouse gas carbon dioxide in ever increasing concentration. Recently, the world's leading climate scientists reached an agreement that human activities, such as burning fossil fuels for energy and transport, are causing the world's temperature to rise. The Intergovernmental Panel on Climate Change has concluded that "the balance of evidence suggests a discernible human influence on global climate". It predicts a rate of warming greater than any one seen in the last 10,000 years, in other words, throughout human history. The exact impact of climate change is difficult to predict and will vary regionally. It could, however, include sea level rise, disrupted agriculture and food supplies and the possibility of more freak weather events such as hurricanes and droughts. Indeed, people already are waking up to the financial and social, as well as the environmental, risks of unsustainable energy generation methods that represent the costs of the impacts of climate change, acid rain and oil spills. The insurance industry, for example, concerned about the billion dollar costs of hurricanes and floods, has joined sides with environmentalists to lobby for greenhouse gas emissions reduction. Friends of the earth are campaigning for a more sustainable energy policy, guided by the principle of environmental protection and with the objectives of sound natural resource management and long-term energy security. The key priorities of such an energy policy must be to reduce fossil fuel use, move away from nuclear power, improve the efficiency with which energy is used and increase the amount of energy obtainable from sustainable, renewable sources. Efficient energy use has never been more crucial than it is today, particularly with the prospect of the imminent introduction of the climate change levy (CCL). Establishing an energy use action plan is the essential foundation to the elimination of energy waste. A logical starting point is to carry out an energy audit that enables the assessment of the energy use and determine what actions to take. The actions are best categorized by splitting measures into the following three general groups:

(1) High priority/low cost: These are normally measures, which require minimal investment and can be implemented quickly. The followings are some examples of such measures:

- Good housekeeping, monitoring energy use and targeting waste-fuel practices.
- Adjusting controls to match requirements.
- Improved greenhouse space utilization.
- Small capital item time switches, thermostats, etc.
- Carrying out minor maintenance and repairs.
- Staff education and training.
- Ensuring that energy is being purchased through the most suitable tariff or contract arrangements.

(2) Medium priority/medium cost: Measures, which, although involve little or no design, involve greater expenditure and can take longer to implement. Examples of such measures are listed below:

- New or replacement controls.
- Greenhouse component alteration e.g., insulation, sealing glass joints, etc.
- Alternative equipment components e.g., energy efficient lamps in light fittings, etc.

(3) Long term/high cost: These measures require detailed study and design. They can be best represented by the followings:

- Replacing or upgrading of plant and equipment.
- Fundamental redesign of systems e.g., CHP installations.

This process can often be a complex experience and therefore the most cost-effective approach is to employ an energy specialist to help.

Energy efficiency brings health, productivity, safety, comfort and savings to homeowners, as well as local and global environmental benefits. The use of renewable energy resources could play an important role in this context, especially with regard to responsible and sustainable development.

POLICY RECOMMENDATIONS FOR A SUSTAINABLE ENERGY FUTURE

Sustainability is regarded as a major consideration for both urban and rural development. People have been exploiting the natural resources with no consideration to the effects, both short-term (environmental) and long-term (resources crunch). It is also felt that knowledge and technology have not been used effectively in utilising energy resources. Energy is the vital input for economic and social development of any country. Its sustainability is an important factor to be considered. The urban areas depend, to a large extent, on commercial energy sources. The rural areas use non-commercial sources like firewood and agricultural wastes. With the present day trends for improving the quality of life and sustenance of mankind, environmental issues are considered highly important. In this context, the term energy loss has no significant technical meaning. Instead, the energy loss has to be considered, as destruction of energy is possible. Hence, energy loss minimization will help in sustainability. In the process of developing, there are two options to manage energy resources: (1) End use matching/demand side management, which focuses on the utilities. The mode of obtaining this is decided based on economic terms. It is, therefore, a quantitative approach. (2) Supply side management, which focuses on the renewable energy resource and methods of utilizing it. This is decided based on thermodynamic consideration having the resource-user temperature or energy destruction as the objective criteria. It is, therefore, a qualitative approach. The two options are explained schematically in **FIGURE 8**. The energy-based energy, developed with supply side perspective is shown in **FIGURE 9**.

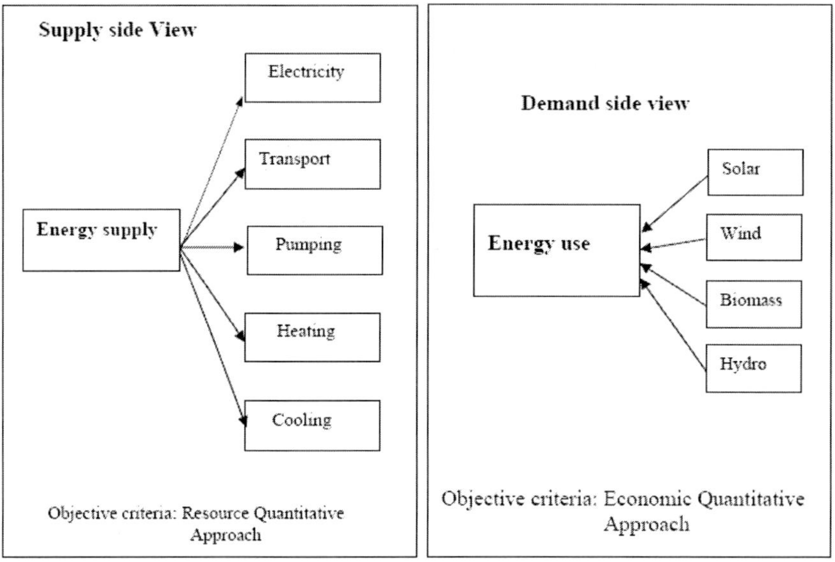

FIGURE 8: Supply side and demand side management approach for energy.

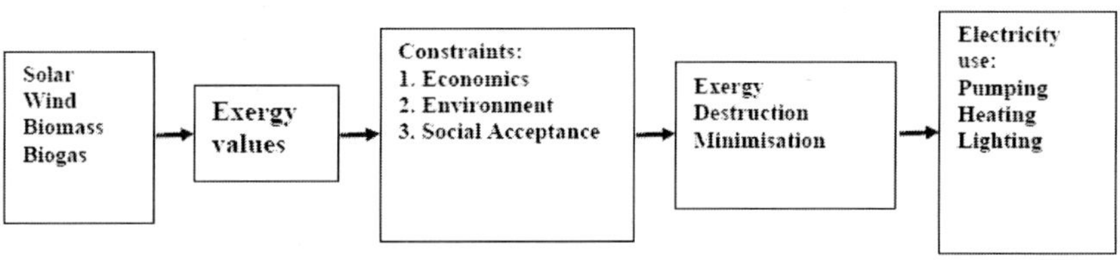

FIGURE 9: Energy based optimal energy model.

The following policy measures had been identified:

- Clear environmental and social objectives for energy market liberalization, including a commitment to energy efficiency and renewable.
- Economic, institutional and regulatory frameworks, which encourage the transition to total energy services, and economic measures to encourage utility investment in energy efficiency (e.g., levies on fuel bills).

- Incentives for demand side management, including grants for low-income households, expert advice and training, standards for appliances and buildings and tax incentives.
- Research and development funding for renewable energy technologies not yet commercially viable.
- Continued institutional support for new renewables (such as standard cost-reflective payments and obligation on utilities to buy).
- Ecological tax reform to internalize external environmental and social costs within energy prices.
- Planning for sensitive development and public acceptability for renewable energy.

Energy resources are needed for societal development. Their sustainable development requires a supply of energy resources that are sustainably available at a reasonable cost and can cause no negative societal impacts. Energy resources such as fossil fuels are finite and lack sustainability, while renewable energy sources are sustainable over a relatively longer term. Environmental concerns are also a major factor in sustainable development, as activities, which degrade the environment, are not sustainable. Hence, as much as environmental impact is associated with energy, sustainable development requires the use of energy resources, which cause as little environmental impact as possible. One way to reduce the resource depletion associated with cycling is to reduce the losses that accompany the transfer of energy to consume resources by increasing the efficiency of energy transfer between resources i.e., increasing the fraction of energy removed from one resource that is transferred to another [9].

As explained above, energy efficiency may be thought of as a more accurate measure of energy efficiency that accounts for quantity and quality aspects of energy flows. Improved energy efficiency leads to reduced energy losses. Most efficiency improvements produce direct environmental benefits in two ways. First, operating energy input requirements are reduced per unit output, and pollutants generated are correspondingly reduced. Second, consideration of the entire life cycle for energy resources and technologies suggests that improved efficiency reduces environmental impact during most stages of the life cycle. Quite often, the main concept of sustainability, which often inspires local and national authorities to incorporate environmental consideration into setting up energy programmes have different meanings in different contexts though it usually embodies a long-term perspective. Future energy systems will largely be shaped by broad and powerful trends that have their roots in basic human needs. Combined with increasing world population, the need will become more apparent for successful implementation of sustainable development.

Heat has a lower energy, or quality of energy, compared with work. Therefore, heat cannot be converted into work by 100% efficiency. Some examples of the difference between energy and energy are shown in **TABLE 3**.

Table 3: Qualities of various energy sources.

Source	Energy (J)	Energy (J)	CQF
Water at 80°C	100	16	0.16
Steam at 120°C	100	24	0.24
Natural gas	100	99	0.99
Electricity/work	100	100	1.00

The terms used in **TABLE 3** have the following meanings:

Carnot Quality Factor (CQF)=(1-To/Ts) (2)

Energy=Energy (transferred) x CQF (3)

Where To is the environment temperature (°K) and Ts is the temperature of the stream (°K).

Various parameters are essential to achieving sustainable development in a society. Some of them are as follows:

- Public awareness
- Information
- Environmental education and training
- Innovative energy strategies
- Renewable energy sources and cleaner technologies
- Financing
- Monitoring and evaluation tools.

Implementation of greenhouses offers a chance for maintenance and repair services. It is expected that the pace of implementation will increase and the quality of work improve in addition to building the capacity of the private and district staff in contracting procedures. The financial accountability is important and should be made transparent. The development of a renewable energy in a country depends on many factors. Those important to success are listed below:

Motivation of the population

The population should be motivated towards awareness of high environmental issues, rational use of energy in order to reduce cost. Subsidy programme should be implemented as incentives to install renewable energy plants. In addition, image campaigns to raise awareness of renewable technology.

(2) Technical product development

To achieve technical development of renewable energy technologies the following should be addressed:

- Increasing the longevity and reliability of renewable technology
- Adapting renewable technology to household technology (hot water supply)
- Integration of renewable technology in heating technology
- Integration of renewable technology in architecture, e.g., in the roof or façade
- Development of new applications, e.g., solar cooling
- Cost reduction.

Distribution and sales

Commercialization of renewable energy technology requires:

- Inclusion of renewable technology in the product range of heating trades at all levels of the distribution process (wholesale, retail)
- Building distribution nets for renewable technology
- Training of personnel in distribution and sales
- Training of field sales force

Consumer consultation and installation

To encourage all sectors of the population to participate in adoption of renewable energy technologies, the following has to be realized:

- Acceptance by craftspeople, marketing by them
- Technical training of craftspeople, initial and follow-up training programmes
- Sales training for craftspeople

- Information material to be made available to craftspeople for consumer consultation

Projecting and planning

Successful application of renewable technologies also requires:

- Acceptance by decision makers in the building sector (architects, house technology planners, etc.)
- Integration of renewable technology in training
- Demonstration projects/architecture competitions

Renewable energy project developers should prepare to participate in the carbon market by:

- Ensuring that renewable energy projects comply with Kyoto Protocol requirements.
- Quantifying the expected avoided emissions.
- Registering the project with the required offices.
- Contractually allocating the right to this revenue stream.

Other ecological measures employed on the development include:

- Simplified building details
- Reduced number of materials
- Materials that can be recycled or reused
- Materials easily maintained and repaired
- Materials that do not have a bad influence on the indoor climate (i.e., non-toxic)
- Local cleaning of grey water
- Collecting and use of rainwater for outdoor purposes and park elements
- Building volumes designed to give maximum access to neighboring park areas
- All apartments have visual access to both backyard and park

Energy saving measures

The following energy saving measures should also be considered:

- Building integrated solar PV system

- Day-lighting
- Ecological insulation materials
- Natural/hybrid ventilation
- Passive cooling
- Passive solar heating
- Solar heating of domestic hot water
- Utilization of rainwater for flushing

Improving access for rural and urban low-income areas in developing countries through energy efficiency and renewable energies. Sustainable energy is a prerequisite for development. Energy-based living standards in developing countries, however, are clearly below standards in developed countries. Low levels of access to affordable and environmentally sound energy in both rural and urban low-income areas are therefore a predominant issue in developing countries. In recent years many programmes for development aid or technical assistance have been focusing on improving access to sustainable energy, many of them with impressive results.

Apart from success stories, however, experience also shows that positive appraisals of many projects evaporate after completion and vanishing of the implementation expert team. Altogether, the diffusion of sustainable technologies such as energy efficiency and renewable energies for cooking, heating, lighting, electrical appliances and building insulation in developing countries has been slow.

Energy efficiency and renewable energy programmes could be more sustainable and pilot studies more effective and pulse releasing if the entire policy and implementation process was considered and redesigned from the outset. New financing and implementation processes are needed which allow reallocating financial resources and thus enabling countries themselves to achieve a sustainable energy infrastructure. The links between the energy policy framework, financing and implementation of renewable energy and energy efficiency projects have to be strengthened and capacity building efforts are required.

GREENHOUSE ENVIRONMENT

The comfort in a greenhouse depends on many environmental parameters. These include temperature, relative humidity, air quality and lighting. Although greenhouse and conservatory originally both meant a place to house or conserve greens (variegated hollies, cirrus, myrtles and oleanders), a greenhouse today implies a place in which plants are raised while conservatory usually describes a glazed room where plants may or may not play a significant role. Indeed, a greenhouse can be used for so many different purposes. It is, therefore, difficult to decide how to group the information about the plants that can be grown inside it. Whereas heat loss in winter a problem, it can be a positive advantage when greenhouse temperatures soar considerably above outside temperatures in summer. Indoor relative humidity control is one of the most effective long-term mite control measures. There are many ways in which the internal relative humidity can be controlled including the use of appropriate ventilation, the reduction of internal moisture production and maintenance of adequate internal temperatures through the use of efficient heating and insulation.

The introduction of a reflecting wall at the back of a greenhouse considerably enhances the solar radiation that reaches the ground level at any particular time of the day. The energy yield of the greenhouse with any type of reflecting wall was also significantly increased. The increase in energy efficiency was obtained by calculating the ratio between the total energy received during the day in greenhouse with a reflecting wall, compared to that in a classical greenhouse. Hence, the energy balance was significantly shifted towards conservation of classical energy for heating or lighting. The four-fold greater amount of energy that can be captured by virtue of using a reflecting wall with an adjustable inclination and louvers during winter attracts special attention. When sky (diffuse) radiation that was received by the ground in amounts shown in **FIGURE 10**, were taken into account, the values of the enhancement coefficients were reduced to some extent: this was due to the fact that they added up to the direct radiation from the sun in both new and classical greenhouses. However, this is a useful effect as further increases overall energy gain. There is also an ironing out effect expressed in terms of the ratios between peak and average insolations.

Finally, the presented theory can be used to calculate the expected effects of the reflecting wall at any particular latitude, under different weather conditions, and when the average numbers of clear days are taken into account. Thereby an assessment of the cost of a particular setup can be obtained. Under circumstances of a few clear days, it may still be worthwhile from a financial point of view to turn a classical greenhouse into one with a reflecting wall by simply covering the glass wall on the north-facing side with aluminum foil with virtually negligible expenditure.

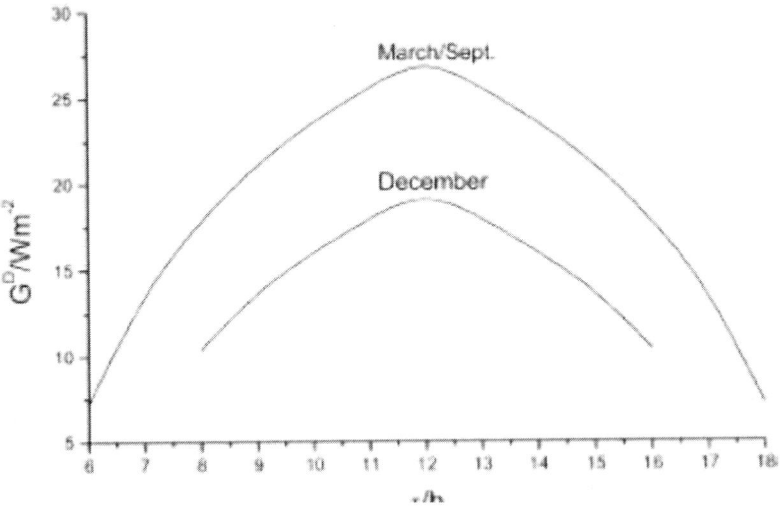

FIGURE 10: Ground irradiance from diffuse (sky) radiation from a clear sky at the shortest winter day and at equinox.

RELATIVE HUMIDITY

Air humidity is measured as a percentage of water vapor in the air on a scale from 0% to 100%, where 0% being dry and 100% being full saturation level. The main environmental control factor for dust mites is relative humidity. The followings are the practical methods of controlling measures available for reducing dust mite populations:

- Chemical control.
- Cleaning and vacuuming.
- Use of electric blankets, and
- Indoor humidity.

CONCLUSIONS

Thermal comfort is an important aspect of human life. Buildings where people work require more light than buildings where people live. In buildings where people live the energy is used for maintaining both the temperature and lighting. Hence, natural ventilation is rapidly becoming a significant part in the design strategy for non-domestic buildings because of its potential to reduce the environmental impact of building operation, due to lower energy demand for cooling. A traditional, naturally ventilated building can readily provide a high ventilation rate. On the other hand, the mechanical ventilation systems are very expensive. However, a comprehensive ecological concept can be developed to achieve a reduction of electrical and heating energy consumption, optimize natural air condition and ventilation, improve the use of daylight and choose environmentally adequate building materials. Plants, like human beings, need tender loving care in the form of optimum settings of light, sunshine, nourishment, and water. Hence, the control of sunlight, air humidity and temperatures in greenhouses are the key to successful greenhouse gardening. The mop fan is a simple and novel air humidifier; which is capable of removing particulate and gaseous pollutants while providing ventilation. It is a device ideally suited to greenhouse applications, which require robustness, low cost, minimum maintenance and high efficiency. A device meeting these requirements is not yet available to the farming community. Hence, implementing mop fans aids sustainable development through using a clean, environmentally friendly device that decreases load in the greenhouse and reduces energy consumption.

REFERENCES

1. Jeremy L (2005) The energy crisis, global warming and the role of renewables. Renewable Energy World.
2. Omer A (2010) Low energy building materials: an overview. In: Proceedings of the Environment: Situation and Perspectives for the European UnionPorto Portugal.
3. Brack D, Graber M, Seki M, Batten R, Mutisya G (2003) Handbook for the International Treaties for the Protection of the Ozone Layer. United Nations Environment Programme Nairobi, Kenya.
4. Viktor D (2002) Ventilation concepts for sustainable buildings. In Proceedings of the World Renewable Energy Congress VII.
5. Lam JC (2000) Shading effects due to nearby buildings and energy implications. Energy Conservation and Management47: 647-659.
6. Raja J, Nichol F, McCartney K (1998) Natural ventilated buildings use of controls for changing indoor climate. In Proceedings of the 5th World Renewable Energy Congress 15:391-394.
7. Limb MJ (1995) Air intake positioning to avoid contamination of ventilation AIVC.
8. Miller G (1990) Resource conservation and management. Wadsworth Publishers California USA.
9. Erlich P (1991) Forward Facing up to Climate Change, in Global Climate Change and Life on Earth.
10. ASHRAE (1993) Energy efficient design of new building except new low-rise residential buildings. BSRIASHRAE proposed standards 90-2P-1993 alternative GAAmerican Society of Heating, Refrigerating, and Air Conditioning Engineers Inc USA 1993.
11. Molla M (1997) Air pollutants and its probable transmutation in the ionosphere. Renewable Energy 10: 327-329.

12. Bahadori M (1988) A passive cooling/heating system for hot arid regions. In: Proceedings of the American Solar Energy Society Conference. Cambridge. Massachusetts.
13. Dieng A, Wang R (2001) Literature review on solar absorption technologies for ice making and air conditioning purposes and recent development in solar technology. Renewable and Sustainable Energy Review 5: 313-342.
14. Lobo C (1998) Defining a sustainable building. In Proceedings of the 23rd National Passive Conference. American Solar Energy Society (ASES'98) Albuquerque USA.
15. Crisp V, Cooper I, McKennan G (1988) Daylighting as a passive solar energy option: an assessment of its potential in non-domestic buildings. Report BR129-BRE. Garston UK
16. Horning M, Skeffington R (1993) Critical loads: concept and applications. Institute of Terrestrial Ecology. HMSO Publishers Ltd London UK.
17. Humphreys M (1978) Outdoor temperatures and comfort indoor. Building Research and Practice.
18. Givoni B (1998) Climate consideration in building and urban design. New York: Van Nostrand Reinhold.
19. Koenigsberger O, Ingersoll T, Mayhew A, Szokolay S (1973) Manual of tropical housing and building. Part 1: Climate design Longmas London UK.
20. Boulet T (1987) Controlling air movement: a manual for architects and builders. McGraw-Hill, New York, USA.
21. Szokolay S (1990) Design and research issues: passive control in the tropic. Proceedings First World Renewable Energy Congress Reading: UK.
22. Borda-Daiz N, Mosconi P, Vazquez J (1989) Passive cooling strategies for a building prototype design in a warm-humid tropical climate. Solar and Wind Technology 6: 389-400.
23. Givoni B (1965) Laboratory study of the effect of window sizes and location on indoor air motion. Architectural Science Review 8: 42-46.
24. Fanger P (1970) Thermal comfort: analysis and applications in environmental engineering. Danish Technical Press.
25. Fordham M (2000) Natural ventilation. Renewable Energy 19: 17-37.
26. Awbi H (1991) Ventilation of buildings. Spon Publisher. London: UK.
27. Givoni B (1976) Man climate and architecture. Applied Science Publisher Ltd London UK.
28. BS 5454 (1989) Storage and exhibition archive documents. British Standard Institute. London.
29. Lazzarin RD, Ascanio A, Gaspaella A (2002) Utilization of a green roof in reducing the cooling load of a new industrial building.
30. David E (2003) Sustainable energy: choices, problems and opportunities. The Royal Society of Chemistry 19: 19-47.
31. Zuatori A (2005) An overview on the national strategy for improving the efficiency of energy use. Jordanian Energy Abstracts 9: 31-32.
32. Anne G, Michael S (2005) Building and land management. (5thedn). Oxford: UK.
33. Randal G, Goyal R (1998) Greenhouse technology. New Delhi: Narosa Publishing House.
34. Yadav I, Chauadhari M (1997) Progressive floriculture. Bangalore: The house of Sarpan.
35. EIBI (Energy in Building and Industry) (1999) Constructive thoughts on efficiency, building regulations, inside committee limited, Inside Energy: magazine for energy professional UKKOPASS.

CHAPTER 2

Economic Analysis of Greenhouse Lighting: Light Emitting Diodes vs. High Intensity Discharge Fixtures

Jacob A. Nelson, Bruce Bugbee

Crop Physiology Laboratory, Department of Plant Soils and Climate, Utah State University, Logan, Utah, United States of America

ABSTRACT

Lighting technologies for plant growth are improving rapidly, providing numerous options for supplemental lighting in greenhouses. Here we report the photosynthetic (400–700 nm) photon efficiency and photon distribution pattern of two double-ended HPS fixtures, five mogul-base HPS fixtures, ten LED fixtures, three ceramic metal halide fixtures, and two fluorescent fixtures. The two most efficient LED and the two most efficient double-ended HPS fixtures had nearly identical efficiencies at 1.66 to 1.70 micromoles per joule. These four fixtures represent a dramatic improvement over the 1.02 micromoles per joule efficiency of the mogul-base HPS fixtures that are in common use. The best ceramic metal halide and fluorescent fixtures had efficiencies of 1.46 and 0.95 micromoles per joule, respectively. We also calculated the initial capital cost of fixtures *per photon delivered* and determined that LED fixtures cost five to ten times more than HPS fixtures. The five-year electric plus fixture cost per mole of photons is thus 2.3 times higher for LED fixtures, due to high capital costs. Compared to electric costs, our analysis indicates that the long-term maintenance costs are small for both technologies. If widely spaced benches are a necessary part of a production system, the unique ability of LED fixtures to efficiently focus photons on specific areas can be used to improve the photon capture by plant canopies. Our analysis demonstrates, however, that the cost per photon delivered is higher in these systems, regardless of fixture category. The lowest lighting system costs are realized when an efficient fixture is coupled with effective canopy photon capture.

INTRODUCTION

Rapid advances in lighting technology and fixture efficiency provide an expanding number of options for supplemental lighting in greenhouses, including numerous LED fixtures (light emitting diode, see [1], [2] for a history of LED lighting in horticulture). Significant improvements have been made in all three high intensity discharge (HID, which includes high pressure sodium, HPS, and ceramic metal halide, CMH) fixture components: the lamp (often referred to as the bulb), the luminaire (often referred to as the reflector) and the ballast. High pressure sodium fixtures with electronic ballasts and double-ended lamps are now 1.7 times more efficient than older mogul-base HPS fixtures.

Lighting technologies vary widely in how radiation is distributed (Figure 1). There is no ideal pattern of radiation distribution for every application. In large greenhouses with small aisles and uniformly spaced plants, the broad, even output pattern typically emitted from HPS fixtures provides uniform (little variation over a large area) light distribution and increased capture of photosynthetic photons. In smaller greenhouses with spaced benches, the more focused pattern typically found in LED fixtures can maximize radiation transfer to plant leaves. As the area (height of width) covered by plants increases, the need for more focused radiation decreases (Figure 2).

Each line represents a cross section of the photon intensity below the fixture. The LED fixture (Lighting Sciences Group) uses optics to achieve a narrow distribution, with the majority of the photons falling in a concentrated pattern directly below the fixture. Conversely, the Cycloptics ceramic metal halide fixture is designed for even light distribution, and therefore casts uniform radiation over a large surface area. Since the area increases exponentially as the distance from the center increases, an equal photon flux farther from the center represents a larger quantity of total photons.

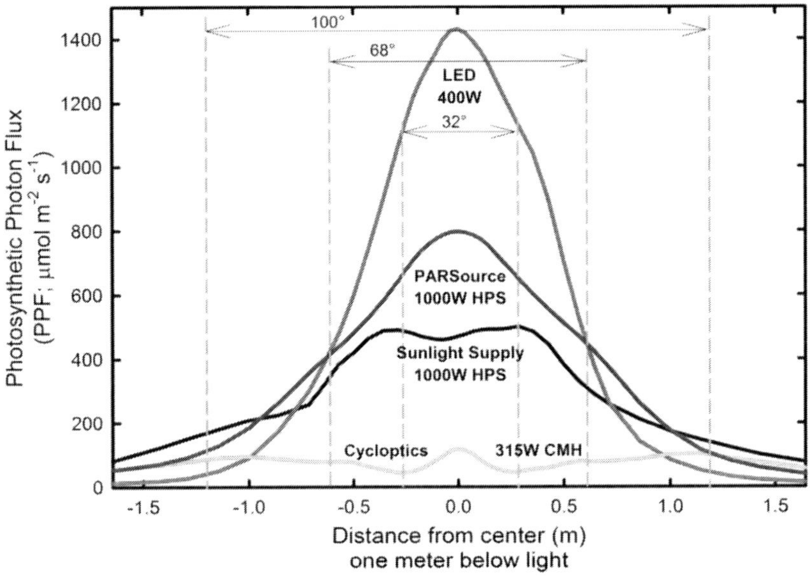

Figure 1. The photon distribution of four fixtures with similar photon efficiency.

As the plant growth area under the fixture gets smaller, wasted radiation often increases. This figure illustrates the concept of canopy photon capture efficiency. Two meters was chosen as a typical mounting height, but this can be scaled as a unit-less ratio. Multiple overlapping fixtures are typically used to minimize PPF variation over a large area.

In greenhouse applications, selection among lighting options should primarily be made based on the cost to deliver photons to the plant canopy surface. This analysis includes two parameters: 1) the fundamental fixture efficiency, measured as micromoles of photosynthetic photons per joule of energy input, and 2) the canopy photosynthetic (400–700 nm) photon flux (PPF) capture efficiency, which is the fraction of photons transferred to the plant leaves.

Electrical efficiency for plant growth is best measured as µmoles per Joule

The electrical efficiency of lamps is often expressed using units for human light perception (efficacy; lumens or foot-candles out per watt in) or energy efficiency (radiant watts out per electrical watt in). Photosynthesis and plant growth, however, is determined by moles of photons. It is thus

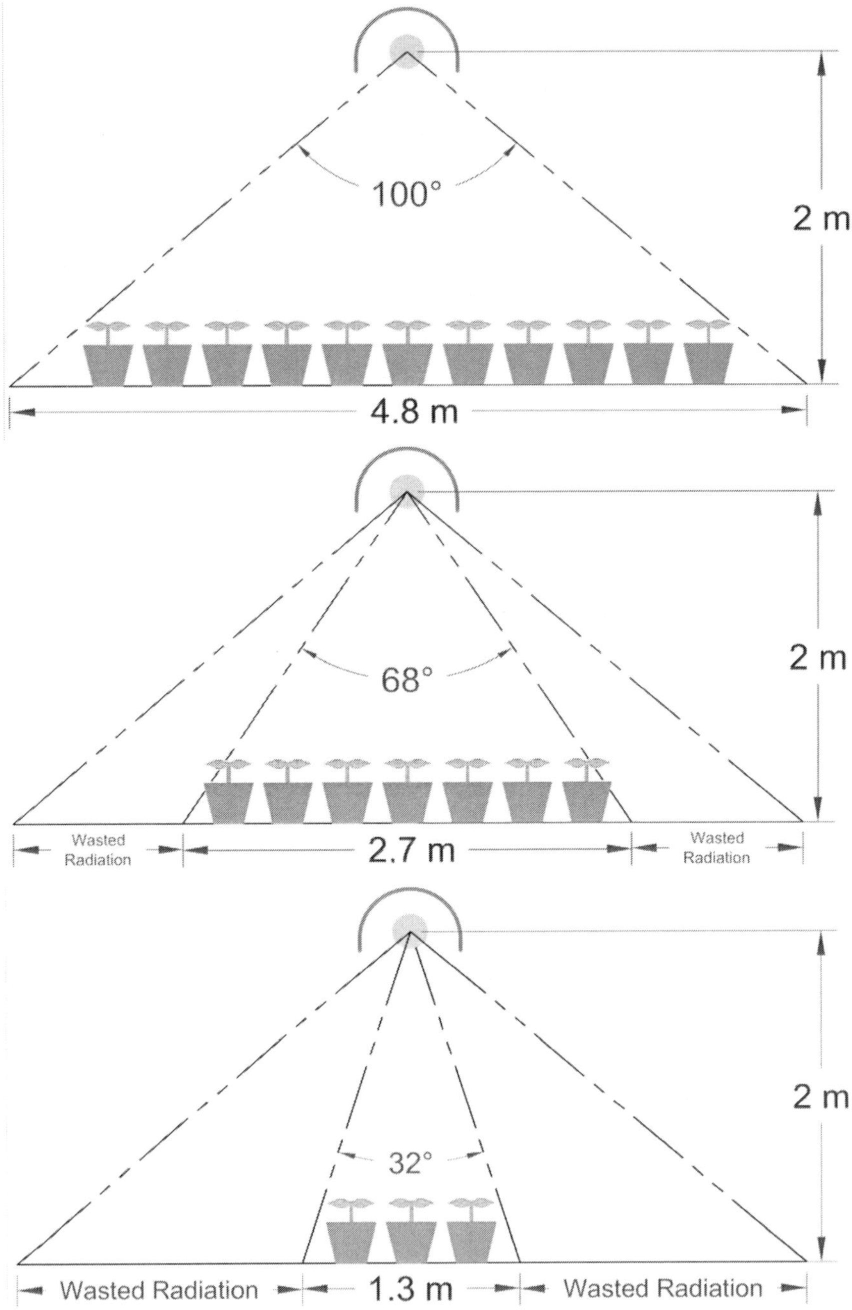

Figure 2. Canopy photon capture efficiency.

important to compare lighting efficiency based on photon efficiency, with units of micromoles of photosynthetic photons per joule of energy input. This is especially important with LEDs where the most electrically efficient colors are in the deep red and blue wavelengths. A dramatic example of this is the comparison of red, blue, and cool white LEDs (Table 1). The lower radiant energy content of red photons allows more photons to be delivered per unit of input energy (radiant energy is inversely proportional to wavelength, Planck's Equation). Conversely, blue LEDs can have a 53% higher energy efficiency (49% vs. 32%) but only a 9% higher photon efficiency (1.87 vs. 1.72).

Table 1. Efficiency of individual LEDs at a drive current of 700

LED Color	Peak wavelength or color temperature	Photon efficiency[z] (μmol/J)	Electrical efficiency[y] (%)	Luminous efficiency[x] (lm/W)
Cool white	5650 Kelvin	1.52	33	111
Red	655 nm	1.72	32	47
Blue	455 nm	1.87	49	17

[z] Photon efficiency is the most appropriate measure for photosynthesis.
[y] The relationship between electrical efficiency and photon efficiency is dependent on wavelength (Plank's equation $E=hc/\lambda$).
[x] Luminous efficiency is shown to demonstrate how inappropriate it is as an indicator of lighting efficiency for plants.

Effect of light quality

There is considerable misunderstanding over the effect of light quality on plant growth. Many manufacturers claim significantly increased plant growth due to light quality (spectral distribution or the ratio of the colors). A widely used estimate of the effect of light quality on photosynthesis comes from the Yield Photon Flux (YPF) curve, which indicates that orange and red photons between 600 to 630 nm can result in 20 to 30% more photosynthesis than blue or cyan photons between 400 and 540 nm (Figure 3)[3], [4]. When light quality is analyzed based on the YPF curve, HPS lamps are equal to or better than the best LED fixtures because they have a high photon output near 600 nm and a low output of blue, cyan, and green light [5].

Effect of wavelength on relative photosynthesis per incident photon for a single leaf in low light (less than 150 μmol m^{-2} s^{-1}) [4].

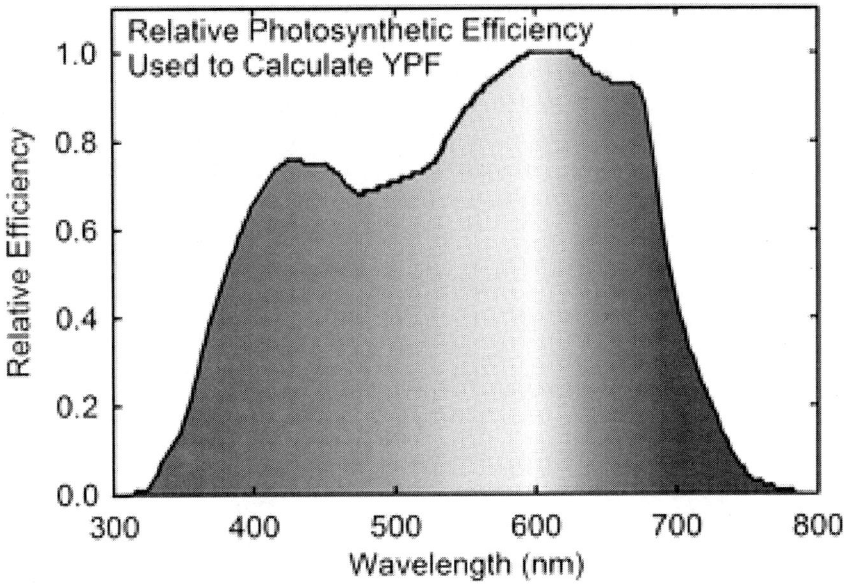

Figure 3. Yield photon flux curve.

The YPF curve, however, was developed from short-term measurements made on single leaves in low light. Over the past 30 years, numerous longer-term studies with whole plants in higher light indicate that light *quality* has a much smaller effect on plant growth rate than light *quantity* [6], [7]. Light quality, especially the fraction of blue light, has been shown to alter cell expansion rate, leaf expansion rate[8], plant height and plant shape in several species [9]–[11], but it has only a small direct effect on photosynthesis. The effects of light quality on fresh or dry mass in whole plants typically occur under low or no sunlight conditions, and are caused by changes in leaf expansion and radiation capture during early growth [6].

Unique aspects of LED fixtures

The most electrically efficient colors of LEDs, based on moles of photosynthetic photons per joule, are blue, red, and cool white, respectively (Figure 4), so LED fixtures generally come in combinations of these colors. LEDs of other colors can be used to dose specific wavelengths of light to control aspects of plant growth [12], due to their monochromatic nature (see [13] for a review of unique LED applications). Ultraviolet (UV) radiation is typically absent in LED fixtures because UV LEDs significantly reduce fixture efficiency. Sunlight has 9% UV (percent of PPF), and standard electric lights have 0.3 to 8% UV radiation (percent of PPF)[5]. A lack of UV causes disorders in some plant species (e.g. Intumescence; [14]) and this is a concern with LED fixtures when used without sunlight. LED fixtures for supplemental photosynthetic lighting also have minimal far-red radiation (710 to 740 nm), which decreases the time to flowering in several photoperiodic species [15]. Green light (530 to 580 nm) is low or absent in most LED fixtures and these wavelengths better penetrate through the canopy and are more effectively transmitted to lower plant leaves [16]. The lack of UV, green, and far-red wavelengths, however, should be minimal when LEDs are used in greenhouses, because most of the radiation comes from broad spectrum sunlight.

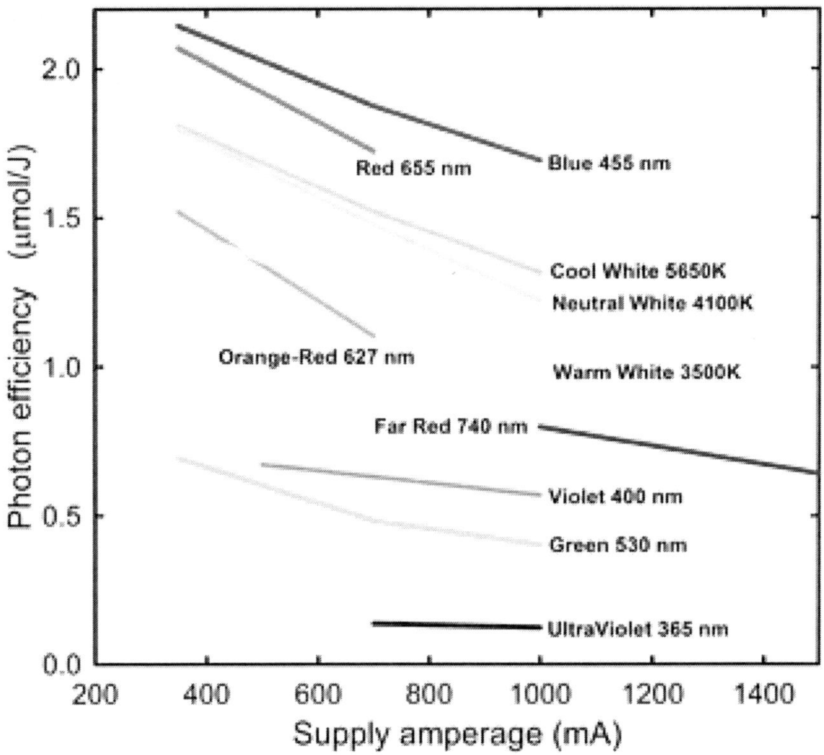

Figure 4. Effect of drive amperage and color on photon efficiency of LEDs.

Data for Philips Lumileds LEDs (May 2014), courtesy of Mike Bourget, Orbitec.

Our objective is to help growers and researchers select the most cost effective fixture options for supplemental lighting in greenhouses. To achieve this goal we measured two fundamental components of each fixture: 1) the efficiency of conversion of electricity to photosynthetic photons that are delivered to a horizontal surface below the lamp, and 2) the distribution pattern of these photons below the fixture.

Materials and Methods

Fixture efficiency

Measurements of fixture efficiency (lamp, luminaire, and ballast) were made by integrating sphere and flat-plane integration techniques. The integrating sphere measurements were made by a certified testing laboratory (TÜV SÜD America) that specializes in the measurement of the efficiency of lighting fixtures using the IES LM79-08 measurement standard [17]. Radiometric output was converted to photon output at each nanometer interval using Plank's Equation and then integrated from 400 to 700 nanometers. The radiation measurements were calibrated to NIST reference standards. These measurements of fixture efficiency are considered repeatable to within 1%.

Flat plane integration

Measurements were made in a dark room with flat black walls using a quantum sensor (LI-COR model LI-190, Lincoln, NE, USA), that was calibrated for each fixture with an NIST-traceable calibrated spectroradiometer (model PS-200, Apogee Instruments, Logan, UT, USA). This calibration is necessary to correct for small spectral errors (±3%) in the quantum sensor that occur because of imperfect matching of the ideal quantum response [18]. Measurements were made in three radial, straight lines below a level fixture and spatially integrated over a flat plane below the fixture to determine total photon output. Measurements were made 2.5 cm apart near the center, increasing to 10 cm near the perimeter as PPF variation decreased (121 measurements total). Fixtures were mounted 0.7 meters above the surface and measurements were made up to a 1.5 meter radius from the center and extrapolated to infinity using an exponential decay function. Fixture height is optional, depending on the size of the room and measurement area as long as measurement resolution captures the spatial variation in fixture output. The flat-plane integration measurements were used to quantify the pattern of photon distribution from the fixture. Total fixture output from these measurements was similar to measurements made using an integrating sphere (Table 2). When redundant measurements were available, the integrating sphere measurements were used to quantify fixture efficiency. Power draw and electrical characteristics were measured using a multimeter and a current clamp (Fluke model 289, Everett, WA, USA).

Table 2. Efficiency of fixtures using integrating sphere measurements compared with flat-plane integration.

	TÜV SÜD America integrating sphere			USU[z] flat plane integration			flat plane/ integrating sphere[y]
Fixture	Elec. input (W or J/s)	Photon output (μmol/s)	Photon efficiency (μmol/J)[x]	Elec. input (W or J/s)	Photon output (μmol/s)	Photon efficiency (μmol/J)[x]	(μmol/J)/(μmol/J)
Gavita Pro 1000DE	1033	1751	1.70	1041	1814	1.74	2.7%
ePapillion 1000W	1041	1767	1.70	1037	1937	1.87	9.1%
LSG violet	384	653	1.70	391	626	1.61	−6.0%
SPYDR 600	326	541	1.66	332	575	1.73	4.4%
LSG red/white	390	634	1.63	397	601	1.51	−7.5%
Illumitex NeoSol	279	390	1.40	281	386	1.38	−1.8%
ParSource GLXII	1026	1334	1.30	1008	1433	1.42	8.6%
Lumigrow Pro 325	304	390	1.29	304	355	1.17	−10.1%
California Light Works SOLARSTORM	337	350	1.04	343	331	0.96	−7.7%
Black Dog BD360U	339	339	1.00	346	323	0.93	−7.2%
Apache AT120WR	169	163	0.96	167	150	0.90	−7.2%
iGrow 400W	394	374	0.95	397	354	0.89	−6.5%
Lumigrow es330	318	264	0.90	317	270	0.85	−5.1%
Hydrogrow Sol 9	423	378	0.89	430	396	0.92	2.9%

[z] Utah State University
[y] The flat-plane integration may have made an inadequate number of measurements to fully characterize the output of some of the lamps. The electric consumption (watts) by the fixture was nearly identical among test sites and did not likely have a significant effect of efficiency.
[x] Photon Output per Electrical Input (μmol per second divided by joules per second).

Cost of electricity

In the United States, commercial electric rates vary widely by region, ranging from $0.07 in Idaho to $0.17 in New York, with residential rates averaging $0.02 higher, and industrial rates $0.02 lower. Electric rates in Europe, and many other countries, can be more than double the rates in the United States. As electricity becomes more expensive, improved lighting becomes more valuable. See U.S. Energy Information Administration for a summary of current electric rates by state and region (accessed April 2014). We used a discounted cash flow model assuming a 5% per year cost of capital on future electrical costs.

RESULTS

The photon efficiency (micromoles per joule) and cost per mole of photons for four categories of lighting technologies (HPS, LED, ceramic metal halide, and fluorescent), in 22 fixtures, are shown in Table 3. One fixture of each model was tested. This table also shows the five-year electric plus fixture costs per mole of photons. Most fixtures (lamp, luminaire and ballast) are now more efficient than the common 1000-W magnetic-ballast, mogul-base HPS fixtures (i.e. Sunlight Supply, 1.02 μmol per joule). If photons coming out of the fixture at all downward angles are considered (180°), the capital cost of the most efficient 400-W LED fixtures we tested is five to seven times more per photon than the 1000-W, double-ended, electronic ballast HPS fixtures (Gavita, ePapillion, Table 3). The high capital cost of LEDs makes the five year cost per mole of photons more than twice that of HPS fixtures (Table 3 and Figure 5A).

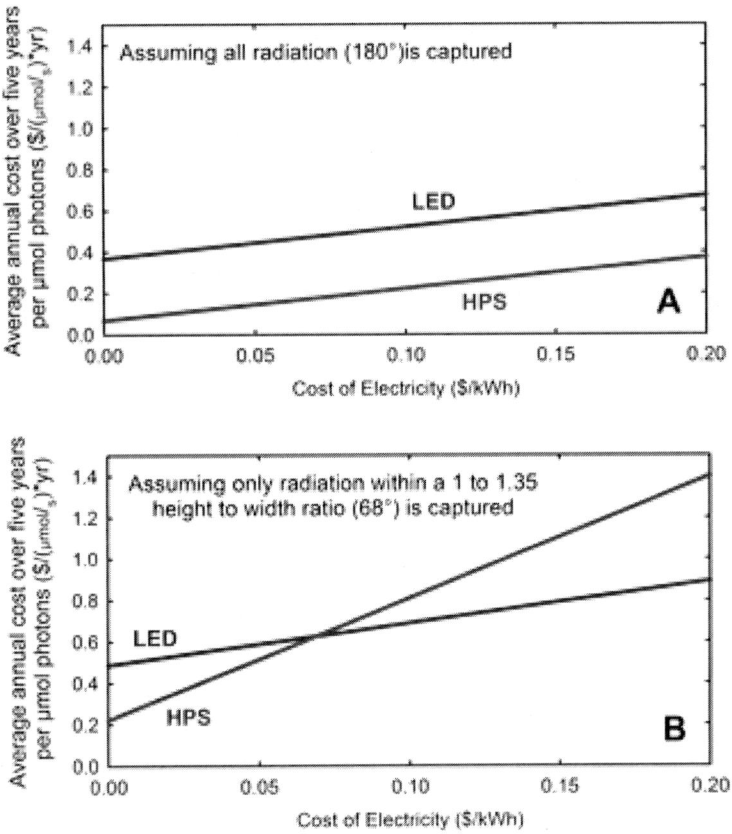

Figure 5. Effect of electricity price on average annual cost over five years for two capture scenarios.

(A) When all radiation is assumed captured, the most efficient HPS fixture (Gavita) has a lower average annual five-year cost per photon than the most efficient LED fixture (Red/Blue fixture, Lighting Sciences Group). (B) When only a narrow region below the fixture (68°) is considered to be captured (e.g. on benches), the LEDs can have a lower cost per photon then HPS fixtures, but the cost per photon increases for both fixtures.

Table 3. Photon efficiency and cost per mole of photons, assuming all photons (180°) are captured by plants.

Lamp type and Ballast	Fixture producer[z]	Electrical input (J/s or watts)	Photon output[y] (µmol/s)	Photon efficiency[x] (µmol/J)	Cost of one fixture[w] ($)	Fixtures needed per millimol/s[v]	Fixture cost per mol/s $/(mol/s)	Electric cost per µmol photons[u] $/(µmol/s)yr	Five year electric cost per µmol photons[t] $/(µmol/s)yr
High Pressure Sodium									
400 W magnetic	Sunlight Supply	443	416	0.94	$200	2.40	$0.48	$0.35	$0.40
1000 W magnetic	Sunlight Supply	1067	1090	1.02	$275	0.92	$0.25	$0.32	$0.33
1000 W magnetic	PARsource GLXI	1004	1161	1.16	$350	0.86	$0.30	$0.29	$0.31
1000 W electronic	PARsource GLXI	1024	1333	1.30	$380	0.75	$0.29	$0.25	$0.28
1000 W electronic	PARsource GLXII	1026	1334	1.30	$310	0.75	$0.23	$0.25	$0.27
1000 W electronic	Gavita	1033	1751	1.70	$500	0.57	$0.29	$0.19	$0.23
1000 W electronic	ePapillon	1041	1767	1.70	$600	0.57	$0.34	$0.19	$0.24
LED									
red/blue	LSG	384	653	1.70	$1,200	1.53	$1.84	$0.19	$0.54
red/white	BML	326	541	1.66	$1,000	1.85	$1.85	$0.20	$0.54
red/white	LSG	390	634	1.63	$1,200	1.58	$1.89	$0.20	$0.55
red/white	Illumitex	279	390	1.40	$1,400	2.56	$3.59	$0.24	$0.92
red/white/blue	Lumigrow (Pro 325)	304	390	1.29	$1,000	2.56	$2.56	$0.26	$0.73
red/white	California Lightworks	337	350	1.04	$1,000	2.85	$2.85	$0.32	$0.85
multiple	Black Dog	339	339	1.00	$950	2.95	$2.80	$0.33	$0.85
red/white	Apache	169	163	0.96	$860	6.14	$5.28	$0.34	$1.35
red/blue	Lumigrow (ES330)	318	284	0.90	$1,200	3.52	$4.22	$0.37	$1.16
red/white	Hydrogrow	423	378	0.89	$1,300	2.64	$3.44	$0.37	$1.01
Ceramic Metal Halide									
315 W 3100 K	Cycloptics	337	491	1.46	$640	2.04	$1.30	$0.23	$0.46
315 W 4200 K	Cycloptics	340	468	1.38	$640	2.14	$1.37	$0.24	$0.48
2@315 W 3100 K	Boulderlamp	651	817	1.25	$1,000	1.22	$1.22	$0.26	$0.47
Fluorescent									
400 W induction	iGrow	394	374	0.95	$1,200	2.68	$3.21	$0.35	$0.94
60 W	T8	56	48	0.84	$40	20.77	$0.83	$0.40	$0.51

[z]- See Table S1 for a list of fixture manufacturers and model numbers.
[y]- Integrated total photon output of fixture.
[x]- Photon Output per Electrical Input (µmol per second divided by joules per second).
[w]-Cost of fixtures as of April 2014.
[v]-The number of fixtures to get a total photon output of one millimol (1000 µmol) of photons per second.
[u]-Assumes 3000 hours per year operation and $0.11/kWh.
[t]-Cost of fixture (multiplied by fixtures needed) plus cost of electricity over 5 years. We used a discounted cash flow model assuming a 5% per year cost of capital. Installation and maintenance costs were assumed to be similar for all lamp types and were not included in this calculation.

Table 3 assumes that all of the photons emitted from the fixture are absorbed by plant leaves. In Table 4, the area under the fixture in which the photons are considered captured by plants is progressively reduced, and the cost per mole of photons increases as more photons are lost around the perimeter. When only highly focused radiation is considered useful (34°), some LED fixtures have a lower cost per photon than the best HPS fixtures (Table 4, Figure 1, Figure 5B and Figure 6), but because photons are lost around the perimeter at this narrow angle, the cost per photon absorbed by plants is much greater. The lowest cost per photon is realized when a large canopy can be arranged to capture the photons.

The cost per mole of photons for LEDs (Red/Blue LED from Lighting Sciences Group) becomes more favorable than the best HPS fixtures (Gavita) when the lighting area is less than 68° from center, assuming $0.11 per kWh cost of electricity and 3000 hours per year use (approximate cumulative operation time at latitudes from 40 to 50 degrees).

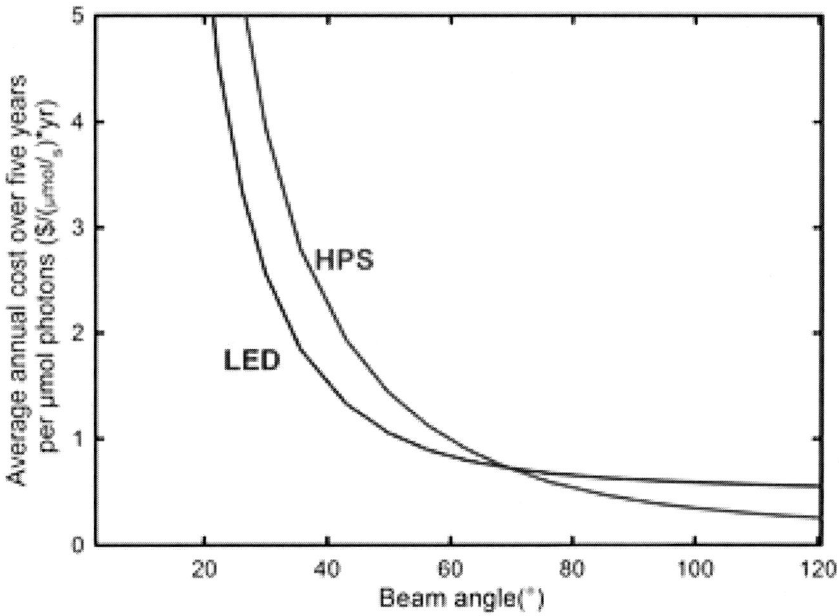

Figure 6. Effect of canopy capture efficiency on average annual cost over five years.

Table 4. Cost per mole photons for four capture assumptions.

		Assuming all radiation (180°) is captured		Assuming radiation within a 1 to 2.38 height to width ratio (100°) is captured		Assuming radiation within a 1 to 1.35 height to width ratio (68°) is captured	
Lamp type and Ballast	Fixture producer[x]	Fixtures needed per mmol/s[y]	Five year electric cost per µmol photons[*] $/(µmol/s)yr	Fixtures needed per mmol/s[y]	Five year electric cost per µmol photons[*] $/(µmol/s)yr	Fixtures needed per mmol/s[y]	Five year electric cost per µmol photons[*] $/(µmol/s)yr
High Pressure Sodium							
400 W magnetic	Sunlight Supply	2.40	$0.40	3.99	$0.66	8.51	$1.42
1000 W magnetic	Sunlight Supply	0.92	$0.33	1.71	$0.61	3.60	$1.30
1000 W magnetic	PARsource GLXI	0.86	$0.31	1.31	$0.47	2.82	$1.01
1000 W electronic	PARsource GLXI	0.75	$0.28	1.14	$0.42	2.49	$0.92
1000 W electronic	PARsource GLXII	0.75	$0.27	1.33	$0.47	2.81	$1.00
1000 W electronic	Gavita	0.57	$0.23	0.96	$0.35	2.12	$0.84
1000 W electronic	ePapillon	0.57	$0.24	1.46	$0.61	3.47	$1.45
LED							
red/blue	LSG	1.53	$0.54	1.62	$0.57	2.03	$0.71
red/white	BML	1.85	$0.54	2.13	$0.62	3.17	$0.93
red/white	LSG	1.58	$0.55	1.67	$0.59	2.09	$0.73
red/white	Illumitex	2.56	$0.92	2.66	$0.96	3.92	$1.37
red/white/blue	Lumigrow (Pro 325)	2.58	$0.73	3.05	$0.87	4.95	$1.42
red/white	California Lightworks	2.85	$0.85	3.09	$0.92	4.92	$1.46
multiple	Black Dog	2.95	$0.85	4.43	$1.27	8.64	$2.48
red/white	Apache	6.14	$1.35	6.58	$1.45	8.21	$1.81
red/blue	Lumigrow (ES330)	2.64	$1.01	2.82	$1.07	4.33	$1.65
red/white	Hydrogrow	3.52	$1.16	5.05	$1.67	10.70	$3.54
Ceramic Metal Halide							
315 W 3100 K	Cycloptics	2.04	$0.46	5.43	$1.22	19.55	$4.38
315 W 4200 K	Cycloptics	2.14	$0.48	5.72	$1.29	20.71	$4.66
2@315 W 3100 K	Boulderlamp	1.22	$0.47	1.56	$0.60	2.90	$1.12
Fluorescent							
400 W induction	iGrow	2.68	$0.94	4.69	$1.65	10.17	$3.58
60 W	T8	20.77	$0.51	38.03	$0.93	83.81	$2.05

[x]- See Table S1 for a list of fixture manufacturers and model numbers.
[y]-The number of fixtures to get 1 millimol (1000 µmol) of photons per second.
[*]-Cost of fixture (multiplied by fixtures needed) plus cost of electricity over 5 years. We used a discounted cash flow model assuming a 5% per year cost of capital. Installation and maintenance costs were assumed to be similar for all lamp types and were not included in this calculation.

DISCUSSION

Importance of photon capture

As reviewed in the introduction, lighting system efficiency is the combined effect of efficient fixtures and efficient canopy photon capture efficiency. Precision luminaires, lenses (e.g. model vivid white, Lighting Sciences Group inc.), or adjustable angle LEDs (e.g. model SPYDR 600, BML inc.) can be used to apply highly focused lighting specifically to the plant growth areas. This is valuable in small greenhouses with widely spaced benches. Canopy photon capture efficiency can be maximized, to above 90%, for large greenhouses with narrow aisles regardless of fixture type. The use of LED intracanopy lighting can increase capture rates to near 100%, and may have other beneficial effects such as increased light sharing with intracanopy leaves [19], [20]. The concentration of heat from HID fixtures makes intracanopy lighting infeasible with high wattage HPS fixtures. Just as precision irrigation can improve water efficiency, precision lighting can improve electrical efficiency.

Effect of fixture shadow

All fixtures block radiation from the sun, and the shadow is proportional to the size of the fixture. For the same photon output, 400-W HPS, ceramic metal halide, fluorescent, and LED fixtures block significantly more sunlight than 1000-W HPS fixtures. We did not include the effect of the shadow in this analysis, but this effect significantly favors the more energy dense, higher wattage HPS fixtures. In the long-term, LEDs can take advantage of innovative design options like mounting along greenhouse support structures, which could provide light without extra shading. Longer, narrower LED fixtures may be preferable to rectangular fixtures because the duration of the shadow is shorter. Fluorescent fixtures, including induction fluorescent, have large shadows relative to their photon output (and have low photon efficiencies) and are therefore generally not economical for greenhouse lighting.

Installation, annual maintenance costs, and life expectancy

Installation costs include wiring for fixtures and physically hanging the fixture. In our experience, the cost of installation is similar for both fixture types, although installation costs can be reduced by fewer, higher wattage fixtures. The annual maintenance costs are small relative to the cost of the electricity, and these costs are better established for HPS fixtures than for

LED fixtures. Maintenance costs are largely determined by the life expectancy of the fixture.

Double-ended HPS lamps (1000-W) have a life expectancy of 10,000 hours to 90% survival (based on manufacturer literature), or 3.3 years when used an average of 8 hours per day or 3,000 hours per year (traditional mogul-base lamps have industry reported life expectancies of 10,000 to 17,000 hours, to 90% survival, and cost approximately $40). The cost of a 1000-W, double-ended replacement lamp is about $140, which averages to $28 per year if we assume a lamp will be replaced once in the first five years. This lamp replacement cost can increase to $30 to $35 per year when the labor to replace the bulb is included, but this is a small amount compared to the approximately $600 per year annual electric cost to operate the fixture. Adding the cost of lamp replacement increases the five-year cost of operation by approximately 5%.

When operated at favorable temperatures, individual LEDs generally have a predicted lifetime (to 70% of the initial light output) of up to 50,000 hours, about 16.7 years when used an average 8 hours per day or 3000 hours per year. The economic life for LED fixtures for plant lighting has not been established, but it depends on the value of the product being produced. The high capital cost of replacement means that LED fixtures would be operated longer, in spite of diminished photon output. Replacement of individual LEDs is more expensive than replacing an HID lamp. The life expectancy of LEDs is reduced if they are driven by higher amperage to achieve a higher output, or exposed to high temperatures. Fixtures may be warmed by radiation from sunlight. The cooler the LED temperature, the longer they last. Power supplies, fans, and other components in LED fixtures can fail well before the LEDs themselves. Fan failure would increase LED temperature and may not be immediately noticed by the user. These components are replaceable, but the labor costs to change fixture components increases operating costs.

For these reasons we have not included a differential operating cost between LED and HPS fixtures. We assumed that maintenance costs will be minimal during the first five years for all types of fixtures. Electronic ballasts for 1000-W HPS lamps are still a relatively new technology, and fixtures vary in quality. We have experienced premature failures of LED power supplies, LED circuit boards, HPS lamps, and electronic HPS ballasts in our greenhouse operations. LED fixtures with improved power supplies and optimized operating amperages are available from reputable manufacturers. Improvements in these new technologies are occurring rapidly.

Importance of PPF uniformity

PPF uniformity is critical in many greenhouse applications, especially in floriculture. It is easier to achieve uniformity with fixtures that have broad distribution of photons. Economically, the value of uniform plants may outweigh the cost of wasted photons. Uniformity has been well characterized and modeled with HID lights [21], [22], but these techniques have not yet been rigorously applied to LED fixtures. Ciolkosz et al. [23] showed that uniform light on the perimeter of a greenhouse requires higher fixture densities in the outer rows, and consequentially may increase the amount of radiation lost beyond the edge of the growing area, decreasing canopy photon capture. HPS fixtures with narrower focus luminaires tend to have lower photon efficiencies.

Effect of fixture efficiency on heating and cooling costs

Improved electrical efficiency reduces the cooling load in a greenhouse, which increases the value of efficient fixtures when cooling is required. The best HPS and LED fixtures have nearly identical efficiency, so cooling costs are similar for both fixture categories. The ability to cycle LED fixtures, which prematurely ages other fixture types, could be used to stabilize the heating and cooling load in a greenhouse during partly cloudy days, which could improve temperature control and increase the lifetime of cooling system equipment.

Additional thermal radiation is useful in warming the plant canopy during the heating season, but is detrimental if the canopy is too warm. When sunlight supplies adequate PPF, supplemental lighting is usually turned off.

CONCLUSIONS

The most efficient HPS and LED fixtures have equal efficiencies, but the initial capital cost per photon delivered from LED fixtures is five to ten times higher than HPS fixtures. The high capital cost means that the five-year cost of LED fixtures is more than double that of HPS fixtures. If widely spaced benches are a necessary part of a production system, LED fixtures can provide precision delivery of photons and our data indicate that they can be a more cost effective option for supplemental greenhouse lighting.

Manufacturers are working to improve all types of lighting technologies and the cost per photon will likely continue to decrease as new technologies, reduced prices, and improved reliability become available.

ACKNOWLEDGMENTS

Disclaimer: Mention of products or vendors does not imply endorsement by Utah State University to the exclusion of other products or vendors that also may be suitable.

We thank Peter Nelson and Alec Hay for their dedicated technical work, A.J. Both and Erik Runkle for conscientious technical review, and Paul Jakus and John F. Burr for analysis of the economics.

AUTHOR CONTRIBUTIONS

Conceived and designed the experiments: JAN BB. Performed the experiments: JAN. Analyzed the data: JAN. Contributed reagents/materials/analysis tools: JAN BB. Contributed to the writing of the manuscript: JAN BB.

REFERENCES

1. Bourget CM (2008) An introduction to light-emitting diodes. HortScience 43: 1944–1946.
2. Morrow RC (2008) LED lighting in horticulture. HortScience 43: 1947–1950.
3. Inada K (1976) Action spectra for photosynthesis in higher plants. Plant Cell Physiol 17: 355–365.
4. McCree KJ (1972) The action spectrum, absorptance and quantum yield of photosynthesis in crop plants. Agric Meteorol 9: 191–216.
5. Nelson JA, Bugbee B (2013) Spectral characteristics of lamp types for plant biology West Lafayette, IN. Available: http://cpl.usu.edu/files/publications /poster /pub_6740181.pdf. Accessed 2014 MAr 27.
6. Cope KR, Snowden MC, Bugbee B (2014) Photobiological Interactions of Blue Light and Photosynthetic Photon Flux: Effects of Monochromatic and Broad-Spectrum Light Sources. Photochem Photobio.
7. Johkan M, Shoji K, Goto F, Hahida S, Yoshihara T (2012) Effect of green light wavelength and intensity on photomorphogenesis and photosynthesis in Lactuca sativa. Environ Exp Bot 75: 128–133
8. Dougher TA, Bugbee B (2004) Long-term blue light effects on the histology of lettuce and soybean leaves and stems. J Am Soc Hortic Sci 129: 467–472.
9. Cope KR, Bugbee B (2013) Spectral effects of three types of white light-emitting diodes on plant growth and development: Absolute versus relative amounts of blue light. HortScience 48: 504–509.
10. Dougher TA, Bugbee B (2001) Differences in the Response of Wheat, Soybean and Lettuce to Reduced Blue Radiation¶. Photochem Photobiol 73: 199–207.
11. Yorio NC, Goins GD, Kagie HR, Wheeler RM, Sager JC (2001) Improving spinach, radish, and lettuce growth under red light-emitting diodes (LEDs) with blue light supplementation. HortScience 36: 380–383.

12. Yang Z-C, Kubota C, Chia P-L, Kacira M (2012) Effect of end-of-day far-red light from a movable LED fixture on squash rootstock hypocotyl elongation. Sci Hortic 136: 81–86
13. Massa GD, Kim H-H, Wheeler RM, Mitchell CA (2008) Plant productivity in response to LED lighting. HortScience 43: 1951–1956.
14. Morrow RC, Tibbitts TW (1988) Evidence for involvement of phytochrome in tumor development on plants. Plant Physiol 88: 1110–1114.
15. Craig DS, Runkle ES (2013) A Moderate to High Red to Far-red Light Ratio from Light-emitting Diodes Controls Flowering of Short-day Plants. J Am Soc Hortic Sci 138: 167–172.
16. Kim H-H, Goins GD, Wheeler RM, Sager JC (2004) Green-light supplementation for enhanced lettuce growth under red-and blue-light-emitting diodes. HortScience 39: 1617–1622.
17. IESNA Testing Procedures Committee (2008) IES Approved Method for the Electrical and Photometric Measurements of Solid-state Lighting Products. Illuminating Engineering Society.
18. Blonquist M, Bugbee B, (2013) Analysis of spectral and cosine errors in quantum sensors West Lafayette, IN. Available: http://www.apogeeinstruments.com /content/Quantum-Sensor-Poster-Park-City-April-2009.pdf. Accessed 2014 Mar 27.
19. Frantz JM, Joly RJ, Mitchell CA (2000) Intracanopy lighting influences radiation capture, productivity, and leaf senescence in cowpea canopies. J Am Soc Hortic Sci 125: 694–701.
20. Gómez C, Morrow RC, Bourget CM, Massa GD, Mitchell CA (2013) Comparison of intracanopy light-emitting diode towers and overhead high-pressure sodium lamps for supplemental lighting of greenhouse-grown tomatoes. HortTechnology 23: 93–98.
21. Both AJ, Ciolkosz DE, Albright LD (2000) Evaluation of light uniformity underneath supplemental lighting systems. Acta Hort ISHS: 183–190.
22. Ferentinos KP, Albright LD (2005) Optimal design of plant lighting system by genetic algorithms. Eng Appl Artif Intell 18: 473–484
23. Ciolkosz DE, Both AJ, Albright LD (2001) Selection and placement of greenhouse luminaires for uniformity. Appl Eng Agric 17: 875–882.

CHAPTER 3

Technology and Studies for Greenhouse Cooling

Hugang Li, Shuangxi Wang*

Engineering College, Shanxi Agricultural University, Taigu, China

ABSTRACT

Greenhouse technology is an efficient and viable option, especially for the sustainable crop production in the regions of adverse climatic conditions. High summer temperature is one of the worst effects on greenhouse crop production throughout the year. The main purpose of this paper is to present some technologies and studies for greenhouse cooling in summer. In the paper, some applicable and practical cooling technologies have been discussed. The choice of efficient cooling method depends on many aspects, such as local climate, agronomic condition, design and covering materials. To achieve desirable benefits, the combination of different cooling methods is necessarily used. Analysis of earlier studies revealed that a naturally ventilated greenhouse with larger ventilation areas (15% - 30%), provided at the ridge and side covered with insect-proof nets of 20 - 40 mesh size with covering material properties of NIR (near infrared radiation) reflection during the day and FIR (far infrared radiation) reflection during night was suitable for greenhouse production throughout year in some special regions. Evaporation cooling is the most effective cooling method for controlling the temperature and humidity inside a greenhouse. However, its suitability is restricted to the respective region and climate when the humidity level is high. The entry of un- wanted radiation or light can be controlled by the use of shading. Researches show that shade net application with different perforated mesh size and their evaluation with respect to local climate and region are necessary to get cooling benefits in summer.

Keywords: Greenhouse Cooling Technologies and Studies, Natural Ventilation, Evaporation Cooling, Shading

1. INTRODUCTION

Greenhouse technology is a breakthrough in the agriculture production technology, it integrates market driven quality parameters with production system profits. Although greenhouse protects crops from external bad weather, high temperature and humidity during summer months cause adverse effect on crop production. Thus, in such regions, reduction of air temperature inside the greenhouse or the regulation of temperature close to the ambient temperature during summer is necessary for successful crop production.

Cooling is considered as the basic necessity for greenhouse crop production to overcome the problems of high temperature during summer months. Development of suitable cooling system that provides optimized micro climate for crop growth is a difficult task as the design is closely related to the local environmental condition. Besides, the selection of appropriate technology for cooling depends on the crop to be grown, maintenance, ease of operation and economic viability [1]. Hence, understanding of greenhouse, its size, shape, covering material and external weather helps in the development of suitable cooling system. Evaluating micro climate in different designs of the greenhouse and establishing physical and physiological relationship of crops is necessary for greenhouse designers to improve cooling system that is suitable for crop growth.

Earlier studies on greenhouse cooling revealed that the present techniques used for cooling were not satisfactory. There is a necessity to find suitable cooling method for greenhouse cooling in peak summer. Thus, development of applicable cooling technologies is an important research endeavor. And appropriate greenhouse design (shape, size and roof configuration), right choice of cladding material coupled with suitable ventilation and cooling techniques address the problems of summer greenhouse production of high temperature regions. So, there is a necessity to carry out in deep investigation on the design aspects of greenhouse and its functional characteristics influence on micro climate. Established cooling technologies like natural ventilation, evaporation cooling, shading and composite systems and their application implications have been discussed with a note in greenhouse technologies. And the information is useful for the researchers' work on the engineering aspects of greenhouse technology.

2. GREENHOUSE COOLING TECHNOLOGIES

2.1. Natural Ventilation

Natural ventilation is the cheapest method of cooling a greenhouse. And its driven force is the pressure differences, which is created and maintained by wind or temperature gradients. It depends on the crop. In this section, research on greenhouse natural ventilation application to high temperature climate in summer has been reviewed and presented in detail. Wang Shuangxi and Wang Xu [2] carried out the experimental studies in plastic covered muti-span greenhouse. In north china, they investigate the ventilation rate for the performances of various vents in it. And found that the colling-up vent had a higher ventilation rate (up to 2.32 times) than the continuous roof one. The better ventilation (more than 60% at the wind speed of 3.0 m/s) could be achieved with vents near the ridge of the roof instead of near the gutter, and the obvious reduction (18% - 22% at the wind speed of 3.0 m/s) in the rate values when an insect-proof screen was fitted to it. Teitel and Tanny [3] carried out theoretical and experimental studies in a four-span greenhouse of 960 m^2, floor area in the south of Isreal.

Continuous vertical roof windows resulted in the decrease of air temperature and humidity ratio with time and approached to a steady state. Teitel et al. [4] studied the effect of continuous screened side vents on the ventilation rate, temperature, humidity and air velocity distributions in a naturally ventilated greenhouse. Results indicated that the ventilation rate increased linearly with wind speed. The temperature and humidity ratios were larger near the roof than near the crop while the air velocity was higher near the crop than near the roof. Teitel et al. [5] investigated the air flow patterns and air temperature distributions in a naturally ventilation greenhouse with vertical roof openings using computational fluid dynamics technique. The results show significant effect of the flow patterns, both inside and at the roof openings. Wind direction, significantly affected the ventilation rate, airflow and crop temperature distributions. Measured ventilation rates are in reasonable agreement with estimated ventilation rates predicted by a model. However, air velocities at the greenhouse openings differed from the measured values. Impron et al. [6] developed a simple model to predict the micro climate of greenhouse appropriately, optimizing the cover properties and ventilation as main parameters for a naturally ventilated greenhouse under tropical low lands of Indonesia. The measured and predicted values are in good agreement.

From the available review on natural ventilation as means of cooling, it can be concluded that wind and buoyancy effect are the main controlling agents for getting a favorable micro climate and air exchange rate in tropics and subtropics. Buoyancy effect cannot be neglected completely, though its effect on natural ventilation is not prominent when external wind speed exceeds 2 - 2.5 m/s. Relative larger ventilation openings provided at the roof and ridge can realize acceptable natural ventilation cooling in tropical and subtropical regions.

2.2. Evaporation Cooling

Evaporation cooling is the most effective cooling method for controlling the temperature and humidity inside a greenhouse. However, its suitability is restricted to the respective region and climate as humid tropics seldom suits for its application due to high humidity levels.

2.2.1. Fan-Pad System

Kozai and Sase [7] and Landsberg et al [8] studied the efficiency of fan-pad system in subtropical region. They reported that for a freely transpiring crop in the glasshouse, inside air temperatures can be reduced by 8°C - 12°C under high ambient temperatures and radiation intensities. Chandra et al. [9] conducted an experimental study in a 24 m^2 plastic covered greenhouse using negative pressure fan and pad system for cooling. Landsberge's model was used to predict air temperature inside greenhouse. Decrease in inside air temperature to a range of 4°C - 5°C compared to outside condition was achieved. Jain and Tiwari [10] conducted theoretical experimental studies of fan-pad evaporation cooling system in a 24 m^2 greenhouse. They reported that the greenhouse air temperature was lowered by 4°C - 5°C against ambient condition by using fan and pad system. The results of the parametric studies to optimize the fan-pad cooling parameters revealed that length of greenhouse and height of the cooling pad were found sensitive to cooling. Ahmed et al. [11] conducted experimental studies of three different types of local evaporation cooling pads on effective growth factors, such as temperature, relative humidity, the results obtained for crop parameters showed that there were significant differences between all parameters inside the greenhouses; however, the greenhouses with sliced wood pads gave the highest yield and the greenhouses with straw pads gave the least and conditions outside gave the lowest. This study indicated that the sliced wood pads are better than the other evaporation cooling pads.

2.2.2. Fog/Mist System

This system is based on spraving water a small droplets (droplet diameter of 2 - 60 μm) with high pressure nozzles. Cooling is achieved by evaporation of droplet. Meanwhile, fogging can also be used to increase the relative humidity apart from cooling the greenhouse.

Montero et al. [12] used an air water fogging system to cool the greenhouse with shade screen of 45% perforations. It was reported that maximum temperature reduction during sunny days was 5°C. Arbel et al. [13] tested the efficiency of the fog system with a droplet size of 2 - 60 μm in 16 m × 24 m greenhouse under Isreal climatic conditions. The results were compared with fan and pad system. They concluded that performance of fog system was better than fan-pad system as temperature and relative humidity variation were <5°C and 20%, respectively. A combination of forced ventilation and fogging system for cooling greenhouse was presented by Arbel et al. [14]. High pressure nozzles of more uniformity distribution coupled with fans placed at both ends were being the treatments imposed to achieve the required cooling demand. It was reported that air temperature and relative humidity of 28°C and 80%, respectively were maintained during the summer at midday with the combination of forced ventilation and fogging.

2.2.3. Roof Evaporation Cooling

Roof evaporation cooling is sprinkling of water onto a surface of the roof so as to form a thin layer of the free water surface causing increase in the evaporation rate, and to fall to the wet bulb temperature of the closely surrounded air.

Cohen et al. [15] experimentally investigated the cooling efficiency of greenhouse by wetting the outer roof and inner crop soil crop soil surfaces where tomatoes were grown. They reported that wetting of roof had a smaller effect on reducing air and canopy temperature than wetting the canopy. However, combination of both wetting treatments reduced inside air temperature by about 5°C and canopy temperature by nearly 7°C below the ambient condition. Sutar and Tiwari studied the effect of flowing water film over the roof on greenhouse air temperature in a low cost plastic greenhouse of Delhi climatic conditions, India. They concluded that inside air temperature was observed to be 4°C - 5°C lower than control greenhouse, however, when a shad cloth on the roof of the greenhouse was used with water film, inside air temperature drop was increased to 10°C. Willits and Preet [17] conducted an experiment with intermittent application of water over externally mounted shad net on greenhouse. The results revealed that rise of greenhouse air temperature was reduced by 41% under wet cloth as compared to 18% under dry cloth.

2.3. Shading

The entry of unwanted radiation or light can be controlled by the use of shading. Shading does not allow important photo-synthetically active radiation (PAR), which is crucial for photosynthesis. However, it is a viable option for crop production during peak summer. Shading paints can be applied to the outside greenhouse cover with some degree of success but paints get washed away during the rainy season. Shade net application with different perforated mesh size and their evaluation with respect to local climate and region is necessary to get cooling benefits in summer. Ali et al. [18] developed an efficient greenhouse design for hot climatic conditions of Kuwait. Entry of solar radiation was allowed only from the roof of the even span greenhouse of 250 m^2 floor area where tomatoes were grown. The roof was used during peak hours to control the inside air temperature. The results showed that combination of shading and fan and pad during peak summer maintained temperature of 30°C during the day and 22°C during night in greenhouse. Hu [19] experimentally investigated the cooling effect of external sun-shading. Two kinds of methods is tested: one is under the condition of closing the greenhouse, unfolding the external sun-shading to collect temperature and humidity date indoors and outdoors. Another one is to open the top window to ventilate to make the indoor and outdoor temperature and humidity in a relatively steady state, and then, unfold the external sun-shading and collect the temperature and humidity data indoors and outdoors. The result showed that, as the sun-radiation enhancing, temperature indoor presents a rise tendency, causing the effect of the external sun-shading, the humidity indoors and outdoors maintain 57% basically. In the subsequent time, both of them present alternate change trend, but the difference is small.

3. CONCLUSION

Test and investigation are respectively conducted with these three cooling measures of natural ventilation, evaporate-cooling and shading cooling. There are some respectively differences among them. All of the methods have disadvantages and advantages. Choosing applicable and practical cooling methods depends on many aspects, such as local climate, agronomic condition, design and covering materials. To achieve desirable benefits, the combination of different cooling methods is necessarily used.

REFERENCES

1. Sethi, V.P. and Sharma, S.K. (2007) Experimental and Economic Study of a Greenhouse Thermal Control System Using Aquifer Water. Energy Conversion and Management, 48, 306-319.
2. Wang S.X. and Wang, X. (2009) Ventilation Rate of Various Vents in Plastic Covered Multi-Span Greenhouse. Transaction of the CSAE, 25, 248-251.
3. Teitel, M. and Tanny, J. (1999) Natural Ventilation of Greenhouses: Experiments and Model. Agricultural and Forest Meteorology, 96, 59-71.
4. Teitel, M., Liran, O., Tanny, J. and Barak, M. (2008) Wind Driven Ventilation of a Mono-Span Greenhouse with a Rose Crop and Continuous Screened Side Vents and Its Effect on Flow Patterns and Micro Climate. Biosystems Engineering, 101, 111-122.
5. Teitel, M., Tanny, J. and Barak, M. (2007) Wind Driven Ventilation of a Mono-Span Greenhouse with a Rose Crop and Continuous Screened Side Vents and Its Effect on Flow Patterns and Micro Climate. Biosystems Engineering, 101, 111-122.
6. Impron, I., Hemming, S. and Bot, G.P.A. (2007) Simple Greenhouse Climate Model as a Design Tool for Greenhouses in Tropical Low Land. Biosystems Engineering, 98, 79-89.
7. Kozai, T. and Sase, S. (1978) A Simulation of Natural Ventilation for a Multi-Span Green-house. Acta Horticulturae, 87, 39-49.
8. Landsberg, J.J., White, B. and Thorpe, M.R. (1979) Computer Analysis of the Efficacy of Evaporative Cooling for Glasshouse in High Energy Environments. Journal of Agricultural Engineering Research, 24, 29-39.
9. Chandra, P., Singh, J.K. and Majumdar, G. (1989) Some Re-sults of Evaporative Cooling of a Plastic Greenhouse. Journal of Agricultural Engineering, 26, 274-280.
10. Jain, D. and Tiwari, G.N. (2002) Modeling and Optimal Design of Evaporative Cooling System in Controlled Environment Greenhouse. Energy Conversion and Management, 43, 2235-2250.
11. Ahmed, E.M., Abaas, O., Ahmed, M. and Ismail, M.R. (2011) Performance Evaluation of Three Different Types of Local Evaporative Cooling Pads in Greenhouses in Sudan. Saudi Journal of Biological Sciences, 18, 45-51
12. Montero, J.I., Anton, A., Beil, A. and Franquet, C. (1994) Cooling of Greenhouse with Compressed air Fogging Nozzles. Acta Horticulturae, 281,199-209.
13. Arbel, A., Yekutieli, O. and Barak, M. (1999) Performance of a Fog System for Cooling Greenhouses. Journal of Agricultural Engineering Research, 72,129-136.
14. Arbel, A., Barak, M. and Shklyar, A. (2003) Combination of Forced Ventilation and Fogging Systems for Cooling Greenhouses. Biosystems Engineering, 84, 45-55.
15. Cohen, Y., Stanhill, G. and Fuchs, M. (1983) An Experi-mental Comparison of Evaporative Cooling in a Naturally Ventilated Glasshouse Due to Wetting the Outer Roof and Inner Crop Soil Surfaces. Agricultural and Forest Meteorology, 28, 239-251.
16. Sutar. and Tiwari (1995) Analytical and Numerical Study of a Controlled Environment Agricultural System for Hot and Dry Climatic Conditions. Energy and Building, 23, 9-18.
17. Willits, D.H. and Preet, M.M. (2000) Intermittent Application of Water to an Externally Mounted Greenhouse Shade Cloth to an Modify Cooling Performance. Transactions of ASAE, 43, 1247-1252.
18. Ali, H.M., Moustafa, S. and El-Mansy, H. (1990) An Efficient Greenhouse Design for Hot Climates. Energy Conversion Management, 30, 433-437.
19. Hu, J. (2013) Experimental Research on Monomial Cooling Measure of Greenhouse in Summer. Smart Grid and Renewable Energy, 4, 48-52.

CHAPTER 4

The Choice between Economic Policies to Face Greenhouse Consequences

Donatella Porrini[1]

[1] *Università del Salento, Lecce, Italia*

ABSTRACT

In the past few years, unstable and extreme weather patterns are increasingly occurring as phenomena of climate change, and the link to greenhouse gas (GHG) emissions is scientifically accepted. From an economic point of view, extreme weather patterns are causing major damage to health, property, and business.In this chapter, following an economic analysis of law (EAL) approach, the issue of the comparison between the alternative environmental economic policies is analyzed starting from the consideration that the emissions of GHGs originate market failures: the environment appears as a "public good" that may not be appropriated and has no market price; the damage to the environment is a case of "externality," where it is fully or partly a social cost that is not internalized into the accounts of the parties causing it. In the EAL literature, an environmental policy instrument has been seen as it may play a role in correcting malfunction and subsequent inefficiencies.In the first part of the chapter, we intend to revise the traditional analysis of the choice of environmental policies. The following part deals with the comparison between tax and tradable permit systems. Then the role that can be played by the insurance sector is considered. The different policy instruments are considered in the framework of climate as an economic global public good. And, finally, some conclusive remarks are presented in relation to the COP 21 conference in Paris in terms of the future policies against GHG effects.

Keywords: climate change, environmental policy choice, insurance, greenhouse gasses, COP 21

1. INTRODUCTION

Over the last centuries, climate change has been raised as a very important issue all over the world. The change in climate results from an increase in the earth's average atmospheric temperature, which is usually referred to as global warming. It may be due to both natural and human causes, especially greenhouse gas (GHG) emissions.

In response to increasing scientific evidence that human activities are contributing significantly to global climate change [1], decision makers are devoting considerable attention to public policies to reduce GHG emissions and thereby prevent or reduce such change.

The policies span a range of regulatory approaches. The main alternative is between command and control (CAC) and market-based (MB) instruments, and a relevant role can also be played by the insurance sector.

This chapter aims to describe the traditional theory on the choice of environmental policies following an economic analysis of law (EAL) approach (Section 2) to analyze the comparison between tax and tradable permit systems (Section 3), to outline the role of the insurance sector (Section 4), and to consider the different policy instruments in a context of economic global public goods. The final objective is to take into account the future COP 21 conference in Paris in terms of the choice of policy instruments against GHG effects.

2. THE TRADITIONAL ISSUE OF THE CHOICE OF ENVIRONMENTAL POLICIES

The problem of the choice of environmental policy instruments has been an issue since Pigou [2] analyzed the need for state intervention when private costs diverge from social costs and suggested that the solution would be to internalize the externalities through taxation.[1] - Coase [4] criticized the proposed state intervention and affirmed that there is no reason to suppose that governmental regulation is called for simply because the problem is not very well handled by the market or the firm.

The ensuing debate has been conducted along these two opposite views: on the one hand, the supporters of the idea that the choice of policy instruments is to be applied as a public matter and the state, as policy designer, should select the optimal instruments and take responsibility for their imposition in the public interest, but, on the other, the supporters of MB instruments are trying to fight a battle against a sort of "anti-market" mentality based on a reluctance to apply MB instruments [5].

So the problem would be to compare the instruments that can be considered public oriented and those that can be considered market oriented, where the former is characterized by a public agency that defines a conduct rule and provides an enforcement system and the latter is characterized by instruments based on market mechanisms stimulating the conduct of the firm indirectly and by a private administration and enforcement system.

Following an EAL approach, traditionally regulatory systems originate from the presence of market failure: in our specific case, the environment appears as a "public good" that may not be appropriated and has no market price; the damage to the environment is a case of "externality," in that it is fully or partly a social cost that is not internalized into the accounts of the parties causing it.[2] - So the comparison of different instruments can consider how they may play a role in correcting malfunction and subsequent inefficiencies [7].

In this way, we can move from the theoretical definition of the efficiency of different instruments to their practical, and so direct, potential to achieve concrete objectives. In particular, the objectives that emerge as relevant in judging the practical efficiency of environmental policies are the following: first is the prevention in the sense of providing incentives for firms to improve safety standards, and second is the connection with technological change in the sense of encouraging firms to adopt lower risk technologies.

The first kind of environmental instrument is the so-called CAC that is characterized by a public agency that provides a definition of conduct rules and enforcement system. Thus, they could be defined as public-oriented instruments, which require the use of a particular technology or the observation of a performance standard, authorizing for the maximum amount that a source can emit.

CAC is divided into two phases as follows: "command", which sets a standard based on the maximum level of permissible emissions, and "control", which monitors and enforces the standard.

Regarding standards, they can be classified into two: ambient standards, which fix a minimum desired level of air or water quality or a maximum level of emissions that must be maintained, and emission standards, which fix a maximum level of permitted emissions that can be performance based, setting emission limits that each firm is allowed, or technology based, specifying the best technology to be used.[3] -

As to the US experience with this kind of regulation, the activity of the Environmental Protection Agency (EPA) provides a clear example of regulation by an independent environmental authority. This agency performs its tasks through the setting of preventive standards, the enforcement of polluting emission thresholds, and the performance of inspections and, possibly, of actions brought to the federal courts. We cannot mention the European Community experience given that a standard setting system has not been established at a European level (but at national level) and that the European Environmental Agency (EEA) has only a very limited role.[4] -

The choice to develop a CAC regulatory system is based on the advantage of centralized agencies to assure a cost-effectiveness calculation on the base of the expected damage and of the marginal cost of different level of preventive care. The centralized structure presents the advantage to provide a continual oversight of problems and a broad array of regulatory tools.

Following the traditional EAL approach, well-defined standards generate the correct incentive for the firm to act with caution and make the best production and prevention decisions [8, 9].

CAC instruments use to be compared with MB instruments that are characterized by a private administration and enforcement system and stimulate indirectly the behavior of the firm. There are essentially two different types of those instruments: taxes that are fees imposed on emitters proportionate to the total amount of emissions released into the environment (they could be divided into emission charges, product charges, and user charges) and marketable (or tradable) permit systems that provide a fixed number of permits equal to the allowed total emissions, distributing them among polluting firms in a specific area.

The two types of MBI instruments can be seen in the following two different approaches: on one side, taxes follow a price approach because producers adjust the quantity, given a fixed price put on emissions; on the other side, a tradable permit system follows a quantity approach because the price is adjusted according to supply and demand, given a maximum quantity of emissions allowed.

3. CARBON TAX VERSUS TRADABLE PERMIT SYSTEM

A carbon tax is a particular tax based on GHG emissions generated by burning fuels and biofuels, such as coal, oil, and natural gas. It has been introduced with the main goal to level the gap between carbon intensive (firms based on fossil fuels) and low carbon intensive (firms that adopt renewable energies) sectors. Due to the introduction of that form of taxation, the relative prices of goods and services will change; the emissions of intensive goods will be more expensive, whereas the emissions of less intensive goods will be lower. Thus, carbon tax provides a strong incentive for individuals and firms to adjust their conduct, resulting in a reduction of the emissions themselves. Hence, by reducing fuel emissions and adopting new technologies, both consumers and businesses can reduce the entire amount they pay in carbon tax.

A tradable permit system is defined as quantity-based environmental policy instrument. The regulatory authority stipulates the allowable total amount of emissions (cap) and the right to emit becomes a tradable commodity. Under a cap-and-trade system, prices are allowed to fluctuate according to market forces. Thus, the price of emissions is established indirectly. Permits could be allocated to firms through auction or free allocation.

Similarly to other environmental taxes, carbon taxes are defined as priced-based policy instruments for the correlated effects to increase the price of certain goods and services, thereby decreasing the quantity demanded. On the other side, tradable permits are defined as quantity-based environmental policy instrument. Although both policy instruments are MB, their implementation is different: carbon taxes fix the marginal cost for carbon emissions and allow quantities emitted to adjust, whereas tradable permits fix the total amount of carbon emitted and allow price levels to change according to market forces.

Which is better? There is no simple yes or no answer, and the policies are not necessarily mutually exclusive. Several important advantages and drawbacks of the respective policies are outlined later.

A well-functioning emission trading system allows emission reductions to take place wherever abatement costs are lowest, regardless of international borders. As costs associated with climate change have no correlation with the origin of carbon emissions, the rationale for this policy approach is that an emission trading system allows to fix a certain environmental outcome and the companies are called to pay a market price for the rights to pollute. This is the reason why an emission trading system is suitable for international environmental agreements, such as the Kyoto Protocol, specifically for the characteristic that a defined emission reduction level can be easily agreed between states.

Emission trading is more appealing to private industry because, by decreasing emissions, firms can actually profit by selling their excess GHG allowances. Creating such a market for pollution could potentially drive emission reductions below targets.

A carbon tax would offer a broader scope for emission reductions [10]. A system of tradable permits entails significant transaction costs, which include search costs, such as fees paid to brokers or exchange institutions to find trading partners; negotiating costs; approval costs; and insurance costs. Conversely, taxes involve little transaction cost over all stages of their lifetime.

Carbon taxes are dynamic economic instruments that offer a continuum incentive to reduce emissions. In fact, technological and procedural improvements and their subsequent efficient diffusion lead to reductions in tax payment. In addition, trading systems are able to self-adjust because emission goals will be easier to meet; there will be a decrease in permits' demand and in their price but not as rapidly as taxes.

The implementation of an emission trading system is very complicated and requires technical steps, including treatment of sinks, monitoring, and enforcement. On the other hand, taxation is a very well-known instrument by policy makers, not very costly because it does not require monitoring and enforcement organization.

The revenue from carbon taxes can be used into the economy to reduce income taxes or levies on labor or capital investment. This may be part of a national or international reform of the taxation systems with the effects to shift the tax burden from "goods" like labor to "bads" like pollution.

In Table 1, we have summarized the main differences among CAC, carbon tax, and tradable permit system.

4. INSURANCE AS AN ENVIRONMENTAL POLICY INSTRUMENT

A relevant role could be played by the insurance sector in the choice of political economic instruments for climate change.

As mentioned in Section 1, following the EAL approach, the emitters of GHGs externalize the true costs of their contribution to climate change, and this implies the need to recover these costs, which manifest through both the costs of impacts and the costs to prevent impacts. Because the insurance sector is the world's largest industry, the response of insurers to the broader climate change challenge will no doubt be the key to, at least partially, solve this internalization problem.

Table 1. Command and control versus carbon tax versus tradable permit system

	Command and control	Carbon tax	Tradable permit system
Certainty over CO_2 price or cost?	No	Yes. The tax establishes a well-defined price	No. But price volatility can be limited by design features, such as a safety valve (price cap) or borrowing
Certainty over emissions?	No. Regulating the rate of emissions, the level uncertain leaves	No. Emissions vary with prevailing energy demand and fuel prices	Yes, in its traditional form. No, with the use of additional cost containment mechanisms
Efficiently encourages least-cost emission reductions?	No, but tradable standard is more efficient than non-tradable standard	Yes	Yes
Ability to raise revenue?	No	Yes. Results in maximum revenue generation compared with other options	Traditionally no
Incentives for R&D in clean technologies?	Yes and no. Standards encourage specific technologies but not Brod innovation	Yes. Stable CO_2 price is needed to induce innovation	Yes. However, uncertainty over permits' prices could weaken innovation incentives
Harm to competiveness?	Somewhat. Regulations increase the cost of manufacturing, but, unlike taxes of tradable permits, do not raise the price of fossil energy	Yes, though if other taxes are reduced through revenue recycling, competitiveness of the broader economy can be improved	Yes (as with a tax), but giving firms free allowances offsets potentially harmful effects on profitability
Practical or political obstacles to implementation?	Yes. Setting the level of standards is difficult	Yes. New taxes have been very unpopular	Yes. Identifying a reasonable allocation and target is difficult
New institutional requirements?	Minimal (unless tradable)	Minimal	Yes, but experience with existing trading programs suggests that markets arise quickly and relatively inexpensively

Source: Parry and Pizer [11].

Generally, insurers can indirectly stimulate prevention behavior in their customers. In the case of climate change, they can play two primary roles through their insurance products. The first role is played in supplying and pricing traditional insurance coverage for damage deriving from climate change by the promotion of actions of businesses and individuals to align policy holders with climate-positive behaviors. The second is played in providing capital to new ventures and reducing the financial risk to investors by facilitate the creation of new markets and services that will help to solve the climate change problems.

In both the roles, insurance companies deal with the management of risk activities. In fact, climate change risk is managed by the insurers through the risks they accept from clients, given that climate change causes an increase in intensity and spreading the distribution of extreme weather events with resultant effects on property claims that could be catastrophic.[6] -

Traditionally, insurance is the main mechanisms available to individuals and business to manage the financial consequences of risky events, such as natural hazards like windstorms and floods. Insurance companies used to work making each individual or business pay a premium to protect themselves against an uncertain loss. The premium is calculated by pooling risks across a large and diverse population considering the pool's expected losses.

In this way, the insurance industry provides a coverage for climate change consequences because climate experts predict changes in the intensity and the distribution of extreme weather events and of the resulting risk of catastrophic property claims.

But specifically, for what concerns the consequences of climate change, many problems arise to provide insurance coverage [13].

First, climate change is connected to global weather patterns that may increase the potential for losses so large. As more severe weather becomes more common and overall variability of conditions increases, there is a threat for the solvency of insurance companies.

Second, because uncertainties in assessing climate change's impacts are high and affect property, casualty, business interruption, and health; as a result, risk has significant ambiguous components and insurers are both likely to charge a significantly higher premium or avoid insuring the risk entirely.

Third, climate change–related risks may be correlated and create a skewed risk pool that could increase the probability of extremely large losses and also risks not well-distributed across existing insurers.

Finally, as a result of insurers' uncertainty aversion and need to protect against extremely large losses from single or related events, it is not clear that insurers will be willing to insure against some climate change–related risks at a price that policy holders are willing to pay.

Climate change will affect, and in some cases is already affecting, most major types of insurance products. Insurers will feel the impact of climate change on property and casualty insurance, where the insurer bears the risk of a loss suffered directly by the policy holder. These property and casualty claims include not only damage to insured property as a direct result of weather but also claims for business interruptions and other consequences of weather-induced events. Also, health and life insurers are going to face increasing costs.

Sometimes, insurance companies are involved in the systems of compensation fund that have been established in some countries. This is the case of special government disaster funds within the target to promote framework of contingency measures to cover the economic costs of natural disasters [14].

Other insurance products are the so-called "financial responsibility" products. This term includes the tools that require companies to demonstrate to have sufficient financial resources for eventual future environmental damage that may arise through their activities. In its common implementation, financial responsibility implies that a production activity is authorized only if companies can prove that they will be able to financially cover economic claims, for example, using surety bonds, cash accounts, deposit certificates, self-insurances, and corporate guarantees.

In the past few years, the insurance industry has developed financial products suitable for dealing with climate change-related risks in the direction to play a role far beyond simply compensating climate change's victims for their losses ex post. The activity of the insurance has become relevant as a political economic instruments within an ex ante strategy to financially manage large-scale catastrophes, as a complement of ex post instruments for the compensation of disaster losses.

Insurance industry is also developing alternative risk transfer products, given that conventional reinsurance arrangements may in future cover a smaller proportion of total losses, and there may be insufficient capital available to insurance markets to cover these losses [15].

A first kind of these insurance products are called "catastrophe bonds" and consist in securitizing environmental risks in bonds, which could be sold to high-yield investors. The catastrophe bonds are able to transfer risk to investors that receive coupons that are normally a reference rate plus an appropriate risk premium. By these products, insurance companies limit risk exposure transferring natural catastrophe risk into the capital markets. In this way, with the involvement of the financial markets, their global size offers enormous potential for insurers to diversify risks.

Weather derivatives are another kind of financial instrument used by companies to hedge against the risk of weather-related losses. Weather derivatives pay out on a specified trigger, for example, temperature over a specified period rather than proof of loss. The investor providing a weather derivative charges the buyer a premium for access to capital. If nothing happens, then the investor makes a profit.

With this kind of financial products, the insurance industry tries to reach two goals. First, there is the need for extra capital and to spread risks beyond the insurance sector. Particularly, cat bonds are used to spread insurance risk in the financial sector. The second goal is to improve the accuracy and the resolution of hazard data and the likely impacts on climate change with the involvement of financial market forecast ability.

The insurance industry can act to tackle the consequences of climate change by playing its part in climate change mitigation, through the promotion of ways to reduce GHG emissions. Insurers are also well placed to help society to adapt to the impacts of climate change, by promoting the effective limitation and management of risks from extreme weather-related hazards.

Individuals and companies that buy insurance products could be stimulated to address climate change-seeking mechanism to facilitate mitigation of GHG emissions and adaptation to the inevitable impacts of climate change [16]. In addition, the insurance companies themselves are motivated to take significant actions to mitigate GHG emissions and increase adaptive capacity to reduce overall uncertainty and other barriers to insurability and are also motivated to limit the insurers' potential exposure to catastrophic risks in excess of their capacity to avoid the potential for property and liability claims in excess.

The insurance industry is also developing new products that would have the consequences to stimulate the adaptation to climate change. On the one hand, products help to create the conditions for active adaptation to build physically resilient communities; on the other hand, products provide capital and liquidity to help communities to cope with losses caused by climate change catastrophes.

Insurance products can be designed in a way for which simply their price reflects the level of climate change-related risk assumed by the insurer. These products, rewarding behavior that reduces risk of financial losses, encourage adaptive behavior. For example, insurance products that incorporate these features provide a premium discounts on property insurance for climate-resilient commercial or residential buildings.

The recent tendency to supply coverage with differential premiums to customers depending on their level of protection from losses caused by weather-related disasters can be seen as a clear opportunity for insurers to reduce their own overall and maximum possible loss exposure while promoting communities overall resilience in the face of climate change's impacts. So, more often, policies include discounts for businesses or homeowners that have taken specific steps to ensure their buildings resistant to floods or other hazards. Moreover, insurance companies also condition their policies on compliance with laws such as building codes, thus playing a role in enforcing laws that promote climate change resilience.

The second kind of insurance products that stimulate adaptation is based on the availability of capital to cope with catastrophes. These are financial arrangements intended to bring needed capital that will reduce the risk that could derive from future climate-related hazards for those who are most likely to be in peril. These products can be defined as adaptation oriented because they help to build the capacity of nations, communities, and businesses to cope with climate change impacts.

Figure 1 shows the different roles that can be played by insurance sector as a form of environmental policy instruments.

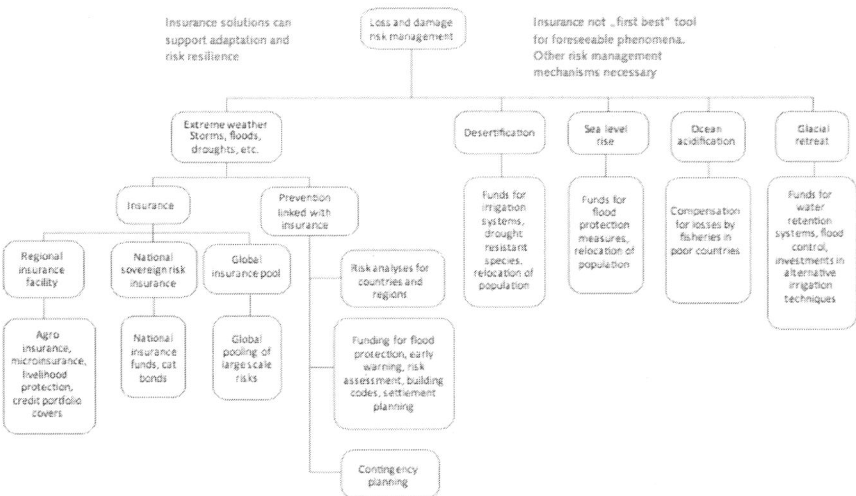

Figure 1. The different roles of the insurance sector in the framework of climate change consequences

5. POLICIES CHOICE IN THE CASE OF CLIMATE CHANGE AND THE CONCEPT OF "ECONOMIC GLOBAL PUBLIC GOODS"

Dealing with the consequences of GHG emissions in terms of climate change implies a definition of "economic global public goods" that can be defined as goods with economic benefits that extend to all countries, people, and generations [18]. Following the EAL approach, they are special case of externalities with a global dimension.

Climate is clearly "global" in both causes and consequences; moreover, the emissions of GHG have effects on global warming independently of their location, and local climatic changes are completely linked with the world climate system.

In addition, global warming is characterized by other important features that imply some difficulties in the implementation of the instruments provided by the standard economic theory of policy choice. First, we cannot determine with certainty both the dimension and the timing of climate change and the costs of the abatement of emissions. Second, the effects of GHG concentration in the atmosphere on climate are intergenerational and persistent across time. Finally, it emerges a relevant equity issue among countries because industrialized countries have produced the majority of GHG emissions, but the effects of global warming will be much more severe on developing countries. In other words, the countries that have more responsibilities will face less consequence in the future and vice versa.

There are major governance issues involved in dealing with global public goods because global coordination is required [19]. With economic public goods, it is difficult to determine and reach agreement on efficient policies because they involve estimating and balancing costs and benefits where neither is easy to measure and both involve major distributional concerns. When dealing with economic public goods like global warming, it is necessary to reach through governments to the multitude of firms and consumers who make the vast number of decisions that affect the ultimate outcomes.

Because global warming is a global public good, the key environmental issue is global emissions, and the key economic issue is how to balance costs and benefits of global emission reductions. Climate change depends only on total GHG emissions and the time path of emissions not on the geographic location of emissions. Moreover, the impacts depend primarily on cumulative emissions that remain in the atmosphere not on the annual flow of emissions.

It is a global issue to decide what the distribution of emission reductions among countries should be and how the costs should be allocated, together with the need for differences among high- and low-income countries, high- and low-emitting countries, and high- and low-vulnerability countries.

Given the global nature of climate change, it is easy to understand the necessity of an action at international level, in order to efficiently implement the different policy instruments that we have analyzed earlier in this chapter.

First, the instruments based on tax mechanism need a method of coordinating policies among countries. In the international environment, it could assume the form of either an international tax or a harmonized domestic tax system. In the case of an international tax, the nations (and not the firms) pay the tax to an international agency, which receives and redistributes the tax revenues. On the other hand, in the case of harmonized domestic tax, the international community should negotiate an agreed level of a domestic emission tax, establishing adequate compensation for the losing countries from the gaining countries.[7] -

Second, it is possible to establish an agreement that sets quantitative limits of emissions and allocates emission permits to firms (or States) but allows to trade among countries, in order to minimize abatement costs. The starting allocation of permits can be set through either an auction or a grandfather allocation. Under an auction, government (or the international community) sells the emission permits, whereas under the grandfather rule, the allocation of emission permits is based on historical records.

In the global-warming context, quantitative limits set targets on the time path of GHG emissions of different countries. Countries then can administer these limits in their own fashion, and the mechanism may allow transfer of emission allowances among countries, as is the case under the Kyoto Protocol.

The European Union Emissions Trading Scheme (EU-ETS) – the world's most extensive carbon pricing market – has now been in operation for 10 years. The EU-ETS was developed as a way of meeting the EU's GHG emission reduction targets in the most efficient and cost-effective manner. To do so, the EU-ETS sets a limit (a cap) on the total emissions; certain EU sectors (mostly heavy industry and aviation) are allowed to use during predefined trading periods. Permits are then distributed among polluters where one permit equals one ton of carbon dioxide equivalent. These permits can then be traded between market participants. As such, the total amount of pollution is set by an external authority, but market participants determine the permit allocation, thereby optimizing efficiency.

In addition, the involvement of the insurance sector as an efficient policy instruments needs a global approach.

Insurance sector can contribute to develop risk management strategy to minimize climate change consequences on an urgent basis to prevent further escalation of global warming. However, insurance needs to be a part of the overall policy of mitigation and adaptation that aims at reducing the severity of many impacts that could result from climate change if current adverse conditions prevail.

In order to organize their own operations to the new challenge, the insurance industry should include climate change risk in its internal governance procedures, in line with the existing financial corporate risk identification. To enable insurance companies to play a responsible role in tackling climate change consequences, they require a reliable, transparent, and international coordinated policy framework as well as long-term, appropriated GHG emission reduction goals coordinated at an international level.

6. FINAL REMARKS: COP 21 AND THE ISSUE OF THE LINKAGE OF DIFFERENT NATIONAL POLICIES

In the COP 21 meeting in Paris at the end of the 2015, global climate policy has faced the tension between the efficiency benefits of uniform global policy and national and regional variation in tastes for differing policies. Although climate negotiations, going back to the framework convention, have had a coordinated global policy as their goal, it could be that we will head toward a less coordinated system of local, national, or regional policies.

In reality, as we have analyzed in the chapter, different countries are undertaking different policies ranging from CAC to MB approaches, such as carbon tax and tradable permit systems.

Variations in policies, although catering to local tastes and preferences, can lead to substantial inefficiencies, and the target will be to reach an optimal degree of policy homogenization.

There is a large literature on the importance of linking economic policies to face GHG consequences [21].

Some authors argue that the basic approach underlying emission reduction credit systems like the Kyoto Clean Development Mechanism (CDM) can be extended to create linkage opportunities in diverse emission control systems in ways that do not necessarily suffer from the shortfalls of the current CDM. Moreover, while emission reduction credit systems are designed to work with MB systems like tradable permits, they describe ways in which it can interact with tax systems as well as certain regulatory systems [22].

By clarifying the opportunities and challenges of insurance as an instrument for adaptation and outlining a practical way forward, it is hoped that this discussion contributes to the opportunities in adopting a comprehensive adaptation strategy that enables risk management and insurance through the funding of a global adaptation strategy [17].

In the near future, the challenge for COP 21 to reduce GHG emissions is to try to link heterogeneous climate policy instruments as a way to reach the solution of climate change issue in the long term.

NOTES

[1] - See the documents of the "Pigou club" [3].
[2] - Economists consider environmental policies within the framework of the category of externalities, as evidenced by Cropper and Oates [6]: "The source of basic economic principles of environmental policy is to be found in the theory of externality."
[3] - As Cropper and Oates [6] explain"...The determination of environmental policy is taken to be a two-step process: first, standards or targets for environmental quality are set, and, second, a regulatory system is designed and put in place to achieve these standards. This is often the environmental decision making proceeds. Under the Clean Air Act, for example, the first task of the EPA was to set standards in the form of maximum permissible concentration of the major air pollutants. The next step was to design a regulatory plan to attain these standards air quality."
[4] - The EEA was set up as a legally independent community body under council regulation (EEC) 1210/90. The EEA's core task is to provide decisionmakers with the information needed for making sound and effective policies to protect the environment and support sustainable development.
[6] - Insurance coverage can be also connected with the liability system. But, for what concerns the consequences of climate change, such as impacts of property damage from extreme weather events, the assignment of liability is a very complex topic. See Reference [12].
[7] - Nordhaus [20] hypothesizes the institution of an harmonized carbon tax (HCM), essentially equivalent to a "dynamic Pigouvian pollution tax for a global public good" and points out 10 different reasons to prefer it to a quantitative approach.

REFERENCES

1. IPCC. Guidelines for national greenhouse gas inventories. Available from www.ipcc-nggip.iges.or.jp/public/2006gl [Accessed: 2015-10-10].
2. Coase R. The problem of social cost. Journal of Law and Economics. 1960;3:1–44.

3. Mankiw G. The Pigou Club Manifesto. Available from http://gregmankiw.blogspot.it/2006/10/pigou-club-manifesto.html [Accessed: 2015-10-10].
4. Pigou AC. The Economics of Welfare. 4th ed. London: McMillan & Co; 1932.
5. Lewis T. Protecting the environment when costs and benefits are privately known. RAND Journal of Economics. 1996;27:819–47.
6. Cropper ML, Oates WE. Environmental economics: A survey. Journal of Economic Literature. 1992;30:675–740.
7. Porrini D. Environmental policies choice as an issue of informational efficiency. In: Backhaus JG, editor. The Elgar Companion to Law and Economics. 2nd ed. Cheltenham: Edwar Elgar; 2005. p. 350–63.
8. Calabresi G. The Cost of Accident. New Haven, CT: Yale University Press; 1970.
9. Shavell S. The Economic Analysis of Accident Law. Cambridge, MA: Harvard University Press; 1987.
10. Kaplow L, Shavell S. On the superiority of corrective taxes to quantity. American Law and Economics Review. 2002;4:1–17. DOI:10.1093/aler/4.1.1
11. Parry WH, Pizer WA. Emissions Trading versus CO2 Taxes versus Standards. Resources for the Future, Issue Brief. 2007;5. p. 80-6.
12. Faure M. Climate Change Liability. Cheltenham: Edwar Elgar Pub; 2011.
13. Charpentier A. Insurability of Climate Risks. The Geneva Papers of Risk and Insurance. 2008;33:91–104. DOI:10.1057/palgrave.gpp.2510155
14. Porrini D, Schwartze R. Insurance models and European climate change policies: An assessment. European Journal of Law and Economics. 2014;38:7–28.
15. Mills E, Lecomte E. From Risk to Opportunity: How Insurers Can Proactively and Profitably Manage Climate Change. Boston, MA: Ceres Report; 2006.
16. Agrawala S, Fankhauser S. Economics Aspects of Adaptation to Climate Change. Paris: OECD; 2008.
17. Munich Climate Insurance Initiative (MCII). Insurance solutions in the context of climate change-related loss and damage. Policy Brief. 2012;6. p. 15.
18. Kaul I, Conceicao P, Le Goulven K, Mendoza RU. How to improve the provision of global public goods. In: UNDP. Providing Global Goods – Managing Globalization. Oxford: Oxford University Press; 2003.
19. Nordhaus DW. To tax or not to tax: Alternative approach to slowing global warming. Review of Environmental Economics and Policy. 2007;1:26–40.
20. Nordhaus DW. After Kyoto: Alternative Mechanisms to Control Global Warming. FPIF Discussion Paper; Washington, DC: Institute for Policy Studies2006.
21. Metcalf GE, Weisbach D. Linking Policies When Tastes Differ: Global Climate Policy in a Heterogeneous World. Harvard: Harvard Kennedy School. Discussion Paper; 2010. p. 10–38.
22. Jaffe J, Ranson M, Stavins R. Linking tradable permit system: A key element of emerging international climate change policy architecture. Ecology Law Quarterly. 2009;36:789–808.

CHAPTER 5

Biodrying under Greenhouse Conditions as Pretreatment for Horticultural Waste

Fabián Robles-Martínez[1], Elizabeth M. Silva-Rodríguez[1], Teodoro Espinosa-Solares[2,3], Belem Piña-Guzmán[1*], Carmen Calixto-Mosqueda[1], Francisco J. Colomer-Mendoza[4], Enrique Durán-Páramo[1]

[1]*Unidad Profesional Interdisciplinaria de Biotecnología, Instituto Politécnico Nacional, Mexico City, Mexico;* [2]*Gus R. Douglass Institute, West Virginia State University, South Charleston, USA;* [3]*Departamento de Ingeniería Agroindustrial, Universidad Autónoma Chapingo, Texcoco, Mexico;* [4]*Departamento de Ingeniería Mecánica y Construcción, Universitat Jaume I, Castellon, Spain.*

ABSTRACT

The biodrying process was studied as an alternative technology to reduce the mass and volume of horticultural waste. Four static piles were prepared inside a greenhouse: two containing whole waste and two consisting of shredded waste. All the piles were compared with a test pile containing whole waste and placed outside the greenhouse. In two cases, ventilation ducts were installed to improve aeration. Each greenhouse was 2.0 m wide, 3.5 m long and 1.16 m high. The air temperature and relative humidity were monitored both inside and outside the greenhouse. Mass, humidity, organic matter and total nitrogen in the waste were measured. Piles inside the greenhouse showed decreases of 80% and 75% in weight and volume, respectively, during the first 16 days. The data obtained in this work suggest that biodrying could improve the handling and transport of horticultural waste while also minimizing the impact of pollutants.

Keywords: Solid Waste; Aerobic Degradation; Stabilization; Greenhouse

1. INTRODUCTION

Nowadays, there are more than 60 distribution centers of fresh food products ("Centrales de Abasto") operating in Mexico. Together, they produce thousands of tons of horticultural waste every day, which rots easily because it is organic matter and has high water content. The biggest distribution center of fresh food in Mexico City generates close to 585 tons of damaged and spoiled fruits, vegetables and flowers daily. Most of this waste is disposed of in landfills without any kind of pretreatment or recovery process [1].

Horticultural waste can be treated by an aerobic process like composting or biodrying. Biodrying, which is a fairly recent development in the field of waste management, is a treatment that uses forced aeration along with the heat generated by natural aerobic bioconversion of some organic matter to dry the waste. Most of the biological heat produced is utilized to evaporate surface and bound water associated with the waste matrix.

The most obvious advantage of biodrying is the reduction in odor, volume, and weight of the waste, which in turn may improve handling, transport and disposal of organic waste when a high volume of waste is an issue. Furthermore, biodrying has also been proposed as an effective means of eliminating some pathogens [2].

As reported previously [3], a controlled air supply can be used to increase the drying rate, which in turn increases the rate of water evaporation. However, the velocity and temperature of the air must be controlled adequately, because the aim of biodrying is to achieve satisfactory water loss in order to obtain a homogeneous, stable and useful product [4]. Biodrying is typically carried out in tunnels or biological reactors, where air is supplied and the ideal conditions of temperature and humidity are maintained to ensure a balance between the drying rate and the aerobic degradation of the organic matter [4-6].

Currently, most studies dealing with the biodrying process focus on reducing the water content of municipal solid waste while preserving its calorific power, in order to obtain a refuse-derived fuel [4,6-8]. Other applications of biodrying have also been researched; for instance, Navaee-Ardeh et al. [9] reported that biodrying is an attractive technique for drying sludge from the paper industry.

In the last few decades, a number of researchers have studied the production and collection of biothermal energy obtained from compost of different kind of organic wastes, to be used in growing plants in greenhouses [10], in heating bedrooms in home [11,12]. Conversely and according to the open literature, no reports have been published on the potential use of the heat of greenhouses, as an efficient collector of solar energy, to improve biodrying of horticultural waste. Thus, in this work, the biodrying process, conducted under both environmental and greenhouse conditions, was studied to evaluate whether it can be considered as an alternative to reduce the mass of horticultural waste in a short period of time in order to enhance handling, transportation and final disposal capacities.

2. MATERIAL AND METHODS

2.1. Feedstock Characteristics

The horticultural waste feedstock was sampled from the biggest market of fresh food products in Mexico City, the "Central de Abasto de la Ciudad de México". It consisted of a mixture of different kinds of vegetable waste, lettuce being the most common product. The trials were performed in small greenhouses constructed as shown in **Figure 1**. Each greenhouse was 2.0 m wide, 3.5 m long and 1.16 m high and was based on a metallic framework. This metal structure of the greenhouse was covered with a white long-life polyethylene film with a thickness of 180 microns. In order to improve the biodrying process, the front and the back of the polyethylene film were removed daily for one hour. In order to provide oxygen and allow air to penetrate and move through the pile by passive convection and diffusion, ventilation ducts were placed in some of the piles as indicated in **Table 1**; the size of the ducts were 0.25 m high and 2.5 m long.

2.2. Sampling Locations

Five static biodrying piles containing between 221 and 325 kg of horticultural waste were prepared as shown in **Table 1**. Air temperature and relative humidity were monitored inside and outside the greenhouse. Pile temperature was measured with a long-stemmed thermometer both in the center (approximately 15 cm below the surface) and in the peripheral zone, and data were recorded daily for 35 days. Samples of 10 g of horticultural waste were taken from the center of the piles and used to evaluate water, organic matter and nitrogen contents.

2.3. Measured Parameters and Measurement Methods

Mass loss, humidity, organic matter and total nitrogen in the waste were measured. Moisture and dry matter content were determined daily by a gravimetric method in compliance with the Mexican standard procedure established by NMX-AA-16-1984 [13]. Briefly, samples weighing between 3 and 5 g were put in an oven at 60°C (CR1-2 Shel Lab Cleanroom Oven) for about 2 hours and allowed to cool to room temperature in desiccators. They were then weighed and heated again for about 2 hours, after which the sample was cooled and weighed a second time. The procedure was repeated until the weight difference between two successive operations was less than 0.01%.

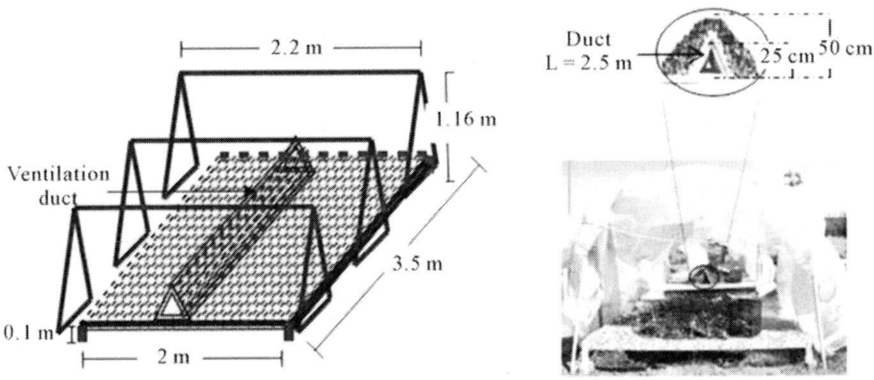

Figure 1. Experimental setup.

Table 1. Operational conditions of biodrying assays.

Pile	Initial mass (kg)	Characteristics (Waste integrity/Ventilation/Surroundings)
1	284.5	Whole waste/ventilation duct/in greenhouse
2	221.1	Shredded waste/ventilation duct/in greenhouse
3	324.8	Whole waste/no ventilation duct/in greenhouse
4	254.5	Shredded waste/no ventilation duct/in greenhouse
5	279.6	Whole waste/no ventilation duct/under environmental conditions

Total Nitrogen was measured by the Kjeldahl method on wet material [14]. Organic matter was quantified, in accordance with the Mexican standard procedure [15], by applying the Watley-Black colorimetric method. Briefly, three 0.1 g samples of horticultural waste were treated with 0.1 N potassium dichromate (99.9% pure, J. T. Baker) as a strong oxidizing agent in sulfuric acid (98% pure, Reasol). The mixture was swirled vigorously for 1 minute and allowed to stand on a heat proof surface for 30 min to allow oxidation of the organic matter to proceed. Distilled water and phosphoric acid (95% pure, Reasol) and diphenylamine (J. T. Baker) were then added as an indicator solution. When the reaction was completed, unreacted dichromate was titrated by adding 0.5 N ferrous sulfate (J. T. Baker).

3. RESULTS AND DISCUSSION

3.1. Evolution of Weight and Water Content during Biodrying

Based on the results presented in **Figure 2**, it was estimated that in the first two days, whole and shredded waste lost 15% and 20% of their water content, respecttively, as leachate. This rapid loss of water during this short period was due to lixiviation by hydrolytic decomposition as a result of partial degradation of the organic matter, since water in horticultural waste can be released as leachate when the wall or membrane cells are degraded, as reported previously [16]. Next, the heat generated in the pile as a by-product of the microbial breakdown of organic material acts as the energy for evaporation. As in composting, in biodrying most of the energy released during microbial activity is used to evaporate water from the waste materials. Then, following the hydrolytic stage, evaporation helped to diminish water content in the horticultural waste; consequently, the weight of the waste had diminished by 82% in the four piles inside the greenhouse on day 16. Nevertheless, the weight reduction observed in the control pile was only 22% over the same period of time, compared with the initial weight.

Our results showed that while drying, waste did not lose water at a constant rate and the rate of water removal under fixed conditions dropped off as drying progressed. At the beginning of biodrying and for some time thereafter, water generally continued to be lost from the waste at a constant rate. This was followed by an inflection in the drying curve, which leads to the period in which the drying rate drops [17]. During the initial period of the drying process (constant velocity), the heat is transferred to the product and to the contained moisture.

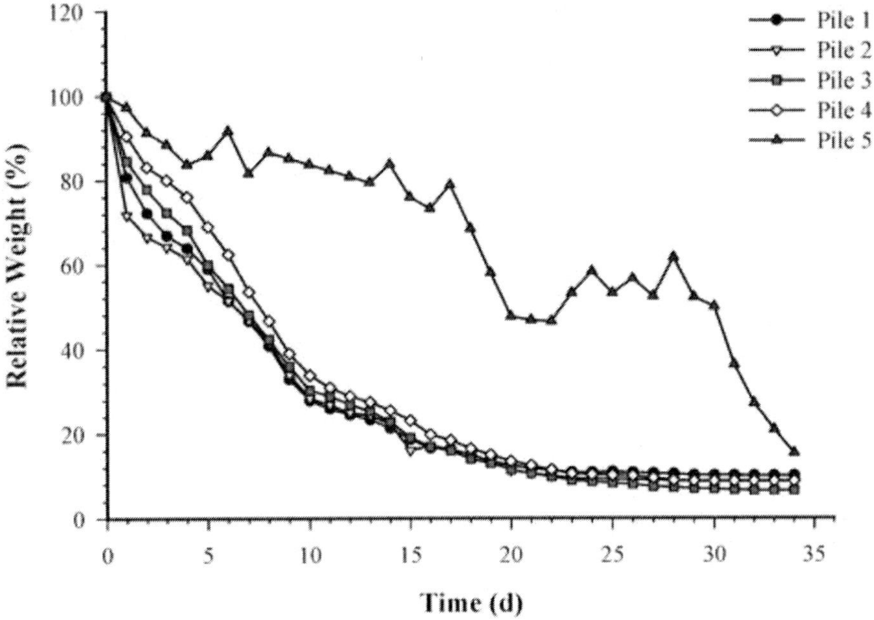

Figure 2. Weight loss of horticultural solid waste during biodrying.

In the next phase of the drying process, free moisture persists on the surfaces and the rate of evaporation alters very little as the moisture content is reduced. There is a gradual and relatively small increase in the product temperature during this period [18]. Initial water content was 95%. According to results, it could be inferred that for all the piles inside the greenhouse, waste humidity is reduced at two drying rates: 1) the constant rate of drying, where the fastest decrease in waste humidity (from 95% to 18%) was attained in a short time (16 days); and 2) the falling rate of drying, where more time (19 days) was required to remove another 9% from the remaining water (from 18% to 9%). The weight loss curve of the piles in the greenhouse became asymptotic on day 24, which means that the material reached the equilibrium state with the surroundings. Thus, zero moisture content was never achieved, which is a similar behavior to that reported by N. K. Thakur and N. S. Thakur [17]. In the control pile, the maximum decrease was reached at the end of the experiment (35 days). The reason for this is that the control pile was often re-humidified by rain and condensing phenomena at dawn, above all on rainy days (days 5 and 21), since this pile was kept under outdoor conditions.

3.2. Loss of Organic Matter during Biodrying

During biodrying, weight loss results from a reduction in water content and, to a lesser extent, from partial degradation of organic matter. As depicted in **Figure 3**, during the first 16 days the organic matter content was reduced from 80% to 50%, which coincides with the period of high microbial activity. The rest of the time the reduction was lower (from 50% to 38%). In the control pile, a bigger reduction was observed in the organic matter content (from 80% to 20%) because the pile was constantly re-humidified (by the rain), thus allowing the microbial activity to continue throughout the experiment.

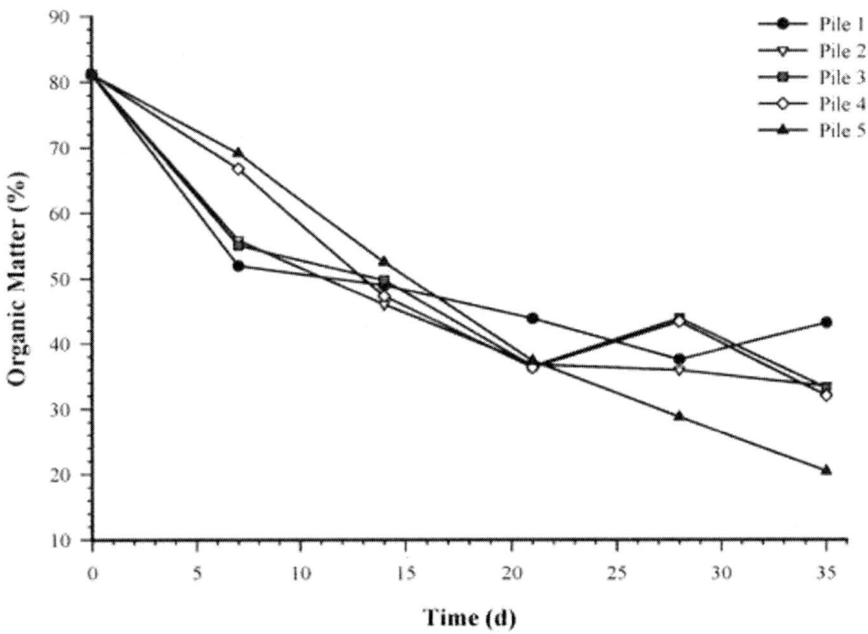

Figure 3. Reduction of organic matter during biodrying.

3.3. Influence of Climate on Biodrying

Due to changes of climate, the air and waste temperatures inside the greenhouse varied during the experiment. Since all the piles in the greenhouse behaved in a similar manner (**Figure 2**), pile 1 was taken as being representative of all of them. The changes in solid waste, together with indoor and outdoor air temperatures are shown in Figures 4 and 5. The highest temperatures of waste inside the greenhouse were observed during the first three days, when they exceeded the greenhouse temperature (**Figure 4**). In this period, microbial activity was more intense. The lowest greenhouse temperatures (17°C) were found on days 5 and 6, when the environmental temperature descended to 10.4°C, which affected the waste temperature and therefore the performance of the process. After that time, waste temperature increased again but never exceeded the air temperature inside the greenhouse, although most of the time it was higher than the environmental temperature.

There was a variable difference in temperature (ΔT) from day 1 to day 15, between the inner and outer layers of the waste piles inside the greenhouse. This suggests that the biodrying process ends on day 16 (case of pile 1) when the surface and center pile temperatures became equal (**Figure 5**).

3.4. Advantages of Biodrying in Greenhouse

The drying process in a greenhouse results in two main effects:

1) Generation of metabolic heat due to the aerobic microbial activity in the waste. Such a metabolic effect is responsible for the process called biodrying, as stated by some authors [6,7]. As a consequence of heat generated by the intense microbial degradation, the waste temperature was higher than the greenhouse temperature during the first three days and on the sixth day (**Figure 4**).

Biodrying piles reached temperatures between 29°C and 37°C during the first four days. The thermophilic phase (above 45°C) was not reached because of the low volume of the piles, and the water content was excessively high. In this case the heat derived from the biodegradation was not enough for water evaporation, as suggested by Zhang et al. [16].

2) Solar energy stored as heat inside the green house that helped to increase mainly the air temperature but also to a lesser extent the waste temperature. On the other hand, the temperature and relative humidity, both of which were higher in the greenhouse than in outdoor conditions, could have a positive effect on microorganisms in waste by improving microbial growth and activeity, mainly in the case of fungi. In addition, the greenhouse protected the piles from the rain, thereby preventing the horticultural waste from being re-humidified.

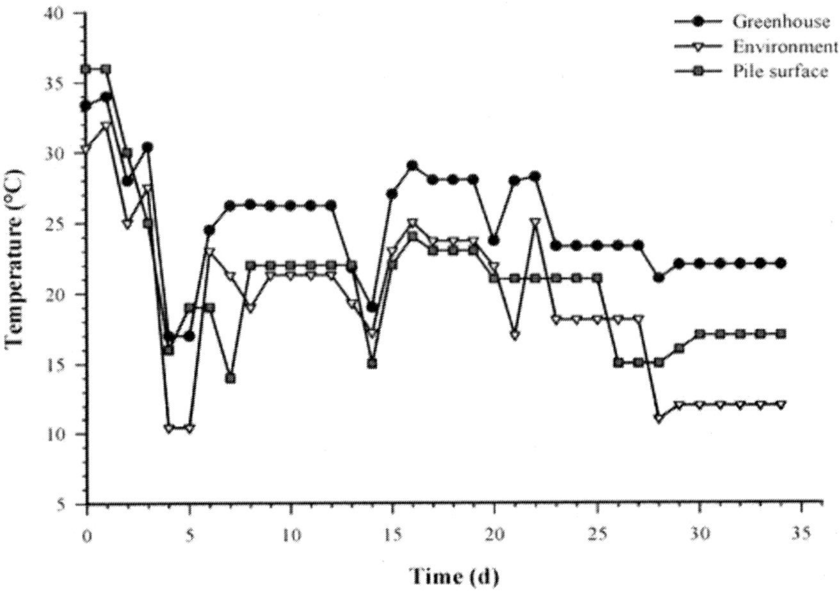

Figure 4. Temperature behavior in greenhouse, environment and pile surface.

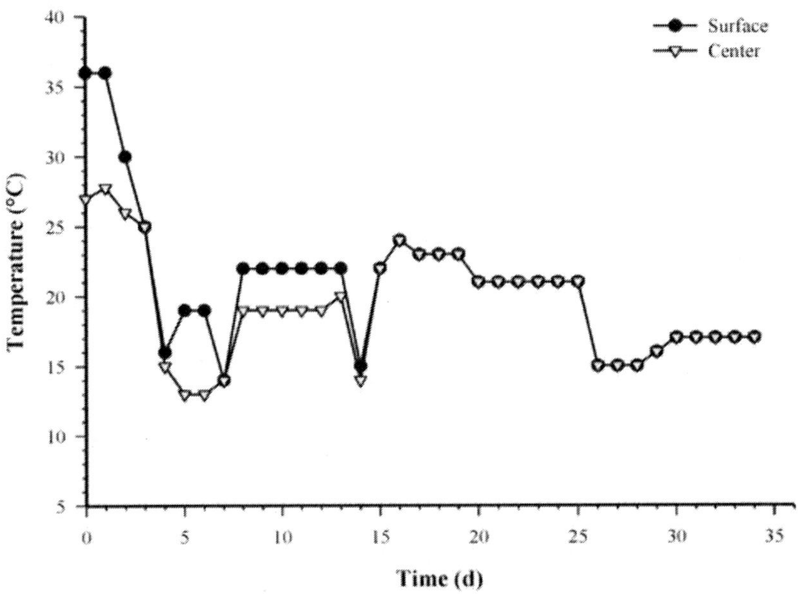

Figure 5. Evolution of internal and surface temperatures of solid waste.

3.5. Physicochemical Characteristics of Biodried Horticultural Waste

The physicochemical characteristics of horticultural waste before and after the biodrying process are shown in Table 2.

The initial humidity of the horticultural waste at the beginning of the process was 95.5%, while at the end of the process it was between 3% and 6%. Such a large reduction in humidity and partial biodegradation of organic matter resulted in a mass reduction above 90% at the end of the process (35 days). Considering that the biodried horticultural waste was partially stabilized, it is possible to infer that the material could be used as organic amendments for soils. Further work is needed to evaluate the influence of this material in crops.

Table 2. Physicochemical parameters of waste during biodrying.

Parameter	Piles in greenhouse		Control pile	
	Initial value (%)	*Final value* (%)	*Initial value* (%)	*Final value* (%)
Humidity	95.48	3 - 6	95.48	14.6
Volume	100	25	100	42
Mass	100	5 - 9	100	18
Organic Matter	81.16	40	81.16	20
N content	1.99	1.25 - 1.75	1.99	1.25

4. CONCLUSION

The final weight, volume and humidity values achieved in this work allow us to suggest biodrying as a pretreatment of horticultural waste in order to improve the capacity for handling, transportation and final disposal of horticultural waste without giving rise to pollution problems. The novelty of carrying out the biodrying process in a greenhouse lies in the fact that the heat of the enclosure and the energy resulting from microbial activity in horticultural waste raise the efficiency of the performance of water removal and water content reduction in waste. Biodrying in greenhouses could be proposed as a technology to stabilize waste with high water content, like the refuse produced by important fresh food markets. Nevertheless, since air moving through the compost pile (either by forced ventilation or passive convection and diffusion) gets hot and evaporates water from the surfaces of particles, further work is needed to evaluate the influence of forced airflow on biodrying kinetics in a greenhouse and performance in piles with a larger volume. Such work will make it possible to lower the time needed to reach target moisture contents and to evaluate whether thermophilic temperatures could be achieved during the processes.

5. ACKNOWLEDGEMENTS

Fabián Robles-Martínez, Elizabeth M. Silva-Rodríguez, Teodoro Espinosa-Solares, Belem Piña-Guzmán, Carmen Calixto-Mosqueda, Francisco J. Colomer-Mendoza, Enrique Durán-Páramo (SIP and COFAA). We particularly wish to thank Ricardo Estrada Núñez from the Mexico City Government (DGSU) and Rogelio Bailón Morales from the IPN compost plant in Mexico.

REFERENCES

1. Secretaria del Medio Ambiente, "Inventario de Residuos Sólidos del Distrito Federal 2008," Secretaría de Medio Ambiente, Gobierno del Distrito Federal, México City, 2010.
2. P. Wright and S. Inglis, "Biodrying Dairy Manure," Proceedings of the 2002 International Symposium on Composting and Compost Utilization, Colombus, 5-7 May 2002, pp. 996-1007.
3. A. Kumar and G. N. Tiwari, "Thermal Modelling of a Natural Convection Greenhouse Drying System for Jaggery: An Experimental Validation," Solar Energy, Vol. 80, No. 9, 2006, pp. 1135-1144.
4. F. Adani, D. Baidoa, E. Calcaterra and P. Genevinia, "The Influence of Biomass Temperature on BiostabilizationBiodrying of Municipal Solid Waste," Bioresource Technology, Vol. 83, No. 3, 2002, pp. 173-179.
5. M. Sugni, E. Calcaterra and F. Adani, "BiostabilizationBiodrying of Municipal Solid Waste by Inverting AirFlow," Bioresource Technology, Vol. 96, No. 12, 2005, pp. 1331-1337.
6. A. C. Velis, P. J. Longhurts, G. H. Drew, R. Smith and S. J. T. Pollard, "Biodrying for Mechanical-Biological Treatment of Wastes: A Review of Process Science and Engineering," Bioresource Technology, Vol. 100, No. 11, 2009, pp. 2747-2761.
7. E. C. Rada, M. Ragazzi, V. Panaitescu and T. Apóstol, "MSW Bio-Drying and Bio-Stabilization: an Experimental Comparison," Proceedings of the International Conference: Towards Integrated Urban Solid Waste Management System, Buenos Aires, 6-10 November 2005, pp. 1-10.

8. J. Dach and S. Bode, "Biological Drying of Biowaste to Generate Biomass," ORBIT 2006 5th International Conference, Biological Waste Management, from Local to Global, Weimar, 13-15 September 2006, pp. 859-866.
9. S. Navaee-Ardeh, F. Bertrand and P. R. Stuart, "Emerging Biodrying Technology for the Drying of Pulp and Paper Mixed Sludges," Drying Technology, Vol. 24, No. 7, 2006, pp. 863-878.
10. S. Diver, "Compost Heated Greenhouses. NCAT Agriculture Specialist. Appropriate Technology Transfer for Rural Areas. U.S. Department of Agriculture," 2001. http://www.mans.edu.eg/projects/heepf/ilppp/cources/12/pdf%20course/45/how/compostheatedgh.pdf
11. I. Pain and J. Pain, "The Methods of Jean Pain or Another Kind of Garden," Committee International Jean Pain, Brussels, 1972.
12. J. W. White, "Letters to the editor. BioThermal Update (BTU)," BioThermal Energy Center, Portland, 1982.
13. NMX-AA-16, "Protección al Ambiente-Determinación de Materia Orgánica. Environmental Protection-Soil Contamination Municipal Solid Residues—Organic Matter Determination," 1984. http://www.pensaryhacer. org.ar/cursosambiente/archivos/norma03.pdf
14. J. B. Jones, B. Wolf and H. A. Mills, "Plant Analysis Handbook," Micro-macro Publishing, Athens, 1991.
15. NMX-AA-021, "Protección al Ambiente-Determinación de Materia Orgánica. Environmental protection-Soil Contamination Municipal Solid Residues—Organic Matter Determination," 1985. http://148.206.53.231/bdcdrom/GAM06/GAMV15/root/docs/NMX-082.pdf
16. D. Q. Zhang, P. He, L. Shao, T. Jin and J. Han, "Biodrying of Municipal Solid Waste with High Water Content by Combined Hydrolytic-Aerobic Technology," Journal of Environmental Sciences, Vol. 20, No. 12, 2008, pp. 1534-1540.
17. N. K. Thakur and N. S. Thakur, "Drying of Fruits and Vegetables," In: L. R. Verma and V. K. Joshi, eds., Postharvest Technology of Fruits and Vegetables: Handling, Processing, Fermentation and Waste Management, Indus Publishing Company, New Delhi, 2000, p. 835.
18. D. A. Traub, "The Drying Curve," 2002. www.process-heating.com/PH/Home/Files/PDFs/0902ph_dryingfiles.pdf

CHAPTER 6

Soil Degradation and Prevention in Greenhouse Production

Y Liang[1,2]*, X Lin[2] , S Yamada3 , M Inoue[4] , K Inosako[3]

[1] *Institute of Soil & Water Conservation, Northwest A&F University, Yangling, Shaanxi, China.*

[2] *Institute of Soil & Water Conservation, the Chinese Academy of Sciences & Ministry of Water Resources, Yangling, Shaanxi, China.*

[3] *Faculty of Agriculture Tottori University, 4-101, Koyama-chominami, Tottori, 680-8553, Japan.*

[4] *Arid Land Research Center, Tottori University, Hamasaka, Tottori, Japan.*

ABSTRACT

Soil degradation has been a very serious problem for sustainable production, especially by a re-cropping of greenhouse-cultivated cucumber (*Cucumis statirus* L.). The aim of this research was to expound the actuality for soil degradation, at the same time, put forward some suggestion for preventing from soil degradation and maintain sustainable production in greenhouse basic on the two experiments conducted in a solar greenhouse during 2001-2008 suburb area of Yan'an, Shaanxi province in North China. The result shown that cucumber fruit productivity increased as the increasing of re-cropping years, but decreased after 5years continuously cropping. As increasing of re-cropping years, the population of fungus and bacteria increased, which was assumingly main factor of soil degradation. There was significant difference in cropping models on soil bio-characteristics and system productivity. The productivity were the highest in cropping model between cucumber and greengrocery, cucumber and cowpea *(Vigna sinensis L)*, the second higher were in cropping model between cucumber and maize (*Zea mays*) for green manure, cucumber and kidney bean *(Phaseolus vulgaris)*. That was the best way to reduce soil bacteria and epiphyte amount to follow lasting three or four months during summer season after cucumber harvest, the better method was planting cowpea or other leguminous crops. Basic on the experiment, the optimums approaches to preventive soil degradation were put forward.

Keywords: Soil degradation greenhouse production cropping model soil bio-characteristics preventing from soil degradation

Solar greenhouse vegetable cultivation has been proven to be a good farming practice in various areas of different countries [1, 2, 3], and has been developed rapidly during recent years because of the comparatively higher economic benefit [4, 5, 6, 7].

Soil degradation is the universal problem in greenhouse soil. It affects the sustainable use of greenhouse soil and also endangers vegetable security.

ACTUALITY OF SOIL DEGRADATION IN GREENHOUSE

Soil salinity

Greenhouse vegetable cultivation often requires a greater degree of management and larger input of nutrients and irrigation. However, over-fertilization and over-irrigation for vegetable cultivation resulted in nutrient accumulation in soil and led to soil salinity [8, 9]. Salinity reduces phosphate uptake and accumulation in crops grown in soils primarily by reducing phosphate availability. Salinity can directly affect nutrient uptake, such as Na^+ reducing K^+uptake or by Cl^- reducing NO_3^- uptake. Salinity can also cause a combination of complex interactions that affect plant metabolism, susceptibility to injury or internal nutrient requirement. Salinity has been described as the 'AIDS of the soil' and its influence is spreading throughout society where crop production has been seriously affected and caused economic hardship.

Continuous cropping

As the development of specialized vegetable production, continuous cropping is inevitable in greenhouses. A lot of vegetable such as tomato (*Lycopersicum esculentum*), pepper (*Capsicum frutescens*), eggplants (*Solanum melongena*), and cucumber have frequently suffered continuous cropping obstacle [10]. At the same time, poorly differentiated plant mass getting into the soil under continuous cropping negatively affects soil organism communities [11]. This negative action leads to the change in metabolism of microbes from the primary to the secondary [12]. Continuous cropping obstacle has became a main problem for sustainable production of vegetable, for example, serious plant diseases and fall of yield, which intimidated facility vegetables production sustainable development, especially in North China [5, 13, 14, 15, 16, 17].

Soil acidification

A number of agricultural practices have expanded the areas of acidification soils. The main causal factor is the growth of plants that use large amounts of basic ions (e.g. legumes); particularly when fertilizers that leaves acidic residues (such as Superphosphate) are used.

Nutrition accumulation

At present, fertilizer application rates in intensive agricultural systems have increased dramatically in recent years, especially in greenhouse vegetable production systems [18]. These high fertilizer inputs and the extremely low crop recoveries of fertilizer nutrients lead to marked deterioration in soil and groundwater quality and the systems are clearly unsustainable [19]. Soil chemical and biological properties in greenhouse vegetable production may change dramatically after several years of continually high inputs of fertilizers and irrigation water and planting of shallow rooting vegetable crops compared with open field cereal production systems. Soil OM, alkalihydrolyzable N, and available P and K have been observed to accumulate significantly in vegetable greenhouses [5]. High concentrations of P and imbalances of N, P and K often occur in the intensively managed soils. High fertilizer application rates may lead to increasing salinity in the surface soils of the plastic film greenhouses in which vegetables are grown [20].

Compaction

Some researcher found that bulk density was higher in greenhouse soils than uncovered soil. In recent years, there has been increasing interests in intensive vegetable cultivation, this caused the soil compacted. Compaction of soils causes a reduction in soil pore space. This reduces the rate at which water can infiltrate and drain through the soil. It also reduces the available space for oxygen in the plant root zones. For this reason, some of the major consequences of compaction are poor drainage, poor aeration, and hard pan surfaces which cause runoff. Repeated cultivation of some soils leads to a breakdown of soil structure and this also increases the likelihood of compaction.

Chemical residue

Although not as large a problem as some of the other types of soil degradation, the presence of chemical residues can be quite a problem on a local scale. These residues derive almost entirely from long term accumulation after repeated use of pesticides, etc., or of use of pesticides or other chemicals with long residual effects. Some problems that result from chemical residues include toxic effects on crop species and contamination of workers, livestock and adjacent streams.

The studies on the soil degradation of greenhouse soil and its affection mechanism on vegetables are highly important for us to maintain high land capacity. For a sustainable management of soil in the greenhouse, it is urgent to improve the fertilizer and irrigation management for vegetable production, especially for the winter season under greenhouse conditions. Excessive nitrogen application can be effectively avoided by site-specific fertilizer application based on soil and plant analysis.

Prevention From Soil Degradation In Greenhouse

MATERIALS AND METHODS

The two experiment were conducted in a solar greenhouse during 2001-2008 suburb area of Yan'an, Shaanxi province in North China, a typical loess hill-gullied region of the Loess Plateau Longitude E109°19'23", Latitude N36°51'30", with an elevation of 1068-1309 m, mean temperature of 8.8°C, the annual precipitation of 500 mm. In the experiment 1, in 2001, the plots were filled with continuously cropped typical sandy soils of winter-spring season cucumber for 0year (hereafter referred to as yr), 1yr, 4yr, or 8yrs, respectively with 4 replications. The plots area was 15.6 m^2. A grafted cucumber with the squash root was transplanted to the continuous cropped soil. The cucumber fruit productivity and species of soil microbes were analyzed. In the experiment 2, the 7 kinds cropping models between winter-spring season cucumber and summer other different crops on the productivity and soil bio-characteristics were studied. The cropping models were: cucumber and greengrocery, cucumber and cowpea, cucumber and maize for green manure, cucumber and kidney bean, cucumber and tomato (*Lycopersicum esculentum*), cucumber and dark bean (*Phaseolus*) for green manure, cucumber and fallow. The plots area was 39.6 m^2 with three replications.

The soil moisture was maintained 75-90% as relative soil water capacity using drip irrigation system. The cucumber fruits were harvested sequentially from the early of December till the next year early of July. Microbe was determined by dilution-plate assay method, PDA broth medium used for soil fungus measured, beef extract peptone broth medium used for soil bacteria, starch-ammonium and martin broth medium used for soil actinamices [21]. To improve the precision of the results, every sample was measured three times, and the average value was calculated.

The experimental data was analyzed using SPSS software, tested with F-test means. Results were compared based on Duncan method (L.S.R).

RESULTS

Cucumber fruit productivity in different re-cropping year

Table 1 showed that there was significant difference in cucumber productivity among re-cropping years, the treatment of 1yr, 2yr, 3yr and 4yr showed significant higher yield, the treatment after 5yr showed lower yield, and this difference had reached the statistics significant level ($p = 0.01$). It means that the productivity increased as re-cropping years increasing before re-cropping 4yr, but the yield decreased obviously after 5 yrs re-cropping, indicting that degradation of soil quality was occurred.

Table 1. The cucumber fruit yield in different re-cropping year (kg/m^2)

Treat.	2002	2003	2004	2005	2006	2007	2008
0yr	12.76 B*	15.31 A	13.83 A	12.41 A	12.39 A	10.03 A	9.36 A
1yr	15.32 A	15.74 A	13.69 A	12.05 A	11.36 A	9.12 B	8.62 AB
4yr	13.95 B	11.23 B	9.65 B	8.32 B	8.16 B	8.28 C	8.23 B
8yr	11.86 C	10.23 B	9.31 B	8.36 B	8.21 B	8.13 C	7.86 B

* Notes: the letters in the same column show significant difference at 0.01 levels.

Soil microbe in different re-cropping year

In the soil, fungus and bacteria increased greatly, but actinamices did not show a big change as re-cropping years increasing (Table 2), which led to a rise of occurrence rate of pathogenic microbe and soil-borne, and lead to a decrease of fruit productivity. Additionally, the change in microbe group affected nutrient conversion in the soil, which possibly became an obstacle for nutrient absorption by crops.

Table 2. Soil fungus, bacteria and actinamices content in different re-cropping year

Treat.	Bacteria $10^6.g^{-1}ds$	Significant	Fungus $10^2.g^{-1}ds$	Significant	Actinamices $10^4.g^{-1}ds$	Significant
0yr	1.57	D*	3.8	D	20.75	A
1yr	1.48	D	1.2	E	11.13	B
3yr	1.03	E	5.4	C	10.49	B
5yr	2.95	C	4.2	D	8.19	C
7yr	3.32	B	6.3	B	3.53	D
9yr	8.90	A	7.2	A	3.43	D

* Notes: the letters in the same column show significant difference at 0.01 levels.

There are some means of soil quality rehabilitation. Reasonable rotation is the best way, for example, use of rotation system of cucumber with maize or dark-bean as green manure, rotation system of cucumber with cowpea. These crops can absorb different nutrients from soil, and abate the occurrence of diseases infected by soil through changing stubble [22]. Maintenance of adequate balance of fertilizers, addition of organic manure, and soil sterilization are secondary recommended for soil quality rehabilitation.

Cucumber fruit productivity in different cropping models

The significant difference in cucumber fruit productivity among cropping models was showed in Table 3, during 2002-2008, the cropping models of cucumber with greengrocery, cucumber with cowpea showed significant higher yield, the second higher yield appeared in the cropping models between cucumber and maize for green manure, cucumber and tomato, cucumber and kidney bean, the cropping models of cucumber with dark-bean for green manure, cucumber with fallow showed lower yield, the difference had reached the significant level ($p = 0.01$).

Table 3.The cucumber fruit yield in different cropping models (kg/m²)

Treat.	2002-2003	2003-2004	2004-2005	2005-2006	2006-2007	2007-2008
Kidney bean	16.24 B*	16.17 AB	13.34 BC	17.30 B	16.70 B	16.40 C
Tomato	15.20 B	15.06 B	14.30 B	17.80 B	15.30 BC	15.10 CD
Fallow	13.40 C	11.91 C	11.63 C	13.90 C	12.50 C	14.80 D
Maize for green manure	16.23 B	13.27 BC	13.84 B	17.60 B	17.80 B	18.50 B
Dark-bean for green manure	14.02 C	10.92 C	12.62 C	15.30 C	14.10 C	16.60 C
Cowpea	17.91 A	17.26 A	14.97 B	18.60 B	20.80 AB	18.60 B
Greengrocery	20.05 A	19.24 A	20.44 A	21.50 A	22.30 A	20.80 A

* Notes: the letters in the same column show significant difference at 0.01 levels.

Soil microbe in different cropping models

In the soil, the number of fungus and bacteria changed greatly when planting different crops during summer season after winter-spring season cucumber harvested (Table 4). The number of bacteria in the soil was the highest after planting tomato during summer season, the second higher number were in the soil after planting greengrocery, maize for green manure and dark-bean for green manure during summer season, the lowest number of bacteria were in the soil after planting kidney bean, cowpea and fallow. The number of fungus in the soil was higher after planting maize for green manure, dark-bean for green manure and kidney bean, the second higher were in the soil after planting tomato and greengrocery, the lowest number were in the soil after planting cowpea and fallow during summer season.

Table 4. Soil fungus and bacteria in different cropping models (18 Oct. 2002)

Treat	Bacteria $10^6.g^{-1}ds$	5% Significant	Fungus $10^2.g^{-1}ds$	1% Significant
Maize for green manure	56.94	b	88.97	A
Tomato	101.36	a	46.97	B
Cowpea	46.43	c	35.71	C
Kidney bean	43.59	c	95.89	A
Greengrocery	62.79	b	40.70	B
Dark-bean for green manure	48.84	bc	79.70	A
Fallow	41.21	c	25.45	C

These results indicated that it was effective method to use different cropping models for decreasing number of microbe in soil, to fallow three or four month or to plant leguminous rotation system during summer season.

CONCLUSIONS

The productivity increased as re-cropping years increasing, but the yield decreased obviously after 5 yrs re-cropping, this indicted that degradation of soil quality was occurred. The increase of fungus and bacteria number became main factors of soil degradation as re-cropping years increasing. The cropping models of cucumber and greengrocery, cucumber and cowpea showed significant higher cucumber yield, the second higher yield appeared in the cropping models between cucumber and maize for green manure, cucumber and tomato, cucumber and kidney bean. Therefore, that was effective method to use different cropping rotation models for decreasing number of microbe in soil, to prevent from soil degradation in greenhouse to fallow three or four month or to plant leguminous during summer season after winter-spring season cucumber harvested.

ACKNOWLEDGEMENTS

The research was supported by the Knowledge Innovation Program of CAS (KZCX2-YW-443) (2009-2011), JSPS Core University Program "Researches on Combating Desertification and Developmental Utilization in Inland China" (2001-2010), and west-action program of the Chinese Academy of Sciences (KZCX2-XB2-05-01).

COMPETING INTERESTS

The authors declare that they have no competing interests.

REFERENCES

1. Li W, Luo HY, Liu JF: **THE MATERIAL AND ENERGY FLUXES IN GREENHOUSE ECOSYSTEM AND THE EVALUATION OF ITS BENEFITS.** *Ecology Agric Res (in Chinese)* 1996, **3:** 53-
2. Nakano A, Yamauchi A, Uehara Y: **EFFECTS OF APPLICATION OF LOW-SULFATE SLOW-RELEASE FERTILIZER (LSR) ON SHOOT AND ROOT GROWTH AND FRUIT YIELD OF TOMATO.** *Jpn Agric Res Qua* 2003, **L37:** 121-127.
3. Darwish T, Atallah T, Moujabber ME, Khatib N: **SALINITY EVOLUTION AND CROP RESPONSE TO SECONDARY SOIL SALINITY IN TWO AGRO-CLIMATIC ZONES IN LEBANON.** *Agric Water Manage* 2005, **78:** 152-164.

4. Liao YC, Wang LX: **FACILITY AGRICULTURE AND THE BUILDING OF CHINESE AGRICULTURAL MODERNIZATION.** *Research of Agricultural Modernization (Chinese)* 1999,**20**(1):5-8.
5. Guo LS, Yang WC: **THE UNDERSTANDING AND FULFILLMENT OF FACILITY AGRICULTURE.** *Sciences Research Management (Chinese)* 2000,**21**(3):56-59.
6. Feng GH: **DISCUSSION ON FACILITY AGRICULTURE.** *Facility Agriculture (Chinese)* 2000, **2:** 5-6.
7. Liang YL, Chen ZJ, Wang ZM: **THE STATUS AND FUNCTION OF FACILITY AGRICULTURE ON ECOLOGICAL ENVIRONMENT CONSTRUCTING.** *Journal of Soil and Water Conservation* 2002,**16**(5):32-35.
8. Stigter TY, Ooijen SPJ, Post VEA, Applello CAJ, Dill AMMC: **A HYDRO GEOLOGICAL AND HYDRO CHEMICAL EXPLANATION UNDER IRRIGATED LAND IN A MEDITERRANEAN ENVIRONMENT.** *Algarve, Portugal J Hydrol (Amst)* 1998, **208:** 262-279.
9. Chen Q, Zhang XS, Zhang HY, Christie P, Li XL, Horlacher D, *et al.*: **EVALUATION OF CURRENT FERTILIZER PRACTICE ANDSOIL FERTILIZER IN VEGETABLE PRODUCTION IN THE BEIJIN GREGION.** *Nutr Cycl Agroecosyst* 2004, **69:** 51-58.
10. Takahashi K: **INJURY BY CONTINUOUS CROPPING IN VEGETABLES: VARIOUS PROBLEMS IN THE CULTIVATION USING GRAFTED PLANTS.** *Yasaishikenjo Kenkyu Shiryo* 1984, **18:** 87-89. (In Japanese)
11. Lech Szajdak: **IMPACT OF CROP ROTATION AND PHONOLOGICAL PERIODS ON RHODANESE ACTIVITY AND FREE SULFURIC AMINO ACIDS CONCENTRATIONS IN SOILS UNDER CONTINUOUS RYE CROPPING AND CROP ROTATION.** *Environment International* 1996,**22**(5):563-569.
12. Bu'Lock JD: *Mycotoxins as secondary metabolism. In: Steyn, P.S.,ed. Biosynthesis of mycotoxins.* New York, NY: Academic Press; 1980:70-85.
13. Liu D, Wu FZ, *et al.*: **EFFECTS OF THE RE-CROPPING ON THE ROOTS ACTIVITY AND THE PHOTOSYNTHESIS RATE OF THE PLASTIC GREENHOUSE CUCUMBERS.** *Journal of Northeast Agricultural University* 1998,**29**(3):219-223.
14. Liang YL, Chen ZJ, Xu FL, *et al.*: **THE EFFECT OF RE-CROPPING YEAR ON PHYSIOLOGICAL CHARACTERISTICS OF CUCUMBER IN SUNLIGHT GREENHOUSE.** *Acta botanical boreali-occidentalia sinica* 2003,**23**(8):1467-1470.
15. Liang YL, Chen ZJ, Xu FL, Zhang CE, Du SN, Yan YG: **SOIL RE-CROPPING OBSTACLES IN FACILITY AGRICULTURE ON LOESS PLATEAU.** *Journal of Soil and Water Conservation* 2004,**18**(4):134-136.
16. Yu JQ, Du YS: **THE RE-CROPPING OBSTACLES PROBLEMS AMONG THE SUSTAINABLE DEVELOPMENT OF VEGETABLE FACILITY GARDENING.** *Journal of Shenyang Agriculture University (Chinese)* 2000,**31**(1):124-126.
17. Wu FZ, Zhao FY, Liu YY: **ANALYSIS OF THE PROBLEMS IN CONTINUES CROPPING SYSTEM OF PROTECTIVE VEGETABLE AND THE CONTROLLING WAYS.** In *Fertilizing for Sustainable Production of High Quality Vegetables.* Edited by: Li, X.L., Zhang, F.S., Mi, G.H. China Agricultural University Publisher, Beijing; 2000:157-163.
18. Ju XT, Wan YJ, Kou CL, Zhang FS: **NITRATE ACCUMULATION IN AGRICULTURAL SOILS AND GROUNDWATER OF THE NORTH CHINA PLAIN.** In *Plant Nutrition for Food Security, Human Health and Environmental Protection.* Edited by: Li. Tsinghua University Press, Beijing; 2005:1056-1057.
19. Zhu JH, Li XL, Christie P, Li JL: **ENVIRONMENTAL IMPLICATIONS OF LOW NITROGEN USE EFFICIENCY IN EXCESSIVELY FERTILIZED HOT PEPPER (CAPSICUM FRUTESCENT L.) CROPPING SYSTEMS. AGRICULTURE.** *Ecosystems and Environment* 2005, **111:** 70-80.
20. Wu FZ, *et al.*: **ON THE REASONS OF RE-CROPPING OBSTACLES IN VEGETABLE FACILITY GARDENING.** *Journal of Northeast Agriculture* 2000, **13:** 63-65.
21. Zhou DQ: **MICROBIOLOGY EXPERIMENT BOOKLET.** *Shanghai, Science and technology press of Shanghai* 1983.

22. Liang YL, You HX, Chen ZJ, Gao JY, Du SN, Zhou MJ, Chen JR, Xiong YM: **SYSTEM PRODUCTIVITY AND SOIL BIO-CHARACTERISTICS OF DIFFERENT CROPPING MODEL IN FACILITY AGRICULTURE.** *Journal of Soil and Water Conservation* 2006,**20**(4):163-165.

CHAPTER 6

Nutrient Analysis of Soil and Soilless Strawberries and Raspberries Grown in a Greenhouse

Chenin Treftz, Stanley T. Omaye

Agriculture, Nutrition and Veterinary Sciences Department and Environmental Sciences and Health Graduate Program, University of Nevada, Reno, USA

ABSTRACT

Soilless (hydroponic) vegetables and fruits grown in greenhouses are gaining popularity and potentially represent a compliment toward sustainable food sources. Only a few studies have looked at the nutrient quality of strawberries (Fragaria × ananassa) and raspberries (Rubus idaeus) grown in soilless systems. Dry weights, content of ascorbic acid, tocopherol, total polyphenolic compounds, glucose, fructose, and soluble solids (BRIX) of strawberries and raspberries grown in soilless systems were compared to their counterpart grown in soil. There was no change in dry weights but BRIX values (28% - 31%), glucose (158% - 175%), and fructose (75% - 102%) content for strawberries and raspberries respectively were significantly higher for the soil grown berries compared to soilless grown berries. Contents of ascorbic acid, tocopherol and total polyphenolic compounds were significantly higher in soilless grown strawberries compared to soil grown strawberries by 74%, 53%, and 22% respectively, and contents of ascorbic acid and total polyphenolic compounds were significantly higher in soil grown raspberries by 83% and 67% respectively compared to soilless grown raspberries. Soilless grown produce warrants future research to strive toward the potential to provide nutrient dense crops and opportunities toward optimized sustainable production.

Keywords: Soilless, Greenhouse, Bioactive Compounds, Strawberry, Raspberry

1. INTRODUCTION

The United States Department of Agriculture estimates that 23 million people live in food deserts with inadequate access to healthy, affordable and fresh food [1] . Inadequate access to food, especially fresh fruits and vegetables, is a public health concern because the consumption of fruits and vegetables is associated with a decreased risk of certain chronic diseases such as cardiovascular disease, type 2 diabetes, obesity and certain types of cancer [2] [3] . Growing fresh produce in soilless systems may be a potential solution to food insecurity issues regardless of soil quality, climate or space [4] . Additionally, soilless growing systems can provide several other environmental benefits such as reduction of water, increased product yields, and less pesticide use. These advantages allow soilless systems to address several environmental issues while providing sustainable systems in food deserts, in arid or urban regions.

Recently, several studies have focused on the nutritional content of soilless produce. As soilless food production grows in popularity, researching the nutritional composition of soilless compared to traditional farming methods will be important because nutrition is one of the main drivers of purchasing and consumption [5] . Some studies indicate that soilless systems provide superior nutrition compared to traditionally grown produce [6] -[9] , while others indicate that no difference or soil grown produce is higher in selected nutritional parameters [10] [11] . The majority of previous research has focused primarily on lettuce, leafy greens and tomato fruit. Additionally, several of the previous studies have limitations on the findings because the comparison was conducted with produce grown in different environments known to affect bioactive compound production in the plant [12] [13] . The comparison of soilless and soil systems must occur in identical environments [14] .

Limited research is available on soilless strawberries and raspberries. The aim of this study was to compare the differences in nutritional quality, as defined by bioactive compounds, Brix and moisture content of strawberries and raspberries grown in soil and soilless conditions. Strawberries (Fragaria × ananassa) and raspberries (Rubus idaeus) are a rich source of bioactive compounds and can provide a plethora of health benefits to the consumer [15] -[17] .

We chose to analyze ascorbic acid, α-tocopherol and total polyphenolics because of their role in antioxidant protection [17] -[19] . Glucose and fructose were analyzed because these two nutrients are the primary sources of sugars in strawberries and raspberries [20] . Research on sugar content is necessary since it can affect the taste of the fruit as well as consumer preferences [21] . We analyzed moisture content and Brix in the fruit, which is an important quality indicator that can influence the texture and flavor of a fruit, as well as shelf life [22] .

2. METHODS

2.1. Chemicals

Thiourea, copper sulfate, and orthophosphoric acid (85%) were purchased from Fischer Scientific (Fair Lawn, NJ). Sulfuric acid, trichloroacedic acid, 2,4-dinitrophenylhydrazine, L-ascorbic acid, ferric chloride, xylene, 4,7-diphenyl-1,10-phanthroline (bathophenanthroline), (±)α-tocopherol, sodium carbonate, Folin-Ciocalteu, trypta- minehypochloride, hydrochloric acid (HCl), fructose, dinitrosalcylic acid, sodium hydroxide, D(+)-glucose, po- tassium sodium tartrate, and sodium sulfite were purchased from Sigma-Aldrich Cooperation (St. Louis, MO). ACS/NSP grade (200 proof) absolute ethanol was purchased from pharmco-AAPER, Kentucky.

2.2. Growing Conditions and Plants

Bare root strawberries and raspberries were ordered from Stark Brothers Nurseries & Orchard Company (Louisiana, MO). In both soilless and soil plants, they were planted on the same day in late Winter 2014. Plants were grown at the University of Nevada, Reno (UNR) Experimental Station. Throughout the growing season, the greenhouse temperature was maintained at 70°F (5:30 AM to 6:30 PM) during the day and 60°F (6:31 PM to 5:29 AM) at night with a relative humidity averaging at 30%. Soil and soilless buckets were numbered and randomized with the available space in the greenhouse at the UNR Experimental Station. The strawberries were placed in 8 rows between two tables, and the raspberry barrels were placed on cinder blocks (Reno, NV), in 3 rows of 4 barrels.

2.2.1. Strawberries

Thirty bare root Ozark Beauty (Fragaria × ananassa) strawberry plants were planted in soil conditions grown in three-gallon nursery pots. Berries were planted in Nevada topsoil mixed with Miracle-Gro potting soil (Maryville, OH) in a 1:1 ratio. The plants were watered by a drip irrigation system for 15 minutes, three times weekly. The plants were fertilized with all-purpose Miracle-Gro fertilizer every six weeks. The pH and parts per million (ppm) of the soil plants was measured with a portable meter before planting and quarterly, averaging at 5.6 and 400 ppm (Oakton Instruments, Vernon Hills, IL).

Thirty bare root strawberry plants were planted in soilless systems. The berries were planted in a bucket system using five-gallon paint buckets from a local hardware store (Reno, NV). The buckets were spray painted black to minimize algae growth. Hydroton, 8-inch netting, a Waterfarm® system pumping column and drip ring for construction of the bucket system was purchased from a local hydroponics store in Reno, Nevada. The plants were aerated using an all-purpose pump (Active Aqua AAPA 15L, Reno, NV). The pH of the plants was maintained between 6.0 - 6.4. The nutrient solution was a commercial General Hydroponics Flora Series, consisting of FloraBloom, FloraGrow and FloraMicro (Sebastopol, CA). Throughout the growing season, the nutrient ratios were changed to match the plant development, as indicated by the manufacturer instructions. The ppm averaged at 400. The pH and ppm were monitored and adjusted three times weekly.

2.2.2. Raspberries

Six bare root Heritage (Rubus idaeus) raspberries were planted in 50 gallon barrels. A combination of Nevada topsoil was mixed with Miracle-Gro potting soil in a 1:1 ratio. The berries were watered one to three times weekly for 15 minutes with a drip irrigation system. The plants were fertilized with all-purpose Miracle-Gro fertilizer every six weeks. The pH and ppm of the soil was checked before planting and quarterly, averaging a pH of 5.6 and ppm averaging at 600.

Six bare root Heritage varieties of raspberries were planted in 19 gallon buckets (United Solutions, TU0014, Reno, NV) using hydroton as the growing medium. Holes were drilled at the bottom of the buckets and were placed on top of the empty fifty-gallon barrel to create a large-scale version of a bucket system described with the strawberries. From the 50 gallon barrel, the water drains into a large water reservoir where a water pump distributes the water to the six soilless buckets via polyvinyl chloride (PVC) and drip tubing (Reno, NV). The water in the reservoir was maintained at a pH between 5.8 - 6.2. This pH was monitored and adjusted if needed three times weekly. The nutrients added to the berries were FloraGrow, FloraBloom and FloraMicro and maintained averaging at 500 ppm.

To support the berries, a T-hedgerow system was built with string and PVC pipe. A T-hedgerow system has been shown to have a comparable yield compared to the V-trellis system.

2.2.3. Harvesting Strawberries and Raspberries

The berries were harvested promptly when they visually reached 100% surface red color. The fruit was harvested between 7 AM and 8 AM for consistency, placed in a plastic laboratory bag, and immediately brought in a −70°C Thermo Scientific™ Revco™ high performance lab freezer (Thermo Fisher Scientific, Waltham, MA). The berries were stored in the freezer until analysis. All berries analyzed for comparisons were harvested on the same day. Nutrients with time sensitive oxidative properties were analyzed within thirty days of harvest, and others (i.e., glucose, fructose) were analyzed within sixty days of harvest.

2.3. Sample Preparation

Samples were randomly selected for analysis by hand. Before analysis, berries were rinsed with deionized water, dried with a paper towel and the stems were manually removed. For all assays, samples were homogenized using a Brinkmann Instruments Polytron homogenizer (Kinematica, Bohemia, NY).

2.4. Brix and Moisture Content

Brix, or soluble solids, is a common measurement of total dissolved solids in the juice, wine and soft drink industry, and can be used to approximate total sugar content. An automated digital refractometer (Milwaukee MA871, Rocky Mount, NC) was used. Procedures have been described previously [23] . Briefly, 10 grams (g) of berry samples were homogenized with a pestle and mortar. A double-dilution with an equal part by weight of distilled water was added to the homogenized berries. The berries were filtered using cheesecloth to remove seeds and pulp. After the samples were filtered, 1000 μL was extracted and the results were read in triplicate.

Moisture content in fruit was estimated by using a modified version of the Official Methods of Analysis of AOAC 934.06 for moisture in dried fruit [24] . The protocols' drying portion was lengthened to 20 hours to account for higher moisture content in fresh fruit compared to dried fruit. Briefly, three 10 g portions of samples were taken and homogenized with a pestle and mortar. The samples were placed in a Lab Line incubator, model 120 (Kerala, India) for 20 hours at 140°F. After drying, the moisture content can be expressed as a percentage of mass determined by the following equation:

$$W = \frac{M_1 - M_2}{M_1 - M_0} \times 100$$

where W is the moisture content, M_0 is the mass of the weight dish, M_1 is the mass of the dish and sample before drying, and M_2 is the mass of the dish and the test portion after drying.

2.5. Ascorbic Acid Analysis

Ascorbic acid content was determined using a modified protocol from measuring ascorbic acid in animal tissues [25] . Ten grams of berries were randomly selected and homogenized with 10 mL cold 20% Trichloroacetic acid (TCA). This mixture was placed into a flask wrapped in aluminum foil with 0.1 grams of activated carbon to remove color intensity and agitated for 15 minutes, and then was allowed to sit overnight [26] . The mixture was then filtered using Whatman no. 2 filter paper. A stock solution was created using L-ascorbic acid and standards were made using 5% TCA with a serial dilution of 0 - 120 μg/mL. After filtering, 100 μL of the liquid was removed and added to new test tubes containing 900 μL of 20% TCA. One mL of a mixture of 2,4-dinitrophenyl- hydrazine (DNPH), thiourea, copper in the presence of sulfuric acid was added to all samples, standards and blank. The copper in the solution oxidized the ascorbic acid to dehydroascorbic acid. The DNPH, thiourea and the sulfuric acid yielded a colored product with minimal interference from other chromogens. The samples, standards and blank was incubated in a 20 L Fischer Water Bath (Fischer Scientific, New Lawn, NJ) at 37°C for three hours. After incubation, 1.5 mL of cold 65% sulfuric acid was added to the samples, standards and blank, and a 30 minute waiting period was observed at 25°C to allow the color to stabilize. The absorbance of the samples, standards and blank were read at 520 nm with a 110 voltage FinstrumentsMicroplate Reader (Model 314, McLean, VA) in triplicate. The samples were compared to a linear regression created from the known standards ($y = 0.0102x + 0.0316$, $R^2 = 0.9957$). The reproducibility was measured by adding a known amount of a standard to a sample and determining the recovery, which was 110% ± 2.1% [27] .

2.6. Tocopherol Analysis

Alpha-tocopherol (α-tocopherol) method was derived from Fabinek et al., 1968, using Fe(III)-bathophenanth- roline spectrophotometry [28] . Ten grams of berries were randomly selected for analysis and homogenized with 10 mL of absolute ethanol. Xylene (1.2 mL) was added to extract the tocopherols from the samples. The samples were then centrifuged for 20 minutes at 3500 rpm in a Sorvall RT6000B refrigerated centrifuge at 7°C. After centrifugation, 100 μL of the organic layer was removed and was added to new test tubes containing 0.4 mL of bathophanthroline. Ferric chloride (0.4 mL) was then added to the tubes and 0.4 mL of orthophorsphic acid (85%) was then added to these test tubes to stabilize the color. In similar fashion, standards were made using a serial dilution between 0 - 50 μg/mL to create a linear regression to estimate α-tocopherol content in the samples ($y = 0.0081x - 0.0053$, $R^2 = 0.9927$). All samples, standards, and blank were read at 530 nm in triplicate with a 110 voltage FinstrumentsMicroplate Reader (Model 314, McLean, VA). The reproducibility was measured by adding a known amount of a standard to a sample and determining the recovery, which was 95% ± 2.5% [6] .

2.7. Total Polyphenolics

Total polyphenolics were determined by using the Folin-Ciocalteu assay. This method has been used in measuring the total reducing capacity in berries by gallic acid equivalents (GAE) [17] [29] [30] . Raspberries (10 g) were homogenized with 10 mL of ethanol. A volume of 1.58 mL of deionized water was added to all samples along with 100 μL of the Folin-Ciocalteu reagent. A series of standards were made using the same method ranging from 0 - 300 mg GAE/L. The solutions were allowed to sit for one minute and then mixed thoroughly. A volume of 300 μL of 25% sodium carbonate solution was added to the samples, standards and blank and was placed into a 40°C 20 L Fischer Water Bath (Fischer Scientific, New Lawn, NJ) for 15 minutes and recorded in triplicate at 690 nm against a 0 GAE mg/L solution FinstrumentsMicroplate Reader, 110 voltage (Model 314, McLean, VA). The GAE in the samples was estimated using the linear regression line created from the standards ($y = 0.0021x - 0.0099$, $R^2 = 0.9988$).

2.8. Fructose

Ten grams of samples were homogenized with 10 mL of deionized water. Samples were then centrifuged for 15 minutes at 3500 rpm in a Sorvall RT6000B centrifuge. Into new test tubes, 100 mL of the supernatant was extracted and 100 mL of a tryptamine reagent (concentration of 10 mM tryptaminehypochloride in 0.1 M HCl) was added to each test tube along with 3 mL 36% HCl. For the standards, a series of fructose solutions were made using serial dilutions ranging from 0 to 1000 mg/mL. All samples and standards were then placed in a 60°C water bath for 15 minutes. The samples and standards were then allowed to stand for forty minutes and then the absorbance was read at 520 nm with FinstrumentsMicroplate Reader, 110 voltage (Model 314, McLean, VA) in triplicate using deionized water as the blank [31] . Fructose was determined from the calibration curve created from the standards ($y = 0.0008x - 0.0525$, $R^2 = 0.98528$).

2.9. Glucose

Glucose was estimated from the reducing sugars assay developed by Miller et al. [32] . Ten grams of berries were homogenized with 10 mL of deionized water. Samples were centrifuged for 15 minutes at 3500 rpm in a Sorvall RT6000B centrifuge. Dinitrosalicylic acid, sodium sulfite, and sodium hydroxide were combined to make the 1% dinitroslicylic acid reagent solution [32] . A series of standards using a serial dilution were made with glucose with concentrations ranging 0 - 1000 mg/mL. All samples, standards and blank were heated in a 90°C water bath for 15 minutes to yield a red-brown color. After heating, 300 mL of 40% potassium sodium tartrate solution was added to all samples, standards and blanks to stabilize the color. The solutions were allowed to stand at room temperature for 30 minutes and then the absorbance was read at 560 nm with a FinstrumentsMicroplate Reader, 110 voltage (Model 314, McLean, VA) in triplicate. Glucose concentration was determined from the linear regression created from the standards ($y = 0.0006x - 0.0315$, $R^2 = 0.9923$).

2.10. Statistical Analysis

Statistical analysis was conducted with Graph Pad Prism Version 6.0f. The independent t-test was used to determine differences in soilless and soil grown berries, with a significance level set at $p < 0.05$. Results are expressed as mean ± standard deviation (SD).

3. RESULTS

3.1. Brix and % Moisture

The results for the moisture content and Brix are shown in Table 1 and illustrated in Figure 1 and Figure 2. Both the soil strawberries and raspberries had a significantly higher Brix value compared to the soilless strawberries and raspberries. The soil grown strawberry had a Brix value 28% higher compared to the soilless strawberry. The raspberry grown in soil had a Brix value 31% higher than the soilless raspberry. There were no significant differences between the percent moisture content between soilless strawberries and raspberries.

Table 1. Brix measurement and % moisture content (mg/100g) of soil and soilless berries (mean ± SD).

	Soilless Strawberries	Soil Strawberries	t	p
Brix	7.5 ± 0.18	9.6 ± 0.23	7.09	<0.0001
% Moisture	90.74 ± 1.06	89.3 ± 0.88	1.05	0.34
	Soilless Raspberries	**Soil Raspberries**	**t**	**p**
Brix	11.7 ± 0.47	8.9 ± 0.25	5.28	<0.0001
% Moisture	86.4 ± 0.61	85.8 ± 0.98	0.55	0.61

N = 3 with 3 replicates.

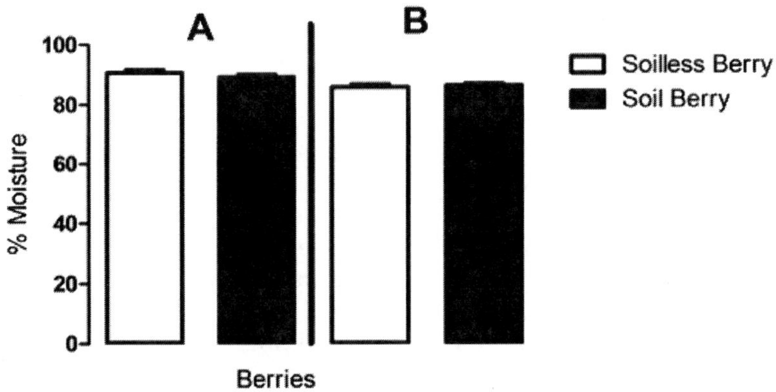

Figure 1. Percent moisture. (A) Indicates soil-grown and soilless- grown strawberries; (B) Indicates soil-grown and soilless-grown raspberries. Star (*) indicates significant differences.

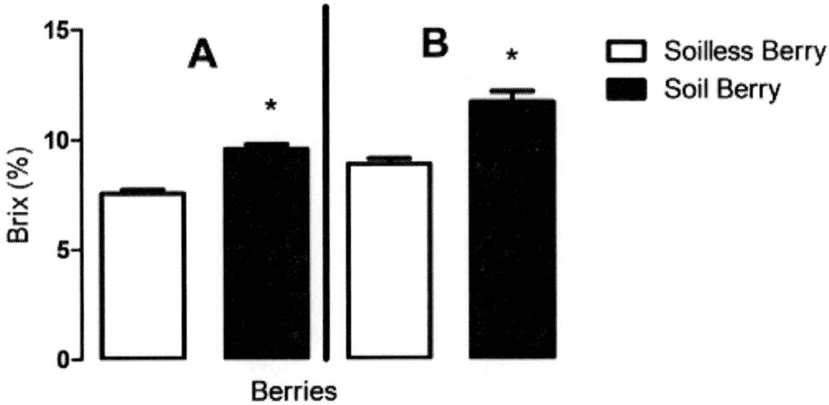

Figure 2. Brix %. (A) Indicates soil-grown and soilless-grown strawberries; (B) Indicates soil-grown and soilless-grown raspberries. Star (*) indicates significant differences.

3.2. Ascorbic Acid, α-Tocopherol, and Total Polyphenolic Compounds

For ascorbic acid, α-tocopherol and total polyphenolic compounds analyzed, soilless grown strawberries were significantly higher compared to soil grown strawberries ($p < 0.05$). For the raspberries, soil grown berries had higher amounts of bioactive compounds compared to the soilless grown raspberries (Table 2, Figures 3-5).

Ascorbic acid content for the soilless grown strawberries contained 74% more compared to the content found in soil grown strawberries. The soilless grown raspberries contained 14% less ascorbic acid content compared to the soil grown raspberries. The α-tocopherol content of soilless grown strawberries was 53% higher compared to the soil grown strawberries. The soil grown ascorbic acid content of raspberries compared to the ascorbic acid content of soilless grown raspberries was a 7% higher amount but was not significant, $p > 0.05$. A significant difference was observed in total polyphenolics, with soilless grown strawberries having significantly higher amounts of total polyphenolics and soilless grown raspberries having significantly less total polyphenolics. The soilless grown strawberries contained 22% higher amounts of total polyphenolic compounds compared to the soil grown strawberries. The opposite trend was seen with the raspberries. The soilless grown raspberries contained 23% less compared to the soil grown raspberries.

3.3. Fructose and Glucose

Fructose and Glucose results are outlined in Table 3 and are illustrated in Figure 6 and Figure 7. The results indicated soil grown strawberries and raspberries contained significantly higher amounts of sugars compared to the soilless grown fruit. The soil grown strawberry contained 75% higher amount of fructose than the soilless grown strawberry. The soil grown raspberry contained 102% higher amount of fructose compared to the soilless grown raspberry. The soil grown strawberries contained 158% higher amount of glucose compared to the soilless grown strawberries. The raspberries showed a similar trend with the soil grown raspberry containing 175% higher amount of glucose compared to the soilless raspberry.

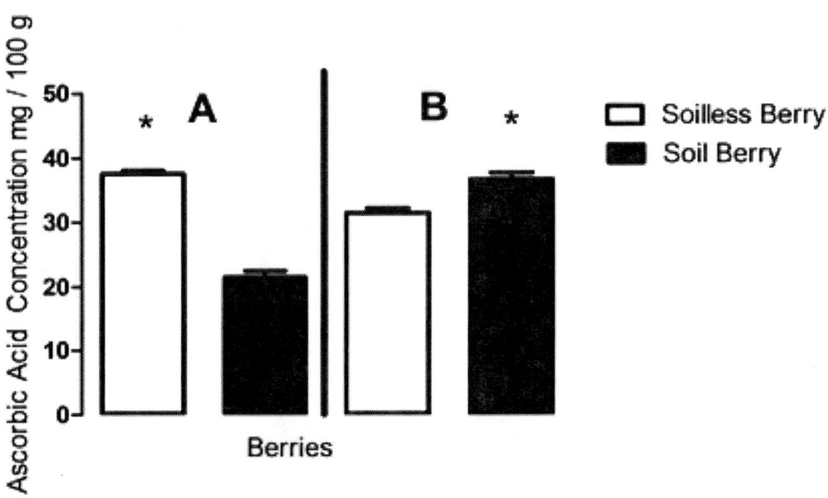

Figure 3. Ascorbic acid concentration. (A) Indicates soil-grown and soilless-grown strawberries; (B) Indicates soil-grown and soilless-grown raspberries. Open bar shows soilless-grown berries and dark bar shows soil-grown berries. Star (*) indicates significant differences.

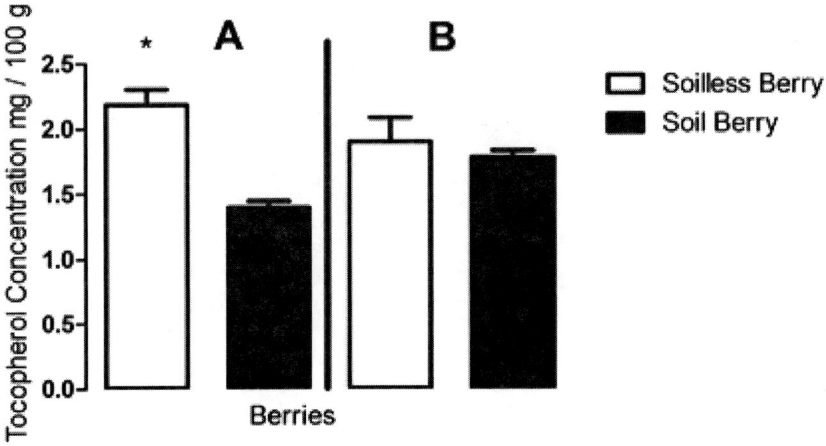

Figure 4. α-Tocopherol concentration. (A) Indicates soil-grown and soilless-grown strawberries; (B) Indicates soil-grown and soilless-grown raspberries. Star (*) indicates significant differences.

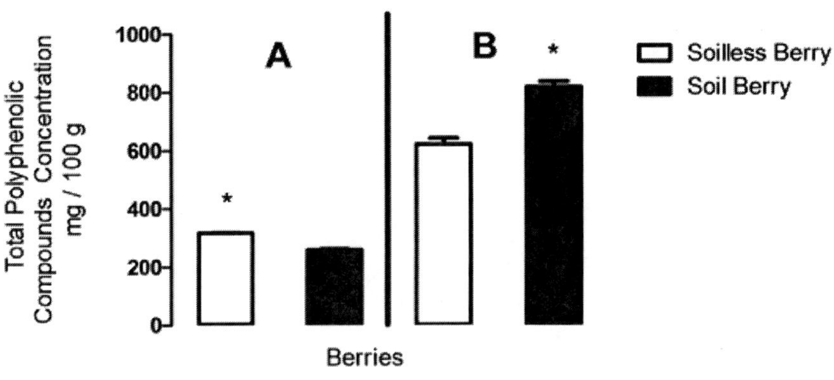

Figure 5. Total polyphenolic compound concentration. (A) Indicates soil-grown and soilless-grown strawberries; (B) Indicates soil-grown and soilless-grown raspberries. Open bar shows soilless-grown berries and dark bar shows soil-grown berries. Star (*) indicates significant differences.

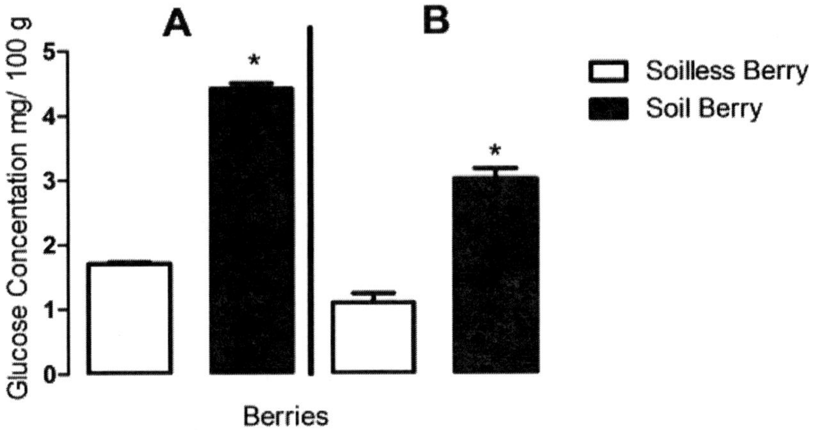

Figure 6. Glucose concentration. (A) Indicates soil-grown and soilless-grown strawberries; (B) Indicates soil-grown and soilless-grown raspberries. Star (*) indicates significant differences.

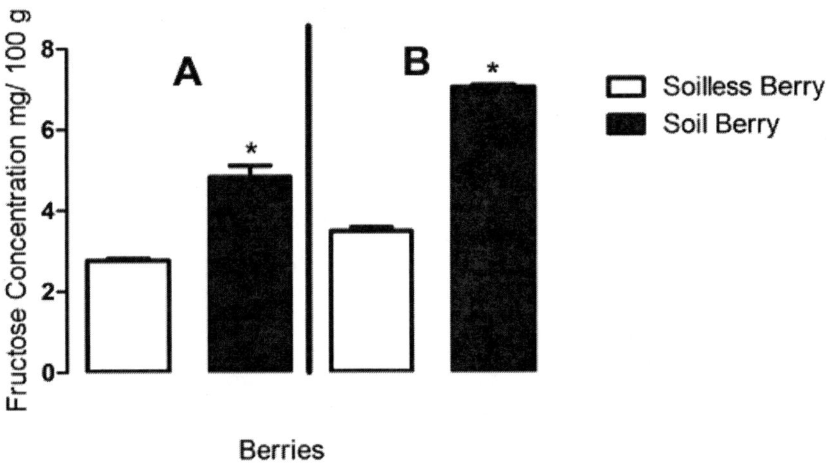

Figure 7. Fructose concentration. (A) Indicates soil-grown and soilless-grown strawberries; (B) Indicates soil-grown and soilless-grown raspberries. Star (*) indicates significant differences.

Table 2. Ascorbic acid, tocopherol and total phenolic content (mg/100g) of soil and soilless berries (mean ± SD).

	Soilless Strawberries	Soil Strawberries	t	p
Ascorbic Acid	37.62± 0.49	21.52 ± 0.95	15	<0.0001
α-tocopherol	2.19 ± 0.12	1.40 ± 0.05	6.05	<0.0001
Total phenolics	317 ± 2.35	259 ± 1.97	18.76	<0.0001
	Soilless Raspberries	Soil Raspberries	t	p
Ascorbic Acid	31.47 ± .074	36.74 ± 0.97	4.3	0.0006
α-tocopherol	1.90 ± 0.85	1.78 ± 0.19	6.05	0.53
Total phenolics	622 ± 20.06	818 ± 19.28	7.03	<0.0001

N = 3 with 3 replicates.

Table 3. Fructose and glucose content (mg/100g) of soil and soilless berries (mean ± SD).

	Soilless Strawberries	Soil Strawberries	t	p
Fructose	2.76 ± 0.43	4.83 ± 0.28	7.29	<0.0001
Glucose	1.71 ± 0.03	4.42 ± 0.07	33.1	<0.0001
	Soilless Raspberries	Soil Raspberries	t	p
Fructose	3.48 ± .096	7.04 ± 0.044	33.8	<0.0001
Glucose	3.02 ± 0.16	1.10 ± 0.15	8.87	<0.0001

N = 3 with 3 replicates.

4. DISCUSSION

Strawberries grown in soilless conditions have higher amounts of bioactive compounds compared to strawberries grown in soil, similar to those who compared bioactive production in produce (Buchanan & Omaye, 2013; Claudia Kiferle, Mariella Lucchesini, Anna Mensuali-Sodi, Rita Maggini & Pardossi, 2011; Palermo, Paradiso, De Pascale, & Fogliano, 2012; Premuzic, Bargiela, Garcia, Rendina, & Iorio, 1998a). However, bioactive compound contents of raspberries were equal to or greater than soil grown raspberries agreeing with others [10] [11] , reiterating that the nutrient density of plants grown by soilless systems is likely highly dependent on the cultivar of interest, environmental conditions (i.e., water stress) and fertilizer bioavailability.

Differences in ascorbic acid may be due to the amount of oxidative stress the plant endures, e.g. ascorbic acid in the biologically active role as an antioxidant. Soilless systems optimize growing conditions, therefore, soilless grown plants are less likely to undergo oxidative stress endured by environmental causes [33] . Ascorbic acid and α-tocopherol work together for antioxidant protection. When tocopherol is oxidized to the tocopheroxyl radical, ascorbic acid can donate electrons to rejuvenate α-tocopherol. Because of the interaction between ascorbic acid and tocopherol, concentration changes in one should be reflective of concentration changes in the other. Lighting (i.e., shading) and fertilizer application can affect ascorbic acid production in plants. Ascorbic acid is created during photosynthesis, however, both of our plant growing systems had the same exposure to light therefore it is more likely the causes were induced by differences in nutrient content. Soilless and soil grown systems are fundamentally different, with soilless having more nutrients bioavailable to the plants all the time.

In agreement with our findings previous research has expressed higher rates of fertilizer increased ascorbic production at the expense of decreasing carbohydrates in the plants [34] . In our soilless system, strawberries, had significantly higher amounts of ascorbic acid but lower amounts of fructose and glucose ($p < 0.05$). Another possible reason for the lower sugar content in the soilless plants compared to soil plants is the potential for higher osmotic pressure in soil plants, increasing the sugar content of the plants. This can commonly occur when plants are drought stressed since plant survival largely depends on carbohydrates [13] . Although our plants were never intentionally drought stressed, it is possible that compared to the soilless plants, which were continuously immersed in water, they may have endured some degree of drought stress with being watered three times weekly. Previous research has indicated a relationship to fertilization and nutritional outcomes in the crop [35] -[39] . Both soil and soilless fertilization concentration was checked using portable ppm meters. The average of the soilless grown strawberries averaged around 400 ppm, and the soil grown plants averaged around 600 ppm. The soilless grown raspberries averaged about 500 ppm and the soil grown plants averaged about 600 ppm, which may influence the differences in nutritional variation within the produce.

5. CONCLUSION

Other research has shown a difference between soilless growing systems and nutritional content of the plant [40] . In order to optimize plant production as well as provide a nutrient dense crop, more research should be conducted to determine the best methods for strawberry and raspberry production. Further research should evaluate feasibility as well as nutritional value of soilless raspberries. We have seen that soilless strawberries have the potential to provide a superior nutrient dense crop compared to soil grown plants. The soilless system has many environmental benefits to provide sustainable food in arid or urban regions. This, added with superior nutrition quality, may contribute significantly to environmental and public health issues that we are currently facing.

ACKNOWLEDGEMENTS

Chenin Treftz, Stanley T. Omaye We thank Eric Horton for greenhouse technical and maintenance assistance. We thank Kevin Hong for laboratory assistance. We acknowledge the Nevada Agricultural Experiment Station, University of Nevada, Reno for the support of this study. HATCH #0745. The research was done and reported to partially fulfill dissertation requirement of C. Treftz.

REFERENCES

1. Agriculture, USDA (2015) Food Deserts. Create Access to Healthy Affordable Food, 1.
2. Steinmetz, K.A. and Potter, J.D. (1996) Vegetables, Fruit, and Cancer Prevention: A Review. Journal of the American Dietetic Association, 96, 1027-1039.
3. Blasa, M., Gennari, L., Angelino, D. and Ninfali, P. (2010) Fruit and Vegetable Antioxidants in Health. In: Bioactive Foods in Promoting Health: Fruits and Vegetables, Academic Press, San Diego, 37-58.
4. Resh, H.M. and Howard, M. (2002) Hydroponic Food Production: A Definitive Guidebook for the Advanced Home Gardener and the Commercial Hydroponic Grower. EUA, St. Bárbara.
5. Glanz, K., Basil, M., Maibach, E., Goldberg, J. and Snyder, D. (1998) Why Americans Eat What They Do: Taste, Nutrition, Cost, Convenience, and Weight Control Concerns as Influences on Food Consumption. Journal of the American Dietetic Association, 98, 1118-1126.
6. Buchanan, D.N. and Omaye, S.T. (2013) Comparative Study of Ascorbic Acid and Tocopherol Concentrations in Hydroponic- and Soil-Grown Lettuces. Food and Nutrition Sciences, 4, 1047-1053.
7. Palermo, M., Paradiso, R., De Pascale, S. and Fogliano, V. (2012) Hydroponic Cultivation Improves the Nutritional Quality of Soybean and Its Products. Journal of Agricultural and Food Chemistry, 60, 250-255.
8. Premuzic, Z., Bargiela, M., Garcia, A., Rendina, A. and Iorio, A. (1998) Calcium, Iron, Potassium, Phosphorus, and Vitamin C Content of Organic and Hydroponic Tomatoes. HortScience, 33, 255-257.
9. Kiferle, C., Lucchesini, M., Mensuali-Sodi, A., Maggini, R., Raffaelli, A. and Pardossi1, A. (2011) Rosmarinic Acid Content in Basil Plants Grown in Vitro and in Hydroponics. Central European Journal of Biology, 6, 946-957.
10. Kimura, M. and Rodriguez-Amaya, D.B. (2003) Carotenoid Composition of Hydroponic Leafy Vegetables. Journal of Agricultural and Food Chemistry, 51, 2603-2607.
11. Selma, M.V., Luna, M.C., Martínez-Sánchez, A., Tudela, J.A., Beltrán, D., Baixauli, C. and Gil, M.I. (2012) Sensory Quality, Bioactive Constituents and Microbiological Quality of Green and Red Fresh-Cut Lettuces (Lactuca sativa L.) Are Influenced by Soil and

Soilless Agricultural Production Systems. Postharvest Biology and Technology, 63, 16-24.

12. Gruda, N. (2005) Impact of Environmental Factors on Product Quality of Greenhouse Vegetables for Fresh Consumption. Critical Reviews in Plant Sciences, 24, 227-247.
13. Brestic, M. and Zivcak, M. (2013) PSII Fluorescence Techniques for Measurement of Drought and High Temperature Stress Signal in Crop Plants: Protocols and Applications. In: Rout, G.R. and Das Bandhu, A., Eds., Molecular Stress Physiology of Plants, Springer India, New Delhi, 87-131.
14. Gruda, N. (2009) Do Soilless Culture Systems Have an Influence on Product Quality of Vegetables ? Journal of Applied Botany and Food Quality, 82, 141-147.
15. Meyers, K.J., Watkins, C.B., Pritts, M.P. and Liu, R.H. (2003) Antioxidant and Antiproliferative Activities of Strawberries. Journal of Agricultural and Food Chemistry, 51, 6887-6892.
16. Hannum, S.M. (2004) Potential Impact of Strawberries on Human Health: A Review of the Science. Critical Reviews in Food Science and Nutrition, 44, 1-17.
17. Liu, M., Li, X.Q., Weber, C., Lee, C.Y., Brown, J. and Liu, R.H. (2002) Antioxidant and Antiproliferative Activities of Raspberries. Journal of Agricultural and Food Chemistry, 50, 2926-2930.
18. Combs, G.F. (2008) The Vitamins: Fundamental Aspects in Nutrition and Health. Elsevier B.V., Burlington.
19. Zhang, Y., Seeram, N.P., Lee, R., Feng, L. and Heber, D. (2008) Isolation and Identification of Strawberry Phenolics with Antioxidant and Human Cancer Cell Antiproliferative Properties. Journal of Agricultural and Food Chemistry, 56, 670-675.
20. US Department of Agriculture, Agricultural Research Service (2011) USDA National Nutrient Database for Standard Reference, Release 24.
21. Crisosto, C.H., Crisosto, G. and Neri, F. (2006) Understanding Tree Fruit Quality Based on Consumer Acceptance. Acta Horticulturae, 712, 183-189.
22. Zude, M., Herold, B., Roger, J.M., Bellon-Maurel, V. and Landahl, S. (2006) Non-Destructive Tests on the Prediction of Apple Fruit Flesh Firmness and Soluble Solids Content on Tree and in Shelf Life. Journal of Food Engineering, 77, 254-260.
23. USDA (2013) Technical Procedures Manual.
24. AOAC (1998) Official Methods of Analysis of AOAC International. Association of Official Analytical Chemists International, CD-ROM.
25. Omaye, S.T., Turnbull, J.D. and Sauberlich, H.E. (1979) Selected Methods for the Determination of Ascorbic Acid in Animal Cells, Tissues, and Fluids. Methods in Enzymology, 62, 3-11.
26. Duygu Ozsoy, H. and Van Leeuwen, J. (2010) Removal of Color from Fruit Candy Waste by Activated Carbon Adsorption. Journal of Food Engineering, 101, 106-112.
27. Kahn, M.M., Rahman, M.M., Islam, S. and Begum, S. (2006) A Simple UV-Spectrophotometric Method for the Determination of Vitamin C Content in Various Fruits and Vegetables at Sylhet Area in Bangladesh. Journal of Biological Sciences, 6, 388-392.
28. Fabianek, J., Apak, E., Gunaydi, E. and Sogen, K. (1968) Micromethod for Tocopherol Determination in Blood Serum. Clinical Chemistry, 14, 456-462.
29. Heinonen, I., Lehtonen, P. and Hopia, A. (1998) Antioxidant Activity of Berry and Fruit Wines and Liquors. Journal of Agricultural and Food Chemistry, 46, 25-31.
30. Marinova, D., Ribarova, F. and Atanassova, M. (2005) Total Phenolics and Total Flavonoids in Bulgarian Fruits and Vegetables. Journal of Chemical Technology and Metallurgy, 40, 255-260.
31. Taylor, K.A.C.C. (1995) A Colorimetric Fructose Assay. Applied Biochemistry and Biotechnology, 53, 215-227.
32. Miller, G.L. (1959) Use of Dinitrosalicyclic Reagent for Determination of Reducing Sugar. Analytical Chemistry, 31, 426-428.

33. Sgherri, C., Cecconami, S., Pinzino, C., Navari-Izzo, F. and Izzo, R. (2010) Levels of Antioxidants and Nutraceuticals in Basil Grown in Hydroponics and Soil. Food Chemistry, 123, 416-422.
34. Premuzic, Z., Bargiela, M., Garcia, A., Rendina, A. and Iorio, A. (1998) Calcium, Iron, Potassium, Phosphorus, and Vitamin C Content of Organic and Hydroponic Tomatoes. HortScience, 33, 255-257.
35. Elia, A., Santamaria, P. and Serio, F. (1998) Nitrogen Nutrition, Yield and Quality of Spinach. Journal of the Science of Food and Agriculture, 76, 341-346.
36. Konstantopoulou, E., Kapotis, G., Salachas, G., Petropoulos, S.A., Karapanos, I.C. and Passam, H.C. (2010) Nutritional Quality of Greenhouse Lettuce at Harvest and after Storage in Relation to N Application and Cultivation Season. Scientia Horticulturae, 125, 93.e1-93.e5.
37. Zhang, Y. (2005) Effects of Nitrogen Supply on Nutritional Quality and Antioxidative Enzyme Activities of Spinach. Chinese Journal of Applied Ecology, 16, 519-523.
38. Randle, W.M. (2000) Increasing Nitrogen Concentration in Hydroponic Solutions Affects Onion Flavor and Bulb Quality. Journal of the American Society for Horticultural Science, 125, 254-259.
39. Sun, Y.D., Luo, W.R. and Liu, H.C. (2014) Effects of Different Nitrogen Forms on the Nutritional Quality and Physiological Characteristics of Chinese Chive Seedlings. Plant, Soil and Environment, 60, 216-220.
40. Gichuhi, P.N., Mortley, D., Bromfield, E. and Bovell-Benjamin, A.C. (2009) Nutritional, Physical, and Sensory Evaluation of Hydroponic Carrots (Daucus carota L.) from Different Nutrient Delivery Systems. Journal of Food Science, 74, S403-S412.

CHAPTER 8

Comparison of Date-Palm Wastes and Perlite as Culture Substrates on Growing Indices in Greenhouse Cucumber

Ahmad Mohamadi Ghehsareh[1]*, Maliheh Hematian[2] and Mahmoud Kalbasi[2]

[1].*Department of Soil Science, Faculty Of AgricultureIslamic Azad University, Khorasgan BranchIsfahanIran*

[2].*Department of Soil Science, Faculty Of AgricultureIslamic Azad University, Khorasgan BranchIsfahanIran*

ABSTRACT

Different substrates have several materials which could have direct and/or indirect effects on plant growth and development. Therefore selecting the best substrate between the various materials is imperative to plant productivity. This research was carried out with using a completely randomized design with three treatments and seven replications. The treatments were including perlite, date-palm waste and soil. During plant growth irrigation rate, temperature, humidity, pest control for all treatments were similar. *Cucumis Stativus* cultivar was used for seeds. During plant growth Papadopolus formula with fertigation method used for nutrient solution. Some physiochemical characteristics of cultivation substrates were measured. The growth indices of cucumber plant were measured at the end of growth period. Results showed that higher amount of yield, biomass weight, SD, plant height, root weight, LAI and fruit TSS related to date-palm waste that had a significant difference with soil at 5% level but generally had not any significant difference at 5% level as compared with perlite substrate.

Keywords: Date-palm waste Perlite Substrate Greenhouse Cucumber

INTRODUCTION

In the recent years a wide range of soilless culture techniques have been developed and commercially introduced for intensive production of horticultural crops, particularly in greenhouses. Reasons for replacing soils as growing media arise from plant protection problems with soil-borne pathogens and environmental regulations against groundwater pollution with nitrate and pesticides. The use of soil in protected agriculture is facing many limitations in this country. Therefore, utilizing substrate-based agriculture is a logical alternative to the current soil-based production approach in the country. The use of different organic and inorganic substrates allows to the plants for best nutrient uptake and sufficient growth and development to optimize water and oxygen holding (Verdonck et al. [1982]). However, different substrates have several materials which could have direct and/or indirect effects on plant growth and development. Therefore, selecting the best substrate among the various materials is imperative to the plant productivity (Olympios [1995]). Replacing soilless growing systems with soil growing for plants especially for cucumber, pepper, tomatoes and other vegetables cause control of plant nutrition and eliminate of plant diseases that caused by soil (Olympios [1995]). Permuzic and et al ([1998]) showed the quality and quantity of tomato fruit in the organic media is better than inorganic media. The results of Inden and Torres ([2004]) on tomato when it cultured in the different substrates showed that the highest amount of total yield and number of fruit were related to Perlite + Rice hull and highest amount of total soluble solids (TSS) related to the Cocopeat substrate. In the other study on tomato plant with perlite and Lica substrates showed that the highest yield and average weight of fruit related to lica media with 12times/days irrigation period that this treatment had not any significant difference with 8 times/day (Shahinrokhsar et al. [2007]). Maximum yield of dry weight was in media include soil + peat (v/v = 50) and sand + peat (25:75) at flowering stage and production fruit (Abdoallahi et al. [2007]). Tomato plants were cultivated on five substrates (rockwool, perlite, and mixtures of perlite to zeolite 1:1, 1:2 and 2:1) in soilless culture with an open system and observed that the highest yield performance obtained by the mixture of perlite and zeolite with 1:1 ratio and highest flowering obtained by perlite substrate. The tomato plants that grown in perlite and zeolite with 2:1 ratio had the best distribution of fruit size, total soluble solid and sensorial quality and so highest dry matter of fruit was found in the perlite substrate (Djedidi et al. [2001]). Alifar et al. ([2010]) investigated the effect of five different growing media including pure Cocopeat, Perlite-Cocopeat(50–50 v/v), Perlite-Cocopeat-Peatmoss (50-20-30 v/v and 50-30-20) and Perlite-Peat moss. Results showed the highest yield of cucumber fruit was obtained from Cocopeat and the lowest one was obtained from Perlite-Cocopeat. The other growth indices such as the largest stem diameter, the highest biomass, the highest fruit's number and the highest fruit's size and diameter were obtained from Cocopeat. The effect of the substrate on yield and fruit quality of tomato in soilless culture studied by Tzortzakis and showed that Plants grown in pumice and perlite substrates obtained lower total yield; and higher yield was obtained from maize substrate. Pumice + 50% maize and 100% maize produced higher total number of fruits per plant. Fruit quality parameters such as mean of fruit weight, fruit firmness, total soluble solid, titrable acidity, ascorbic acid and carotenoids were influenced by substrates , while they had not any effect on EC, pH and dry matter content. The results suggested that addition of maize to perlite and pumice could improve properties of inorganic substrates for tomato soilless culture, leading to higher yields and better of quality fruit (Tzortzakis et al. [2008]). Samiei and et al investigated the effect of peat moss and date-palm wastes as substrates on growing of Aglaonema and his result showed that peat moss and date-palm peat were similar in some characteristics such as CEC, pH, EC and organic carbon but water holding capacity in peat moss was higher than date-palm peat. If this characteristic of date-palm peat improve, it would be a proper substitute in the future (Samiei et al. [2005]). Date-palm extensively exist in the world and Iran and produce a lot of residues and wastes per annum. Currently, appropriate management and optimize procure is not to use this material at now. It seems that residues and wastes of date-palm can used as a substrate in greenhouse cultivation; therefore, it decrease problem of date-palm owners same fire. With attention to much development of greenhouse cultivations and using of substrate in greenhouses, the objective of the present research is study of date-palm waste and residues as a substrate and its effect on yield and quality of greenhouse cucumber in soilless culture.

MATERIALS AND METHODS

This research was carried out in the research greenhouse of Islamic Azad universit, khorasgan branch, using a completely randomized design with three treatments and seven replications. The treatments were including perlite, date-palm waste and soil. Average temperature of day and night were 30°C and 18°C respectively and proportional humidity was 37/1-61/2%. During plant growth irrigation rate, temperature, humidity, pest control for all treatments were similar. The *Cucumis Stativus* cultivar was used for seeds. During plant growth Papadopolus formula with fertigation method used for nutrient solution (Papadopolus [1991] and [1994]; Benton Jones J, [2005]) that adjusted its pH until 5.5 to 6. The Physiochemical characteristics of culture media including bulk density, total porosity, WHC and CEC were measured (Baruah and Barthakur ([1998]), Rhoades ([1982]; 1988)). Amount of pH and electrical conductivity (EC) was measured in verdonck method ([1992]). Some growth indices including stem diameter, height of plant, dry and wet mass of plant, dry and wet mass of root, TSS of fruit and fruit yield were measured at the end of growth period. The analysis of all data was done by SPSS statistical software and the comparison of the averages was done by Excel.

RESULT AND DISCUSSION

Substrates

Some physiochemical properties of substrates presented in Table1. The soil texture was sandy loam and its bulk density was higher than other substrates. The amount of bulk density in Palm peat was minima (0.041%); therefore, root media aeration in this treatment is better than others. Porosity percentage that is an index for root media aeration was high for Palm peat substrate (85%) and it was low for soil (37%). When root media aeration is sufficient, supply of water and nutrient elements for plants is easily. The amount of EC (Electric Conductivity) in soil was more than other substrates and this term in Perlite was lower than other treatments but according to amount of leaching requirement (20%), it seems that differences of EC in substrates had not important effects.

Table 1. COMPARISON OF PHYSIOCHEMICAL PROPERTIES OF SUBSTRATES

Substrate	bulk density (%)	Porosity (%)	EC (ds/m)	pH	CEC (Cmol/kg)
Palm waste	0.041	85	1.28	6.7	95
Perlite	0.32	48	0.38	6.9	13
Soil	1.25	37	2.3	7.1	43.1

Amount of CEC in substrates very different and higher and lower amounts related to Palm peat and Perlit respectively; therefore, the Palm peat media had more capacity for supplement of nutrient elements for plant.

Growing indexes of plant

Amount of cucumber yield in different substrates had significant difference at 5% level. Higher amount of yield related to Palm peat media that had not any difference with Perlite media but they had a significant difference with soil. Amount of plant biomass in wet and dry condition in different substrates had not any significant differences at 5% level. Amount of stem diameter (SD) of cucumber plant in Palm waste media was more than other substrates (18 mm) and had significant differences at 5% level as compared with other media and so comparison of means showed that results of plant height were similar to SD. Sufficient conditions with a view to bulk density and porosity in palm peat media caused to good support of water and nutrient elements for plant and leading to good growth. The results of Alifar and et al. ([2010]) indicated that yield of cucumber fruit, plant stem diameter, biomass, fruit's number and fruit's size and diameter was obtained from Cocopeat media. They connected this results to preference of physiochemical properties of Cocopeat as compared with other media such as Perlite.

Table 2. SOME GROWING INDICES OF CUCUMBER THAT CULTIVATED IN DIFFERENT MEDIA

Substrate	Fruit yield (kg)	ABP(W) gr	ABP(D) gr	SD mm	Height cm	Root weight(w) gr	Root weight(D) gr	LAI Cm2	TSS (Brix%)
Palm	25.68a	550.2a	88.44a	18a	328.1a	30.3a	3.5a	277.2a	3.0a
Perlite	22.72a	350.7b	41.95b	8.1b	234.2b	23.3a	3.3a	181.7c	3.7a
Soil	12.77b	297.9b	41.91b	9.2b	258.9b	11.4b	1.8b	236.2b	3.1a

Definitions : LAI = Leaf Area Index , TSS = Total Solution Solid , ABP = Arial Biomass of Plant , W = Wet , D = Dry, SD = Stem Diameter.

The physiochemical properties in date-palm media were better than others that cause to increasing in growing indexes. The low amount of bulk density and the high amount of porosity related to date-palm media that allowed the plant root penetrate in substrate easily and it could use more volume and space of media, thus available water and nutrient elements were sufficient for plants grow up in this media. Also the amount of CEC in date-palm media were higher than others. This term related to nutrient elements, buffer capacity that have an essential role in plant supplement. The amount of physical indices of perlite were between palm and soil; therefore, physiochemical properties of substrates affected on plant growing indices and tend to above results. The effect of substrates include Perlite, date-palm peat, date-palm peat + Perlite and Coco peat + Perlite on growing indices of tomato studied by Mohammadi Ghehsareh et al. ([2010]) and results showed that the yield of tomato fruits and TSS had not any significant differences between treatments at 5% level. Generally, Coco peat and Palm peat media with aspect to physiochemical characteristics had not any significant differences on growing of Aglaonema plant (Samiei et al. [2005]).

CONCLUSIONS

Results of last investigations showed that coco peat and perlite were sufficient substrates for growing of some plants especially for vegetables and grower use this materials as growing media in greenhouse cultures. Also, the results of this investigation indicated that Perlite and date-palm waste media had similar properties and they had not create any significant difference on qualitative and quantitative indices in cucumber fruit. Thus, with considering to low cost, availability and abundance of date-palm cultivation in Iran (242000 ha), it seems that date-palm wastes can be replaced with perlite media for growing of certain vegetable in Iran.

COMPETING INTERESTS

The authors declare that they have no competing interests.

AUTHORS' CONTRIBUTIONS

All authors read and approved the final manuscript.

REFERENCES

1. Abdoallahi K, Movahedi SAA, Mashayekhi K, Mazaheri M: **INVESTIGATION OF CONCENTERATION AND NUTRIENT UPTAKE AND GROWTH TOMATO IN THE MEDIA CULTURE PROVIDE BY PEAT CAPTIVATE WATER MARSH OF SARI.** *Journal Agricultural Sciences and Natural Resources* 2007,**14**(4):35–42. (in Persian) (in Persian)
2. Alifar N, Mohammadi Ghehsareh A: *The effect of Coco peat, Perlite and Peat moss on some greenhouse cucumber's growth indices in soilless culture. international Soil Science Congress on " management of natural resources to sustain soil health and quality.* Ondokus Mayis University, Sumsun - Turkey; 2010:26–28.
3. Alifar N, Mohammadi Ghehsareh A, Honarjoo N: **THE EFFECT OF GROWTH MEDIA ON CUCUMBER YIELD AND ITS UPTAKE OF SOME NUTRIENT ELEMENTS IN SOILLESS CULTURE.** *J of sci and techn Greenhouse Culture* 2010, **1:** 19–25. Isfahan University of Technology Isfahan University of Technology
4. Baruah TC, Barthakur HB: *A Textbook of Soil Analysis.* Vikas Publishing House PMT Ltd, New Dehli, India; 1998:282.
5. Benton Jones J: **A PRACTICAL GUIDE FOR THE SOILLESS GROWER.** *CRC Press* 2005, **2ND ED.:** 423.
6. Djedidi M, Grasopoulos D, Maloupa E: **THE EFFECT OF DIFFERENT SUBSTRATES ON THE QUALITY OF F. CARMELLO TOMATOES (LYCOPERSICON ESCULLENTUM MILL) GROWN UNDER PROTECTION IN A HYDROPONIC SYSTEM.** *Cahiers Option Mediterranean's* 1997, **31:** 379–83.
7. Inden HA, Torres : **COMPARISON OF FOUR SUBSTRATES ON THE GROWTH AND QUALITY OF TOMATOES.** *Acta Hort. (ISHS).* 2004, **644:** 205–210.
8. Mohammadi Ghehsareh A, Samadi N, Borji H: **COMPARISON OF DATE-PALM WASTES AND PERLITE AS GROWTH SUBSTRATES ON SOME TOMATO GROWING INDEXES.** *Afri J Biote.* 2011,**10**(24):4871–4878.
9. Olympious CM: **SOILLESS MEDIA UNDER PROTECTED CULTIVATION ROCKWOOL, PEAT, PERLITE AND OTHER SUBSTRATES.** *Acta Hort.* 1995, **401:** 443–451.
10. Papadopolus AD: *Growing greenhouse tomatoes in soil and soilless.* Communication Branch, Agriculture, Canada. Ottawa; 1991:79.
11. Papadopolos AP: *Growing greenhouse cucumbers in soil and in soilless media.* , Agriculture and Agri-Food Canada Publication 1902/E.; 1994:108.
12. Permuzic Z, Bargiela M, Garci A, Rendina A: **CALSIUM, IRON, POTASSIUM, PHOSPHOROUS AND VITAMIN C CONTENT OF ORGANIC AND HYDROPONIC TOMATOES.** *Hort Science.* 1998,**33**(2):255–257.
13. Samiei L, KHalighi A, Kafi M, Samavat S, Arghavani M: **AN INVESTIGATION OF SUBSTITUTION OF PEAT MOSS WITH PALM TREE CELLULOID WASTES IN GROWING AGLAONEMA (AGLAONEMA COMMUTATUM CV. SILVER QUEEN).** *Iranian J of Agri Sci.* 2005,**36**(2):503–510. in Persian in Persian
14. Sahinrokhsar P, Davari K, Peyvast GH, GHahraman B, Nemati H: **INFLUENCE OF IRRIGATION INTERVAL AND SUBSTRATES ON PRODUCTION OF TOMATO (CV.HAMRA) IN SOILLESS (BAG)CULTURE.** *J. Agri Eng Res* 2007,**8**(1):31–46. in Persian in Persian
15. Tzortzakis N, Economakis GCD: **IMPACTS OF THE SUBSTRATE MEDIUM ON TOMATO YIELD AND FRUIT QUALITY.** *Hort culture Sci.* 2008, **2:** 83–89.
16. Verdonck O, Gabriels R: **REFERENCE METHOD FOR THE DETERMINATION OF PHYSICAL PROPERTIES OF PLANT SUBSTRATES.II. REFERENCE METHOD FOR THE DETERMINATION OF CHEMICAL PROPERTIES OF PLANT SUBSTRATES.** *Acta Horticulturae* 1992, **126:** 251–258.
17. Verdonck O, Vleeschauwer D, De Boodt M: **THE INFLUENCE OF THE SUBSTRATE TO PLANT GROWTH.** *Acta Hort. (ISHS)* 1982, **126:** 251–258.

CHAPTER 9

Utilization of Hairy Footed Flower Bee Anthophora plumipes (Hymenoptera: Apidae) for Pollination of Greenhouse Strawberry

Radha Devkota Adhikari[1,2], Ryoichi Miyanaga[1*]

[1]*Faculty of Life and Environmental Science, Shimane University, Matsue, Japan*

[2]*The United Graduate School of Agricultural Sciences, Tottori University, Tottori, Japan*

ABSTRACT

Pollination efficiency of hairy footed flower bee Anthophora plumipes for greenhouse strawberry was evaluated in a small greenhouse located in Shimane University, Matsue, Japan (lat. 35°29', long. 133°27' and elevation 170 m). Twenty five female and six male bees were released in a greenhouse on April 15, 2014. Red clay soil blocks were provided as the nesting materials. Two hundred and fifty pots of strawberry (Fragaria annanasa) were allocated for the following three treatments: 1) no supplementary pollination (control), 2) hand pollination, and 3) bee pollination. Both male and female bees effectively foraged flowers throughout the day. Rate of fertilized seed by the bee pollination was significantly greater than no supplementary pollination (control) and similar to that in the hand pollinated flowers. Quality of the fruits (evaluated based on the shape and deformity) was also improved by the bee pollination. A. plumipes effectively pollinated strawberry flowers in the greenhouse and could be developed as an alternative pollinator for this crop.

Keywords: Anthophora plumipes, Greenhouse, Strawberry, Pollination, Fruit Quality

1. INTRODUCTION

Cultivation of strawberry in greenhouses is increasing worldwide for getting better quality and higher value of the fruits. In Japan, about 90% of the total strawberry production is carried out under protected structures [1] . Strawberry flowers are hermaphrodite and self-compatible to a certain extent. Berry deformity is a severe problem in commercial cultivation of strawberries in greenhouses. Strawberry flowers contain many carpels and it is necessary that each carpel should contain a fertilized ovule (seed) in order to produce a well-formed berry. Berries are deformed at parts in which achenes are not fertilized. Therefore, a successful pollination is required for seed formation and better quality of the fruits. Studies have shown that number of fertilized achenes and quality of strawberry fruit are directly related to the pollination [2] -[5] . Since fertilization of strawberry has direct influence on the fruit quality, it is crucial that the flower inside greenhouses needs agents for pollination.

Hand-pollination using special devices such as hand, brushes or electric brushes, is commonly practiced for the pollination of strawberry in greenhouses [6] . Although these methods improve the pollination efficiency, some devices are expensive and have risks of damaging the flowers. Alternatively, bees have been used as pollinators for quality fruit production under the greenhouse.

European honeybees (Apis mellifera) are commonly used as pollinators for strawberries inside the greenhouse [7] -[10] and in the open fields [11] [12] . Bumble bees are an alternative pollinator to honeybees for strawberries [7] [10] [13] . Several species of stingless bees, including Nannotrigona testaceicornis (Lepeletier), Trigona minangkabau and Tetragonisca angustula (Latreille) have been successfully tested for strawberry pollination inside greenhouses in Japan [14] -[17] . Some Megachilidae, such as Osmia rufa and O. cornifrons have also been found to be effective for pollinating this crop [18] [19] .

Hairy footed flower bee Anthophora plumipes (Pallas) (Hymenoptera: Apidae) is a ground-nesting, solitary bee, occurring from Western Europe to East Asia [21] . It is one of the first bee species active during the spring [20] -[22] . In a previous study, we found that A. plumipes females were manageable pollinators with stable flight activities in the greenhouse [23] .

In this study, pollination efficiency of A. plumipes bees was evaluated for developing alternative pollinators for strawberry under greenhouse conditions. This paper describes the foraging activities of A. plumipes bees on greenhouse strawberry and quality of the fruits pollinated by the bee.

2. MATERIALS AND METHODS

2.1. Bees and Nesting Materials

Adult A. plumipes bees were collected from their natural habitat in a forest located in Izumo City, Shimane prefecture, Japan. The collected bees were kept in an incubator for about one month at 6°C before liberating in the greenhouse. Twenty five female and six male bees were released in the greenhouse for the pollination experiment. Soil blocks were provided as nesting substrates of the bees in the greenhouse (Figure 1(B)). The blocks were constructed using red clay soil.

2.2. Greenhouse and Treatments

A small greenhouse (8.1 m long, 4.5 m wide, and 2.9 m high) located at Shimane University, Matsue, Japan (lat. 35°29', long. 133°27', and elevation 170 m) was used in this study. Two hundred and fifty strawberry (Fragaria ananassa) (cv. Hokowase) seedlings at 5-leaf stage were planted in the plastic pots (each pot was 35 cm wide and 40 cm deep) in October 2013 (Figure 1(A)). Twenty, 50, and 180 pots of strawberry were allocated for control (no supplementary pollination), hand-pollination (using a brush), and bee pollination, respectively. Plants receiving hand-pollination and no supplementary pollination (control) were covered with nets before releasing the bees. The bees were released on April 15, 2014. The flowers bloomed before liberating the bees were hand- picked.

During the experiment, temperature inside the greenhouse ranged from 10.7°C to 29.6°C, relative humidity was from 21% to 77%, and solar radiation levels were from 1082 to 1830 lux.

Figure 1. Photographs showing strawberry plants in a greenhouse (A), soil blocks for nesting (B), a A. plumipes bee visiting strawberry flower (C), and a A. plumipes bee digging its nest in soil block (D).

2.3. Fertilizer, Disease and Insect Management

Pots were filled with soil mixed with composted cow manure (2%, w/w). Ammonium sulfate (3 g/pot) was top dressed just before the flower initiation. No pesticide was applied for controlling disease and insects. Insects were hand-picked when observed.

2.4. Observation of Foraging Activities

Foraging and related activities of the bees after liberation in the greenhouse were observed during the flowering period. Number of flowers visited per unit of time and handling time (time spent per flower) by male and female bees were also recorded.

2.5. Measurement of Fruit Quality Parameters

Strawberries were harvested every two to three days from May 8 to May 31, 2014 when fruits were red. Number of fertilized and non-fertilized achenes were recorded to evaluate the pollination efficiency. After separating the seeds from the pulp, it was taken in a beaker filled with water. Submerged and floated seeds in the water were recorded as fertilized and non-fertilized achenes, respectively.

Fruit length and width were measured using Vernier calipers. Five-level deformity ranking was made based on the measurement of shapes and visual observations as follows: quite few unfertilized achenes and not deformed (A), some unfertilized achenes but not deformed (B), many unfertilized achenes and slightly deformed (C), quite unfertilized achenes but slightly deformed (D), and many unfertilized achenes and highly deformed (E).

2.6. Data Analysis

Significance of three pollination treatments was evaluated by analyzing the data using ANOVA (Minitab version 14). Treatment means were compared using the least significant difference test at $p < 0.05$.

3. RESULTS

3.1. Foraging Behavior of A. plumipes

The bees started foraging nectar from the strawberry flowers soon after the liberation. Pollen collection was observed after four days from the liberation in the greenhouse. While visiting the flowers (Figure 1(C)), the bees always landed on the top of the flower with a circular movement either from right to left or left to right. The bees landed at the corner of flower disc for collecting nectar and walked to the whole pistils for collecting pollen. Both females and males successfully visited strawberry flowers (Table 1). The average time (±SD) spent on the flower (handling time) by the male bees was 8.7 sec (±4.9) ($n = 132$) and that of the females was 11.0 sec (±6.9) ($n = 340$). Male bees visited an average of 10.2 (±3.0) ($n = 28$) flowers per minute, while females visited 6.5 (±3.0) ($n = 68$) flowers per min.

3.2. Effect of Pollination on Seed Fertilization

Figure 2 shows the rate of seed fertilization for each treatments. The rate of fertilization in bee-pollinated flowers was 59.9%, which was significantly higher than that in the control (31.0%). No significant difference in seed fertilization rate was observed between the bee- and hand-pollination indicating the A. plumipes bees can effectively pollinate the strawberry flowers in the greenhouse.

Table 1. Hourly flower handling and visiting patterns of males and females of A. plumipes bees on strawberry flowers in a greenhouse.

Time	Handling time (sec/flower) ±SD (number of observations)		Number of flower visited/min ±SD (number of observations)	
	Male	Female	Male	Female
8:00-9:00	6.3 ± 3.0 (8)	12.7 ± 8.4 (6)	13.4 ± 3.0 (9)	5.5 ± 4 (6)
9:00-10:00	7.7 ± 4 (15)	12.6 ± 8.7 (27)	11.1 ± 4.0 (4)	7.1 ± 4 (17)
10:00-11:00	9.0 ± 5.6 (35)	13.2 ± 8.3 (29)	10.5 ± 2.0 (9)	7.5 ± 2.0 (6)
11:00-12:00	7.1 ± 3.7 (25)	9.4 ± 5.3 (68)	11.3 ± 5.0 (3)	8.7 ± 1.0 (9)
12:00-13:00	7.7 ± 4.8 (39)	8.0 ± 2.6 (43)	6.2 ± 1.0 (2)	7.3 ± 5.0 (9)
13:00-14:00	8.7 ± 4.5 (4)	10.77 ± 6.4 (47)	8.5 ± 0.0 (1)	4.9 ± 5.0 (4)
14:00-15:00	7.5 ± 3.9 (2)	11.0 ± 8.9 (61)	ND	5.2 ± 2.0 (8)
15:00-16:00	15.6 ± 9.3 (4)	10.5 ± 6.3 (59)	ND	5.5 ± 1.0 (9)
Mean	8.7 ± 4.9 (132)	11.0 ± 6.9 (340)	10.2 ± 3 (28)	6.5 ± 3 (68)

ND, Not detected.

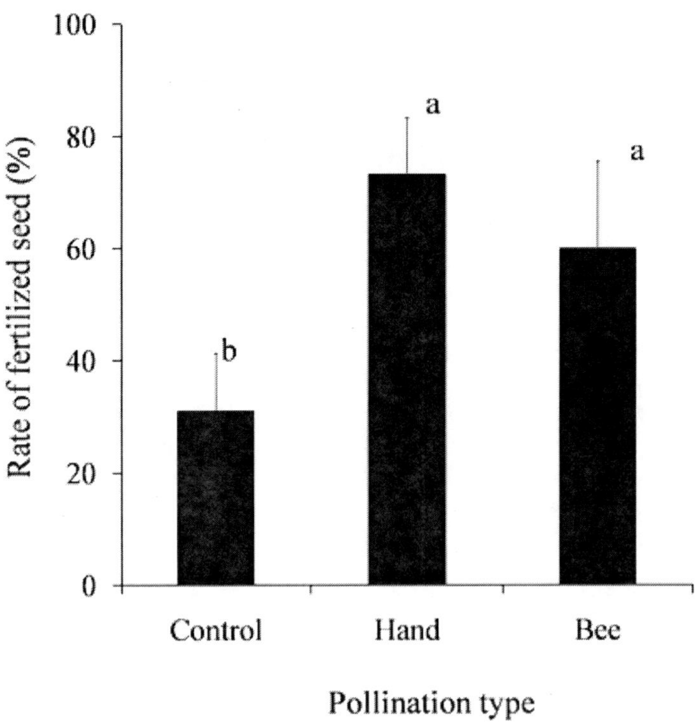

Figure 2. Rate of fertilized seeds in control, hand-pollinated, and A. plumipes bee pollinated strawberry flowers. Values with the same letter are not significant ($p < 0.05$, least significant difference test).

3.3. Fruit Quality

The shape of the fruits varied among the three pollination treatments (Figure 3). The length/width ratio of almost all bee- and hand-pollinated fruits (>95%) was from 1.0 to 1.5, while the length/width ratio in non-pollinated (control) fruits was 40% only. As the length/width ratio of strawberry fruits from 1.0 to 1.5 is highly preferred by the Japanese consumers, pollination seems necessary for not only increasing the fruit production, but also for getting higher profit.

Less deformed fruits were observed in bee pollinated and hand-pollinated fruits (Figure 4). On a 5-scale deformity ranking basis, most of the fruits in non-pollinated control were medium to highly deformed, while a large proportion of the fruits were normally formed with bee-pollination.

4. DISCUSSION

A. plumipes pollinated strawberry flowers in the greenhouse as effectively as hand pollination. Both female and male bees visited the flowers efficiently suggesting this bee as a potential alternative pollinator for greenhouse strawberry. The seed fertility and quality were significantly higher in bee- and hand-pollinated flowers compared with the flowers which received no supplementary pollination. These results are similar to those reported for strawberry pollination by honeybees under greenhouse conditions [24] .

A. plumipes bees showed a high rate of adaptability in closed condition and started foraging and nesting activities soon after liberation [23] , while no such adaptability to the closed condition is reported in honey bees, bumble bees, stingless bees, carpenter bees and Osmia spp. [15] [16] [19] [25] . Results indicated that when the weather condition is adverse, A. plumipes could be a suitable pollinator for strawberry and other flowering plants.

Although, the handling time and foraging duration of A. plumipes was longer than that of Osmia cornifrons in a similar study [19] , the longer handling time in this study might be due to a small number of flowers during the active foraging period.

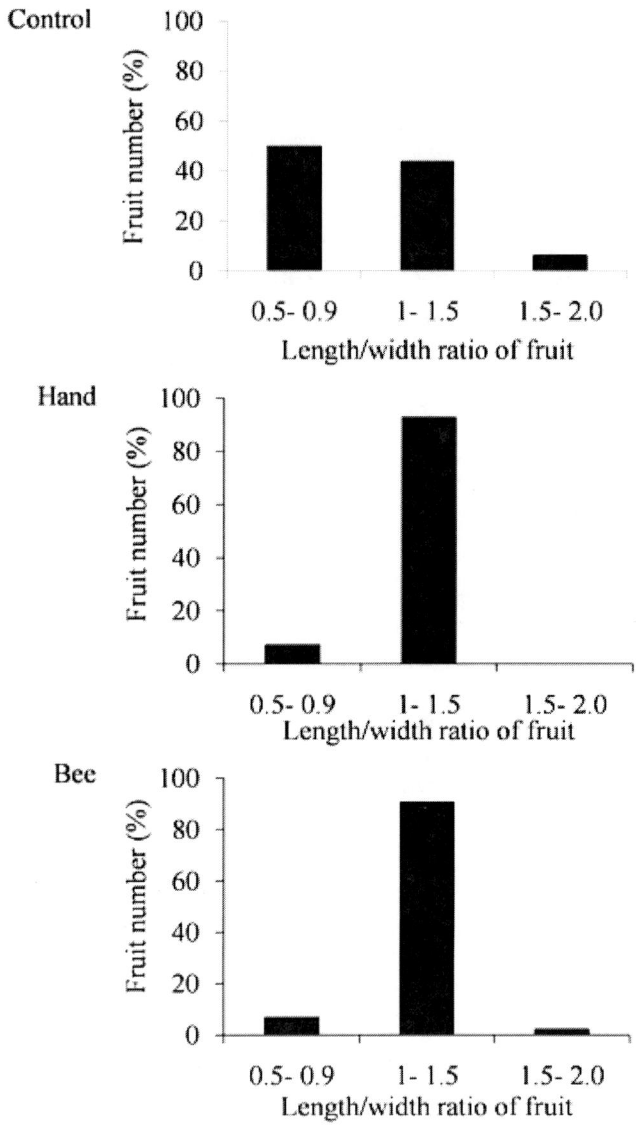

Figure 3. Proportion of fruits with different shapes (length/width ratio) in control, hand-pollinated, and A. plumipes bee pollinated strawberry flowers.

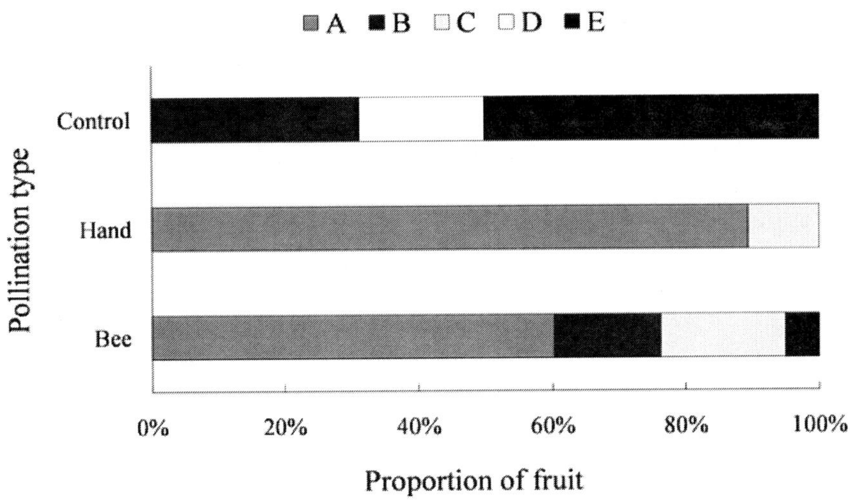

Figure 4. Relative proportion of the fruits of different deformity levels in no supplementary pollination (control), hand pollination and A. plumipes bees pollinated strawberry. Deformity levels: quite few unfertilized achenes and not deformed (A), some unfertilized achenes but not deformed (B), many unfertilized achenes and slightly deformed (C), quite unfertilized achenes but slightly deformed fruits (D), and many unfertilized achenes and highly deformed (E).

Flowers pollinated by bees make a better quality of fruit than self-pollinated and open-pollinated [5] [15] [16] [19] . Also in this study, a higher quality of fruits by the pollination of bees was found. Higher rate of the fertilized seed and shape of the fruits pollinated by the bees suggested that use of A. plumipes could be a good alternative for making higher profit from the greenhouse strawberries.

5. CONCLUSION

A. plumipes bees could be used as an alternative pollinator for greenhouse strawberry. For the sustainability of greenhouse strawberry production using A. plumipes as a pollinator, study on the reproductive ability of this bee under greenhouse condition is necessary when strawberry is the only floral resource.

CITE THIS PAPER

Radha Devkota Adhikari, Ryoichi Miyanaga (2016) Utilization of Hairy Footed Flower Bee *Anthophora plumipes* (Hymenoptera: Apidae) for Pollination of Greenhouse Strawberry. *Advances in Entomology*,**04**,25-31.

REFERENCES

1. Takeda, F. (1999) Out-of-Season Greenhouse Strawberry Production in Soilless Substrate. Advances in Strawberry Research, 18, 4-15.
2. Nitsch, J.P. (1950) Growth and Morphogenesis of Strawberries as Related to Auxin. American Journal of Botany, 37, 211-215.
3. McGregar, S.E. (1976) Insect Pollination of Cultivated Crop Plants. Agricultural Handbook No. 496, Agricultural Research Service, United States Department of Agriculture.
4. Crane, E. (1990) Bees and Beekeeping: Science, Practice and World Resources. Cornell University Press, 640.
5. Pratap, U. (2000) Pollination of Strawberry by the Asian Hive Bee, Apis cerana. Oxford and IBH, New Delhi, 48, 178-182.
6. Wang, S.Y. and Zheng, W. (2001) Effect of Plant Growth Temperature on Antioxidant Capacity in Strawberry. Journal of Agricultural and Food Chemistry, 49, 4977-4982.
7. Free, J.B. (1993) Insect Pollination of Crops. 2nd Enlarged Edition, Academic Press, London, 684.
8. Chang, Y.D., Lee, M.Y. and Mah, Y. (2000) Pollination on Strawberry in the Vinyl House by Apis mellifera L. and Apis cerana Fab. Acta Horticulture, 561, 257-262.
9. Li, J.L., Peng, W.J., Wu, J., An, J.D., Guo, Z.B., Tong, Y.M. and Huang, J.X. (2006) Strawberry Pollination by Bombus lucorum and Apis mellifera in Greenhouses. Acta Entomologica Sinica, 49, 342-348.
10. Zaitoun, S.T., Al-Ghzawia, A.A., Shannag, H.K., Rahman, A. and AL-Tawaha, A.R.M. (2006) Comparative Study on the Pollination of Strawberry by Bumble Bees and Honey Bees under Plastic House Conditions in Jordan Valley. Journal of Food Agriculture and Environment, 4, 237-240.
11. Albano, S., Salvado, E., Duarte, S., Mexia, A. and Borges, P.A.V. (2009) Pollination Effectiveness of Different Strawberry Floral Visitors in Ribatejo, Portugal: Selection of Potential Pollinators. Part 2. Advances in Horticultural Sciences, 23, 246-253.
12. Chagnon, M., Ingras, J. and Oliveira, D. (1993) Complementary Aspects of Strawberry Pollination by Honey and Indigenous Bees (Hymenoptera). Journal of Economic Entomology, 86, 416-420.
13. Payadas, S., Eti, S., Kaftanoglu, O., Yasa, E. and Derin, K. (2000) Effects of Pollination of Strawberries Grown in Plastic Greenhouses by Honeybees and Bumblebees on the Yield and Quality of the Fruits. Acta Horticulturae, 513, 443-451.
14. Maeta, Y., Tezuka, T., Nadano, H. and Suzuki, K. (1992) Utilization of the Brazilian Stingless Bee Nannotrigona testaceicornis as a Pollinator of Strawberries. Honey Bee Science, 13, 71-78. (In Japanese with English Summary)
15. Kakutani, T., Inoue, T., Tezuka, T. and Maeta, Y. (1993) Pollination of Strawberry by the Stingless Bee, Trigona minangkabau, and the Honey Bee, Apis mellifera: An Experimental Study of Fertilization Efficiency. Researches on Population Ecology, 35, 95-111.
16. Roselino, A.C., Santos, S.B., Hrncir, M. and Bego, L.R. (2009) Differences between the Quality of Strawberries (Fragaria × ananassa) Pollinated by the Stingless Bees

Scaptotrigona aff. depilis and Nannotrigona testaceicornis. Genetics and Molecular Research, 8, 539-545.

17. Malagodi-Braga, K.S. and Kleinert, A.D.M.P. (2004) Could Tetragonisca angustula Latreille (Apinae, Meliponini) Be Effective as Strawberry Pollinator in Greenhouses? Crop and Pasture Science, 55, 771-773.
18. Wilkaniec, Z. and Radajewska, B. (1997) Solitary Bee Osmia rufa L. (Apoidea, Megachilidae) as Pollinator of Strawberry Cultivated in an Unheated Plastic Tunnel. Acta Horticulture, 439, 489-493.
19. Maeta, Y., Nadano, H. and Kitamura, K. (2012) Utilization of the Japanese Mason Bee, Osmia cornifrons (Radoszkowski) for Pollination of Strawberries in Greenhouses (Hymenoptera: Megachilidae). Bulletin Hoshizaki Green Found, 15, 19-45.
20. Batra, S.W.T. (1994) Anthophora pilipes villosula Sm. (Hymenoptera: Anthophoridae), a Manageable Japanese Bee That Visits Blueberries and Apples during Cool, Rainy, Spring Weather. Proceedings of Entomological Society of Washington, 96, 98-119.
21. Neeman, G., Shavit, O., Shaltiel, L. and Shmida, A. (2006) Foraging by Male and Female Solitary Bees with Implications for Pollination. Journal of Insect Behavior, 19, 383-401.
22. Stone, G.N. (1995) Female Foraging Response to Sexual Harassment in the Solitary Bee Anthophora plumipes. Animal Behavior, 50, 405-412.
23. Adhikari, R.D., Maeta, Y., Izumi, Y. and Miyanaga, R. Nesting Activity of Anthophora plumipes under the Closed Circumstance (Hymenoptera: Apidae). Japanese Journal of Entomology. (In Japanese with English Summary) (in press)
24. Bond, D.A. and Kirby, E.J.M. (1999) Anthophora plumipes (Hymenoptera: Anthophoridae) as a Pollinator of Broad Bean (Vicia faba Major). Journal of Apicultural Research, 38, 199-203.
25. Sadeh, A., Shmida, A. and Keasar, T. (2007) The Carpenter Bee Xylocopa pubescens as an Agricultural Pollinator in Greenhouses. Apidologie, 38, 508-517.

CHAPTER 10

Effect of Date Palm Wastes and Rice Hull Mixed with Soil on Growth and Yield of Cucumber in Greenhouse Culture

Ahmad Mohammadi Ghehsareh

[1].*Department of Soil Science, Faculty of AgricultureIslamic Azad University, Khorasgan BranchIsfahanIran*

ABSTRACT

Different substrates have several materials which could have direct and/or indirect effects on plant growth and development. The use of different organic and inorganic substrates allows that plants have best nutrient uptake and sufficient growth and development to optimize water and oxygen holding. This work was carried out using a completely randomized design with six treatments and six replications. The treatments were pure palm peat, pure rice hull, soil + 5% (weight) palm peat, soil + 5% (weight) rice hull, soil + 5% (weight) palm peat + 5% (weight) rice hull and pure soil. During plant growth Papadopolus formula with fertigation method was used for nutrient solution. Selected physiochemical properties of culture media and selected growth indices of plants were measured at the end of growth period. Results showed that amount of porosity, water holding capacity (WHC) and cation exchange capacity (CEC) in date-palm peat was higher than soil and rice hull but amount of bulk density in date-Palm peat was lower than the others. Also The results showed that many growth parameters were affected by the culture media. Most amount of yield and plant height in each was related to palm waste (100%) and had significant difference at 5% level as compared with the others. Results showed that plant growing indices for cucumber plant were sufficient when cultured only in date Palm waste and rice hull substrates and when this materials were added to the soil, although it amended physiochemical properties of media but decreased the plant growing indices.

Keywords: Palm peat Rice hull Yield Greenhouse cucumber Soil

INTRODUCTION

When crop residues are returned to the soils, their decomposition can have both positive and negative effects on crop production and the environment (Kumar and Goh 1999). The use of different organic and inorganic substrates allows the plants the best nutrient uptake and sufficient growth and development to optimize water and oxygen holding (Verdonck et al. 1982). Soil organic matter is one of the most important constituents of soils due to its capacity in affecting plant growth indirectly and directly (Bongiovanni and Lobartini 2006). Indirectly, it improves the chemical and physical conditions of soils by increasing cation exchange capacity, termed buffering effect, and enhancing aggregation, aeration and water retention. Improvement of soil biological properties affects soil microbial diversity and population, thereby creating a suitable environment for root growth of plants and soil microbes (Senesi and Loffredo 1999).

Hassan dokht et al. (2001) used tea waste and bark as soil amendment and then cultured lettuce. They reported that the plants cultured in tea waste + bark + soil treatment had more yield as compared with alone soil treatment, but amount of total soluble solids (TSS) and growth indices in tea waste + soil treatment was higher than bark + soil treatment. Investigations showed that addition of rice hull to soil in rice farm caused to increase in yield in long time. The results of use of zeolite, vermiculite and some organic materials as media for hydroponic tomato production showed that the lowest yield was obtained from the rice hull and mixture of rice hull with other materials. The highest of yield, shoot dry weight and number of cluster per pot were obtained from cocopeat and perlite + mica media (Saberi et al. 2006).

Borji et al. (2010) investigated the effect of some media include date-palm peat, perlite, cocopeat and mix of these materials on growing indices of tomato. Their results showed that fruit yield, fruit number, stem length, titrable acidity and ascorbic acid (in fruit) in different media had no significant differences. Samiei et al. (2005) investigated effect of peat moss and date-palm wastes as substrates on growing of Aglaonema and their results showed that peat moss and date-palm peat were similar in some characteristics. The effect of some culture media such as date-palm peat, cocopeat and perlite on some tomato growing indexes was studied by Mohammadi Ghehsareh et al. (2011a). The treatments were perlite, date-palm peat and different ratio of coco peat + perlite and date-palm peat + perlite. Results showed higher amount of total soluble solids (TSS) was related to Coco peat + perlite treatment that had no significant difference with date-palm peat + perlite, perlite and date-palm peat treatments.

According to present evidences original birthplace of date palm tree were in south of Iran and Iraq but today it cultivated in more areas of world. Total numbers of date palms around the world are about 100000000.

More than 50% of date palms in the world are in Iran, Iraq and Saudi Arabia (21% in Iran). Total weight of date palm wastes annually produced in the world is estimated about 4.5 million tons (Barreveld 1993; FAO 2002). Currently, appropriate management and optimized procedure is not in use for this material. It seems that residues and wastes of date palm trees could can be used as culture media for soilless systems in greenhouse or it could be applied to soil as organic fertilizer to improve soil chemical and physical conditions. Also this will decrease the adverse effect of burning these wastes to the environment, the disposal method currently is used for date palm wastes.

The objective of the present study was to compare date palm waste, rice hull, and their mixture with soil as culture media on cucumber production in greenhouse.

METHODS

This work was carried out in a research greenhouse of Islamic Azad University, Khorasgan branch, using a completely randomized design with six treatments and six replications. The treatments were pure palm waste (Pa), pure rice hull (Rh), soil + 5% (weight) palm waste (S + 5%Pa), soil + 5% (weight) rice hull (S + 5%Rh), soil + 5% (weight) palm waste + 5% (weight) rice hull (S + 5%Pa + 5%Rh) and pure soil (S). *Cucumis Stativus* cultivar was used for seed.

Cucumber seeds were sowed in a box of peat moss and young plants at three leaf stages were transferred in 10 liter pots filled with culture media. One cucumber plant was cultivated per pot. Heights of pots were 40 cm and distance between pots was 50 cm. Plant were grown for a period of four months (May to August). Average temperature of day and night were 30 and 18°C respectively in greenhouse. Relative humidity in greenhouse was 37.1- 61.2 percent during the growth period. Plants were fertigated with Papadopoulos nutrient solution (Papadopoulus 1991 and Papadopoulos 1994; Benton Jones 2005). pH of the nutrient solution was kept between 5.5-6. Irrigation was performed by hand, three times a day based on water and leaching requirement (20%).

Some physiochemical characteristics of the culture media including bulk density (Baruah and Barthakur 1998), organic carbon (% OC) (Walkley and Black 1934), total porosity (Baruah and Barthakur 1998), water holding capacity (WHC) (Verdonck and Gabriels 1992), cation exchange capacity (CEC) (Rhoades 1982) were measured. Electrical conductivity (EC) and pH (Asiah et al., 2004) of culture media were determined with 1:2 extraction method. Microorganisms' population in culture media were determined with counting method. Plants were grown for a period of four months (May to August) and selected growing indices including stem diameter (mm) at the plant curb location, plant height (cm), shoot dry mass (g/plant), root dry mass (g/plant), fruit length (cm), fruit number and pedicel length (cm) were measured at the end of growth period. The fruit yield (Kg/m^2) for each plant accumulated for fruitage period. Total soluble solids (TSS) of fruit (Brix%) (by Refractometer), leaf area index (LAI) (by Planimeter set), fruit firmness index (kg/cm^2) (by Penetrometer), leaf chlorophyll content (%) (by CL-10 set) were measured at the end of the growth period. Experimental data normality was verified, and then data were submitted to analysis of variance, using SPSS statistical software. Means were compared using Duncan multiple test ($P < 0.05$).

RESULT AND DISCUSSION

Physiochemical properties of media

Some physiochemical properties of culture media are presented in Table 1. Soil texture was sandy loam and its bulk density was higher than the other media cultures. The lowest bulk density (0.04) and highest porosity was related to palm waste (86%), therefore root media aeration in this treatment was better than the other media. When root media aeration is sufficient, supplying of water and nutrient elements for plants is better. The EC of the soil treatment was higher than the other culture media and this term in palm waste was lower than the other media.

Table 1. SOME PHYSIOCHEMICAL PROPERTIES OF CULTURE MEDIA

Substrate	Porosity (%)	WHC %	Bulk density (gr/cm^3)	OM %	EC (ds/m)	pH	CEC (Cmol/kg)
Palm waste	86	92	0.04	84.48	1.28	6.7	95
Rice hull	73	88	0.09	88.52	2.24	6.2	72.6
Loamy soil	37	78	1.25	3.52	2.34	7.1	43.1

The CEC of the culture media varied between 13 and 59 Cmol/kg. Maximum and minimum amounts of CEC were related to palm waste and soil media respectively, therefore the palm waste media had more capacity to supply of nutrient elements for plant. Mohammadi Ghehsareh et al. (2011b), Mohammad Khiyami et al. (2008) and Borji et al. (2010) had similar results.

Growing indices

Some growing indices of cucumber plant are presented in Table 2. Amount of cucumber yield in different culture media had significant difference at 5% level as compared with other treatments. Highest and lowest amount of fruit yield was related to palm waste and S + Pa5% + Rh5% treatments respectively. Also most amount of plant height, stem diameter and shoot dry mass was related to palm waste media that had significant differences at 5% level as compared with other treatments. Amount of bulk density, Porosity, WHC and CEC in date palm waste was maxima as compared with other media and these physiochemical properties were affected on plant growth indices and fruit yield. Sufficient conditions in the palm waste media caused to good support of water and nutrient elements for plant and leading to good growth (Olympious 1992; Kumar and Goh 1999). Results showed that the highest amount of root weight in dry condition was related to palm waste and rice hull media that had significant difference at 5% level as compared with other treatments. The roots of cucumber plant had more growth in the palm waste and rice hull media because porosity in these cultures media was higher than the others and so bulk density in these media was lower as compared with other treatments therefore resistance to root motion was minimum in these media. A good growing media would provide sufficient anchorage to the plant, serves as reservoir for nutrients and water, allow oxygen diffusion to the roots and permit gaseous exchange between the roots and atmosphere outside the root substrate (Bunt 1988). As result root growth in the date palm waste and rice hull treatments was higher than other treatments.

Table 2. EFFECT OF CULTURE MEDIA ON YIELD AND SOME GROWTH INDICES OF CUCUMBER

Culture media	Fruit yield (Kg)	Plant height (cm)	Stem diameter (cm)	Shoot dry (g) mass	Root dry mass (g)	Chlorophyll %
Palm waste	3.67^{a}	328.1^{a}	1.8^{a}	88.4^{a}	3.5^{b}	50.94^{a}
Rice hull	2.12^{b}	260^{b}	0.9^{b}	55.4^{b}	3.9^{a}	63.24^{a}
Loamy soil	1.82^{b}	258.9^{b}	0.9^{b}	41.9^{c}	1.8^{c}	35.13^{c}
S + Pa5%	1.41^{c}	247.4^{b}	0.9^{b}	42.4^{c}	1.7^{c}	48.12^{b}
S + Rh5%	1.46^{c}	196.6^{c}	0.9^{b}	36.1^{c}	2.0^{c}	47.13^{b}
S + Pa5% + Rh5%	1.04^{d}	234^{bc}	0.9^{b}	40.5^{c}	2.7^{bc}	45.25^{b}
Culture media	**WSF (g)**	**TSS (brix%)**	**Fruit firmness index (Kg/cm^2)**	**Fruit length (cm)**	**Fruit diameter (cm)**	**Pedicel length (cm)**
Palm waste	85.1^{b}	2.67^{b}	1.92^{ab}	12.5^{b}	2.9^{ab}	3.2^{bc}
Rice hull	77.9^{b}	2.75^{ab}	1.83^{ab}	13.5^{a}	2.8^{c}	2.9^{c}
Loamy soil	82.9^{b}	3.58^{a}	1.42^{b}	13.6^{a}	2.9^{ab}	2.6^{c}
S + Pa5%	82.7^{b}	3.58^{a}	2.25^{a}	13.5^{a}	3^{a}	2.7^{c}
S + Rh5%	89.4^{b}	3.33^{ab}	1.83^{ab}	13.8^{a}	3^{a}	3.4^{b}
S + Pa5% + Rh5%	93.6^{a}	3.58^{a}	1.75^{ab}	13.9^{a}	3^{a}	3.9^{a}

* S = soil, Pa = Palm waste, Rh = Rice hull, WSF = weight of single fruit.
**Different letters in each Colum represent significant differences and same letters don't represent significant differences at level 5%.

Highest and lowest chlorophyll percentage was observed in rice hull and soil treatments respectively. The weight of single fruit in S + Pa5% + Rh5% treatment was highest and had significant difference as compared with other substrates. The lowest amount of TSS was observed in date palm waste that had significant difference with other substrates. The lowest fruit firmness index was related to loamy soil treatment and the highest amount of fruit firmness index was related to S + 5%Pa treatment, and it had significant difference as compared with other treatments. The fruit length in the palm waste media culture was lowest and had significant difference with other media culture although other treatments had no significant difference together. Highest amount of fruit diameter was related to S + 5%Pa, S + 5%Rh and S + Pa5% + Rh5% treatments and had significant difference with other media culture so most amount of pedicel length was related to S + Pa5% + Rh5% treatment. These results were similar to Olympious (1992), Kumar and Goh (1999) findings that physiochemical properties of culture media affected on plant growth and yield. Results of Mohammadi Ghehsareh et al. (2011a) on tomato plant showed that fruit yield in date palm and perlite had no significant differences at 5% level. Saberi et al. (2006) reported lowest and highest amounts of tomato yield were obtained from rice hull and perlite media cultures respectively. A lot of these results were similar to Mohammadi Ghehsareh et al. (2011b) and Borji et al. (2010) investigations. Tzortzakis and Economakis (2008) showed that highest yield, highest total number of fruits per plant, fruit weight, fruit firmness and total soluble solid of tomato were obtained from plants grown in organic media (Maize) as compared with inorganic media.

Some researchers reported that when organic wastes were added to the soil, it improved the chemical and physical conditions of soils and thereby creating a suitable environment for root growth of plants (Senesi and Loffredo, 1999; Bongiovanni and Lobartini 2006; Hassan dokht et al. 2001) but in this study some growing indices of cucumber plant were decreased when date palm waste and rice hull were mixed with soil. Addition of palm waste and rice hull to the soil improved the physiochemical properties of mix media but amount of cucumber yield in mix media (soil + 5% (weight) palm waste + 5%(weight) rice hull) was lowest and even it was lower than the soil media. These results may be related to microorganisms' population and their activity in different substrates. Micro organisms population before planting in the soil, palm waste and rice hull were $830*10^7$, $490*10^7$ and $335*10^7$ respectively. Microorganisms' population in the soil was higher and when it was mixed with organic matter (palm waste and rice hull), activity of micro organisms for organic matter mineralization were increased and caused competition between plant roots and microorganisms for nutrient elements uptake. The organic wastes that were used in this study were fresh and have not passed fermentation period therefore mineralization process maybe were lead to decreasing in fruit yield and number of fruit in the soil + 5% (weight) palm waste + 5% (weight) rice hull treatment.

CONCLUSION

Results of this study clearly showed that date palm waste has a great potential to be used as the culture media for plant growth in hydroponic systems. Date palm waste provided better physical and chemical condition for plants growth and plants grown in this media produced highest yield. when fresh date Palm waste and rice hull were mixed with soil, although caused the amendment of the physiochemical properties of soil but the plant growing indices decreased and were even lower than soil treatment. It seems that addition of fresh organic waste to soil caused to increase in microorganisms' activity, immobilization process in rhizosphere and affected on plant activity. Therefore in mixing process with soil, firstly, date palm waste must be fermented and composed and after the completion of the process it would be mixed with soil.

ACKNOWLEDGEMENTS

Sincere gratitude goes to Jahad–e-agriculture and Islamic Azad University of Khorasgan for providing plant materials, experimental sites, and technical assistance.

COMPETING INTERESTS

The authors declare that there is no conflict of interest.

REFERENCES

1. Asiah A, Mohd Razi I, Mohd Khanif Y, Marziah M, Shaharuddin M: **PHYSICAL AND CHEMICAL PROPERTIES OF COCONUT COIR DUST AND OIL PALM EMPTY FRUIT BUNCH AND THE GROWTH OF HYBRID HEAT TOLERANT CAULIFLOWER PLANT.** *Pertanika J. Trap. Agric. Sci.* 2004,**27**(2):121–133.
2. Barreveld WH: *Date palm products*. FAO, Agriculture Services Bulletin No; 1993. 101 101
3. Baruah TC, Barthakur HB: *A textbook of soil analysis*. Vikas Publishing House PMT Ltd, New Dehli, India; 1998.
4. Benton Jones J: *A practical guide for the soilless grower*. 2nd edition. CRC Press; 2005:423.
5. Bongiovanni MD, Lobartini JC: **PARTICULATE ORGANIC MATTER, CARBOHYDRATE, HUMIC ACID CONTENTS IN SOIL MACRO- AND MICROAGGREGATES AS AFFECTED BY CULTIVATION.** *Geoderma* 2006, **136:** 660–665.
6. Borji H, Mohammadi Ghehsareh A, Jafarpour M: **EFFECTS OF THE SUBSTRATE ON TOMATO IN SOILLESS CULTURE.** *Res J of Agri and Biol Sci* 2010,**6**(6):923–927.
7. Bunt AC: **MEDIA AND MIXES FOR CONTAINER-GROWN PLANTS.** *A manual on the preparation and use of growing media for pot plant* 1988, **2:** 309.
8. FAO: *Date palm cultivation, FAO plant production and protection paper vol. 1*. 156th edition. Publishing Management Service, Information Division, FAO, Viale delle Terme di Caracalla, 00100 Rome, Italy; 2002. or by e-mail to copyright@fao.org or by e-mail to copyright@fao.org
9. Hassan dokht M, Mostor IF, Yadasht Dehkaei M: **THE EFFECT OF TEA WASTE AND BARK ON CULTURE MEDIA AND QUALITATIVE AND QUANTITATIVE PROPERTY OF GREENHOUSE LETTUCE.** *J of Hort Sci of Iran* 2001., **40:** In Persia In Persia
10. Khiyami M, Masmali I, Abu-khuraiba M: **COMPOSTING A MIXTURE OF DATE PALM WASTES, DATE PALM PITS, SHRIMP, AND CRAB SHELL WASTES IN VESSEL SYSTEM.** *Saudi J of Biolo Sci* 2008,**15**(2):199–205.
11. Kumar K, Goh KM: **CROP RESIDUES AND MANAGEMENT PRACTICES: EFFECTS ON SOIL QUALITY, SOIL NITROGEN DYNAMICS, CROP YIELD, AND NITROGEN RECOVERY.** *J Advances in Agro* 1999, **68:** 197–319.
12. Mohammadi Ghehsareh A, Borji H, Jafarpour M: **EFFECT OF SOME CULTURE SUBSTRATES (DATE PALM PEAT, COCOPEAT AND PERLITE) ON SOME

GROWING INDEXES IN GREENHOUSE TOMATO.*Afr J Micro Reas* 2011,**5**(12):1437–1442.

13. Mohammadi Ghehsareh A, Samadi N, Borji H: **COMPARISON OF DATE-PALM WASTES AND PERLITE AS GROWTH SUBSTRATES ON SOME TOMATO GROWING INDEXES.** *Afri J of Biotec* 2011,**10**(24):4871–4878.
14. Olympious CM: **SOILLESS MEDIA UNDER PROTECTED CULTIVATION ROCKWOOL, PEAT, PERLITE AND OTHER SUBSTRATES.** *Acta Hort* 1992, **401:** 443–451.
15. Papadopoulos AP: *Growing greenhouse tomatoes in soil and soilless.* Research Station Harrow, Ontario. Communication Branch, Agriculture, Canada. Ottawa; 1991:79.
16. Papadopoulos AP: *Growing Greenhouse Cucumbers in Soil and in Soilless Media.* Agriculture and Agri-Food Canada Publication 1902/E; 1994:108.
17. Rhoades JD: **CATION EXCHANGE CAPACITY. CHEMICAL AND MINERALOGICAL PROPERTIES.** In *Methods of soil analysis, vol 2.* Agron. No. 9 edition. Edited by: Page AL. Am. Soc. Agron, Madison, WI, USA; 1982:149–157.
18. Saberi Z, Khoshgoftarmanesh AH, Kalbasi M, Mobli M: *Usage of Zeolite.* Isfahan University of Technology, Mica and some neutral materials as substrates for tomato cultured in hydroponic method. MS thesis; 2006:116.
19. Samiei L, KHalighi A, Kafi M, Samavat S, Arghavani M: **AN INVESTIGATION OF SUBSTITUTION OF PEAT MOSS WITH PALM TREE CELLULOID WASTES IN GROWING AGLAONEMA (AGLAONEMA COMMUTATUM CV. SILVER QUEEN).** *Iranian J of Agri Sci* 2005,**36**(2):503–510. (in Persian) (in Persian)
20. Senesi N, Loffredo E: **THE CHEMISTRY OF SOIL ORGANIC MATTER.** In *Soil physical chemistry.* Edited by: Spark DL. CRC Press, Boca Raton, FL; 1999:239–370.
21. Tzortzakis N, Economakis GCD: **IMPACTS OF THE SUBSTRATE MEDIUM ON TOMATO YIELD AND FRUIT QUALITY.** *Hort culture Sci* 2008, **2:** 83–89.
22. Verdonck O, Gabriels R: **REFERENCE METHOD FOR THE DETERMINATION OF PHYSICAL PROPERTIES OF PLANT SUBSTRATES.II. REFERENCE METHOD FOR THE DETERMINATION OF CHEMICAL PROPERTIES OF PLANT SUBSTRATES.** *Acta Horticulturae* 1992, **302:** 169–179.
23. Walkley A, Black IA: **AN EXAMINATION OF THE DEGTJAREFF METHOD FOR DETERMINING SOIL ORGANIC MATTER AND A PROPOSED MODIFICATION OF THE CHROMIC ACID TITRATION METHOD.***Soil Sci* 1934, **37:** 29–38.CrossRef

CHAPTER 11

Design of Soft Computing Based Optimal PI Controller for Greenhouse System

A. Manonmani[1], T. Thyagarajan[2], S. Sutha[2], V. Gayathri[2]

[1]*Department of Electronics and Instrumentation Engineering, Saveetha Engineering College, Chennai, India*

[2]*Department of Instrumentation Engineering, MIT Campus, Anna University, Chennai, India*

ABSTRACT

Greenhouse system (GHS) is the worldwide fastest growing phenomenon in agricultural sector. Greenhouse models are essential for improving control efficiencies. The Relative Gain Analysis (RGA) reveals that the GHS control is complex due to 1) high nonlinear interactions between the biological subsystem and the physical subsystem and 2) strong coupling between the process variables such as temperature and humidity. In this paper, a decoupled linear cooling model has been developed using a feedback-feed forward linearization technique. Further, based on the model developed Internal Model Control (IMC) based Proportional Integrator (PI) controller parameters are optimized using Genetic Algorithm (GA) and Particle Swarm Optimization (PSO) to achieve minimum Integral Square Error (ISE). The closed loop control is carried out using the above control schemes for set-point change and disturbance rejection. Finally, closed loop servo and servo-regulatory responses of GHS are compared quantitatively as well as qualitatively. The results implicate that IMC based PI controller using PSO provides better performance than the IMC based PI controller using GA.

Also, it is observed that the disturbance introduced in one loop will not affect the other loop due to feedback-feed forward linearization and decoupling. Such a control scheme used for GHS would result in better yield in production of crops such as tomato, lettuce and broccoli.

Keywords: Greenhouse System, Feedback-Feed Forward Linearization and Decoupling, IMC Based PI Controller, Genetic Algorithm, Particle Swarm Optimization, Nonlinear System

1. INTRODUCTION

The GHS consists of highly coupled subsystems: the greenhouse climate and the greenhouse crop. The greenhouse control problem is to create a favorable environment for the crop in order to reach predetermined results for high yield, high quality and low costs. It is very difficult to control the GHS in practice, due to the complexity of the greenhouse environments such as high non-linearity, strong coupling between Multi-Input Multi-Output (MIMO) systems.

There are several models available for the GHS. Some of them use white box (first principle model) [1] [2] and others use black box model [3] - [5] . The best benchmark climate model is presented by Albright et al. [1] for controlling the temperature and humidity. A model based on feedback-feed forward compensation technique used for linearization, decoupling and disturbance compensation is presented in [1] [2] . Multi-ob- jective optimization of greenhouse system using evolutionary algorithm is given in [6] .

In this paper, a model of nonlinear thermodynamic laws between numerous system variables affecting the greenhouse climate is formulated and a feedback-feed forward approach to system linearization and decoupling is done. A conventional IMC based PI controller is designed based on the model. The IMC-PI controller parameters are optimized using Genetic Algorithm (GA) and Particle Swarm Optimization (PSO). The performance measures are compared for servo and servo-regulatory systems.

This paper is organized as follows: Section 2 briefs about a greenhouse system model. The control schemes are explained in Section 3 followed by closed loop analysis in Section 4. Section 5 gives details about the quantitative comparison. Finally, conclusions are presented in Section 6.

2. GREENHOUSE SYSTEM MODEL

2.1. Description of Greenhouse Model

The three models for the greenhouse system include: cooling model, heating model and ventilating model. Ventilation is one of the most important components for a successful greenhouse. If there is no proper ventilation, greenhouses and their plants become prone to problems. The main purpose of ventilation is to regulate the temperature at the optimum level, to ensure movement of air and also to ensure the supply of fresh air for photosynthesis and plant respiration. Heating model is essential, when the inside temperature of the greenhouse is very low during winter climate and at night time. Whereas, cooling model is required when the outside temperature is very high which may affect the plants during the summer mode at day time. The functional block diagram of GHS is shown in Figure 1.

A simple greenhouse heating-cooling-ventilating model can be obtained (as given in Equation (1)) by considering the diff erential equations, which govern sensible and latent heat, as well as water balances on the interior volume [5] [6] :

$$\begin{cases} \frac{\mathrm{d}}{\mathrm{d}t}T_{in}(t)=\frac{1}{\rho C_p V_T}\left[Q_{heater}(t)+S_i(t)-\lambda Q_{fog}(t)\right]-\left(\frac{V_R(t)}{V_T}+\frac{UA}{\rho C_p V_T}\right)\cdot\left[T_{in}(t)-T_{out}(t)\right] \\ \frac{\mathrm{d}}{\mathrm{d}t}H_{in}(t)=\frac{1}{V_H}Q_{fog}(t)+\frac{1}{V_H}\left[E\left(S_i(t),H_{in}(t)\right)\right]-\frac{V_R(t)}{V_H}\left[H_{in}(t)-H_{out}(t)\right] \\ \left[E\left(S_i(t),H_{in}(t)\right)\right]=\alpha\frac{S_i(t)}{\lambda}-\beta_T H_{in}(t) \end{cases} \tag{1}$$

where

T_{in}/T_{out}: Indoor/outdoor temperature (°C)

H_{in}/H_{out}: Interior/exterior humidity (g[H_2O/kg[dry air])

Q_{heater}: Heat provide by the greenhouse heater (W)

Q_{fog}: Water capacity of fog system (g/s)

V_R: Ventilation rate (m^3/s)

The greenhouse system model parameters are shown in the Table 1.

In general, the operating conditions of the ventilation/cooling process are rather dominated by solar radiation alone (i.e. $\beta_T=0$), hence $\beta_T H_{in}(t)$ can be neglected. Suppose that $C_0=\rho C_p V_T$, $\alpha'=\alpha/(\lambda V_H)$, $\lambda'=\lambda Q_{fog}^{\max}$

and $V' = V_H / Q_{fog}^{max}$. The control variables can be normalized as follows: $V_{R,\%} = V_R / V_R^{max}$, $Q_{fog,\%} = Q_{fog} / Q_{fog}^{max}$ and $Q_{heater,\%} = Q_{heater} / Q_{heater}^{max}$.

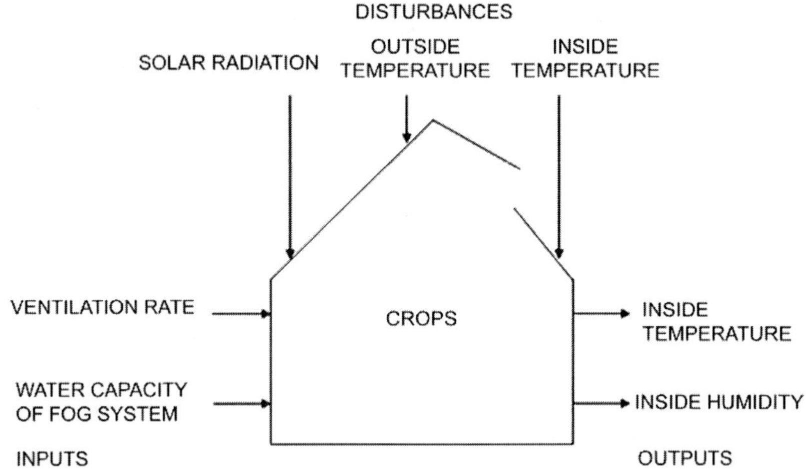

Figure 1. Block diagram representing greenhouse system.

Table 1. Greenhouse system model parameters.

Description	Values
Heat transfer coefficient (UA)	29.81 (W/K)
Air density (ρ)	1.2 (kg/m^3)
Specific heat of air (C_P)	1006 (J·(kg^{-1}·K)$^{-1}$)
Latent heat of vaporization (λ)	2257 (J/g)
Maximum ventilation rate (V_R^{max})	20 air changes per hour
Maximum water capacity of fog system (Q_{fog}^{max})	26 g[H_2O]s^{-1}
Maximum heating energy (Q_{heater}^{max})	150 W·m^{-2}
Solar radiation (S_i)	300 W·m^{-2}
Outside Temperature (T_{OUT})	35 (°C)
Outside Humidity (H_{OUT})	4 g/Kg

In summer mode operation, Q_{heater} is set to zero. The climate model provided have two variables to be regulated, namely the indoor air temperature (T_{in}) and the humidity ratio (H_{in}), through the process of ventilation $V_{R,\%}(t)$ and fogging $Q_{R,\%}(t)$. After normalizing the control variables, the differential equation for cooling model for greenhouse system is re-written as given in Equation (2)

$$\begin{cases} \frac{\mathrm{d}}{\mathrm{d}t}T_{in}(t) = \frac{1}{C_0}\left[S_i(t) - \lambda' Q_{\%,fog}(t)\right] - \left(\frac{V_{R,\%}(t)}{t_v} + \frac{UA}{C_0}\right)\cdot\left[T_{in}(t) - T_{out}(t)\right] \\ \frac{\mathrm{d}}{\mathrm{d}t}H_{in}(t) = \frac{Q_{\%,fog}(t)}{V'} + \alpha' S_i - \frac{V_{R,\%}(t)}{t_v}\cdot\left[H_{in}(t) - H_{out}(t)\right] \end{cases} \quad (2)$$

In order to express the GHS in state-space form, the inside temperature and absolute humidity are defined as the dynamic state variables, $x_1(t)$ and $x_2(t)$, respectively; the ventilation rate and the water capacity of the fog system as the control (actuator) variables, $u_1(t)$ and $u_2(t)$, respectively; the intercepted solar radiant energy, the outside temperature, and the outside absolute humidity as the disturbances, $v_i(t)$, i = 1, 2, 3. The Equation (2) can alternatively be written in the following state-space form:

$$\dot{x}_1(t) = -\frac{UA}{C_0}x_1(t) - \frac{1}{t_v}x_1(t)u_1(t) - \frac{\lambda'}{C_0}u_2(t) + \frac{1}{C_0}v_1(t) + \frac{UA}{C_0}v_2(t) + \frac{1}{t_v}u_1(t)v_2(t) \quad (3)$$

$$\dot{x}_2(t) = -\frac{1}{t_v}x_2(t)u_1(t) + \frac{1}{V'}u_2(t) + \alpha' v_1(t) + \frac{1}{t_v}u_1(t)v_3(t) \quad (4)$$

Relative Gain Array Analysis

From the state-space form the values of A, B and C is given by:

$$A = \begin{bmatrix} -\frac{UA}{C_0} & 0 \\ 0 & -\frac{1}{t_v} \end{bmatrix}$$

$$B = \begin{bmatrix} \left(-\frac{1}{t_v}\right)x_1 + \frac{1}{t_v}v_2 & -\frac{\lambda'}{C_0} \\ \left(-\frac{1}{t_v}\right)x_2 + \frac{1}{t_v}v_3 & \frac{1}{V'} \end{bmatrix}$$

$$C = \begin{bmatrix} 1 & 0 \\ 0 & 1 \end{bmatrix}$$

$$\Lambda = \begin{bmatrix} 0.7764 & 0.2236 \\ 0.2236 & 0.7764 \end{bmatrix}$$

In this case relative gains are positive values. If $0 < \lambda_{11} < 1$, then an interactions exists. From the interaction analysis it is found that y_1 is paired withu_1 and y_2 is paired withu_2 to form two loops with minimum interaction. However, the disturbances at one loop will affect responses of other loop due to coupling. Hence, decoupling is necessary for GHS. The feedback-feed forward linearization technique is used for decoupling is discussed in the subsequent section.

2.2. Feedback-Feed Forward Linearization and Decoupling

One method for linearization and decoupling for systems with external disturbances is presented in [2] [3] which is based on the feedback-feed forward linearization technique proposed by Isidori [4] . In the particular case of GHS model expressed in Equations (3) and (4), the feedback-feed forward linearization and decoupling procedure described above is followed and compensator is designed. The compensator that delivers the process input control variables $u(u_1, u_2)$, has the form given in Equations (5) and (6).

$$u_1 = \left[\frac{1}{V'} \frac{UA}{C_0} x_1 - \frac{1}{V'} \frac{UA}{C_0} v_2 - \frac{1}{V'} \frac{1}{C_0} v_1 + \frac{1}{V'} \hat{u}_1 - \alpha' \frac{\lambda'}{C_0} v_1 + \frac{\lambda'}{C_0} \hat{u}_2 \right] \Big/ Q(t) \tag{5}$$

$$u_2 = \left[\left(-\frac{1}{t_v} v_3 + \frac{1}{t_v} x_2 \right) \left(\frac{UA}{C_0} x_1 - \frac{UA}{C_0} v_2 - \frac{1}{C_0} v_1 + \hat{u}_1 + \left(\frac{1}{t_v} v_2 - \frac{1}{t_v} x_1 \right) (\alpha' v_1 + \hat{u}_2) \right) \Big/ Q(t) \right] \tag{6}$$

where Q(t) must be anon-zero quantity.

$$Q(t) = \frac{1}{V'} \left(\frac{1}{t_v} v_2 - \frac{1}{t_v} x_1 \right) + \frac{\lambda'}{C_0} \left(\frac{1}{t_v} v_3 - \frac{1}{t_v} x_2 \right) \tag{7}$$

where,

$\hat{u}_1 \rightarrow$ temperature change;

$\hat{u}_2 \rightarrow$ humidity change.

The transfer functions derived from the linearized and decoupled GHS are as given in Equations (8) and (9)

$$G_{11} = \frac{5.5988}{1+13.7s} \quad (8)$$

$$G_{22} = \frac{5.574}{1+6.068s} \quad (9)$$

3. DESIGN OF IMC BASED PI CONTROLLER

3.1. Conventional IMC-PI Controller

In the Internal Model Control (IMC) formulation, the controller, q(s), is based directly on that part of the process transfer function which can be inverted. The IMC formulation generally results in only one tuning parameter, the closed loop time constant (λ, the IMC filter factor). The PI tuning parameters are then becomes a function of this closed-loop time constant. The selection of the closed-loop time constant is directly related to the robustness (sensitivity to model error) of the closed-loop system.

The feedback controller $g_c(s)$ contains the internal model $\tilde{g}_p(s)$ and internal model controller, q(s). Now, the IMC design procedure can be used to design a standard feedback controller. The standard feedback controller is a function of the internal model, $g_p(s)$ and internal model controller, q(s) as given in Equation (10).

$$g_c(s) = \frac{q(s)}{1 - g_p(s)q(s)}$$

For a given first-order process in Equation (10), the values of the PI tuning parameters are obtained using Equations (11), (12).

$$\tilde{g}_p(s) = \frac{K_p}{\tau_p s + 1} \quad (10)$$

$$K_c = \frac{\tau_p}{K_p \lambda} \quad (11)$$

$$K_I = \frac{K_C}{\tau_I} = \left[\frac{\tau_P}{K_p \tau}\right]\left[\frac{1}{\tau_I}\right] \quad (12)$$

The desired value for λ is defined as a trade-off between performance and robustness and its optimum value is obtained using genetic algorithm. The objective function considered here is Integral Square Error (ISE) and the decision variables obtained is λ, filter factor.

The controller parameter values for IMC-PI are calculated by using trial and error method and the parameter values are tabulated in Table 2.

Table 2. Controller parameter values for IMC-PI controller.

Controller	K_C	K_I
Temperature	0.8862	0.0987
Humidity	0.9452	0.1176

3.2. IMC-PI Using GA

Although PI parameters obtained using conventional method gives a good response, it is not satisfactory. In order to meet the desired specification PI controller parameters are tuned using optimization technique [7] [8] namely Genetic Algorithm (GA). The various steps involved in the GA based optimization of PI controller are listed below.

Step 1: Choose the string length, population size, probability of crossover, probability of mutation and number of generation.

Step 2: Initialize the population size K_c and K_I.

Step 3: Carry out selection operation. Check for acceptance criteria, if satisfied stop otherwise goto next step (Step 4).

Step 4: Perform crossover operation. Check for acceptance criteria, if satisfied stop otherwise goto next step (Step 5).

Step 5: Perform mutation operation. Check for acceptance criteria, if satisfied stop otherwise goto next step (Step 3).

The GA parameters used for tuning GA based IMC- PI scheme are given in Table 3.

Table 3. GA parameters used for tuning GA based IMC-PI scheme.

Parameters	Values
Population size	10
Number of generations	20
Probability of crossover	0.99
Probability of mutation	0.01
Selection procedure	Ranking
Crossover type	Single point crossover
Objective function	ISE

The desired PI parameters, around an operating point, are generated in random as the initial population. Each of these PI settings is applied to the system and the fitness of each chromosome is calculated. From that, the chromosomes with the best fit value are taken to the next iteration whereas the chromosomes with the worst fit are discarded and are replaced with new chromosomes derived from the mutation and crossover of best parent chromosomes. The function of the crossover operator is to generate new or "child" chromosomes from two "parent" chromosomes by combining the information extracted from the parents. Mutation operates individually on each individual by probilistically perturbing each bit string. This process repeats until the desired performance index is met with.

The objective function considered here is Integral Square Error (ISE) and the decision variables obtained are PI controller parameters namely K_C and K_I. The controller parameter values, thus obtained, for IMC-PI using GA are tabulated in Table 4.

Table 4. PI controller parameter values for IMC-PI using GA ($\lambda = 0.1107$).

Controller	K_C	K_I
Temperature	1.6133	0.0877
Humidity	1.6205	0.2671

The IMC-PI parameters obtained using GA based optimization technique. This is reflected in the value of λ obtained using GA based optimization algorithm gives a good response but the reduction in error is minimum. In order to further minimize the error the PI parameter is tuned using an optimization technique namely Particle Swarm Optimization (PSO).

3.3. Design of IMC-PI Using PSO

In the design of IMC-PI using PSO, the desired PI parameters, around an operating point, are generated through a bird flocking in two-dimension space. The position of each agent is represented by XY axis position and also the velocity is expressed by v_x (velocity of X axis) and v_y (velocity of Y axis). Each of these PI settings is applied to the system; the velocity and position are calculated. From that, the best values are taken and the agent position is realized by the position and velocity information. The objective function considered here is Integral Square Error (ISE). To implement the PSO algorithm the values to the variables of the PSO algorithm is given by sampling number N = 12, the local attractors and global attractors $C_1 = 2$, $C_2 = 2$, [9] . The steps involved in PSO based optimization are listed below.

Step 1: Initialize the population of particles with random positions and velocities.

Step 2: For each particle, evaluate the fitness.

Step 3: Compare the particles current fitness (X_i) with the particle best fitness (P_i). If current is the best then set P_i fitness equal to current values and set P_i location equal to current location otherwise go to step 4.

Step 4: Compare the fitness of the particle (P_i) with the population overall best P_g. If the current value of particle fitness is better than P_g, then P_g fitness equal to current value otherwise go to step 5.

Step 5: Change the velocity and position using the Equation (13).

V_i (new) = $W_iV_i + C_1$ rand $(P_i - X_i) + C_2$ rand $(P_g - X_i)$ (13)

Step 6: Repeat step 2, 3 & 4 till we get the best fitness.

The controller parameter values for IMC-PI controller using PSO are tabulated in Table 5.

Table 5. IMC-PI controller parameters obtained using PSO.

Controller	K_C	K_I
Temperature	1.6099	0.0297
Humidity	1.6097	0.0189

4. CLOSED LOOP ANALYSIS

The closed loop control strategy for Greenhouse system is shown in Figure 2. The GHS is a multivariable system. The inputs to GHS are ventilation rate (r_1) and water capacity of fog system (r_2). The external disturbance include: solar radiation (V_1), outside temperature (V_2) and outside humidity (V_3). The outputs from the GHS are: inside temperature (y_1) and inside humidity (y_2). $\hat{u}_1$ & $\hat{u}_2$ are the temperature change and humidity change. u_1 & u_2 are the control variables. Generally the closed loop responses of multivariable GHS are affected by the disturbance variable namely solar radiation, outside temperature and outside humidity. From the RGA analysis it is found that the system outputs are affected by disturbance variable. Hence the decoupled and linearized is developed from the desired closed loop using feedback-feed forward linearization based on the model IMC controller is designed. The GHS output responses namely the inside temperature and inside humidity obtained using three different control schemes for the linearized and decoupled of the greenhouse are shown below.

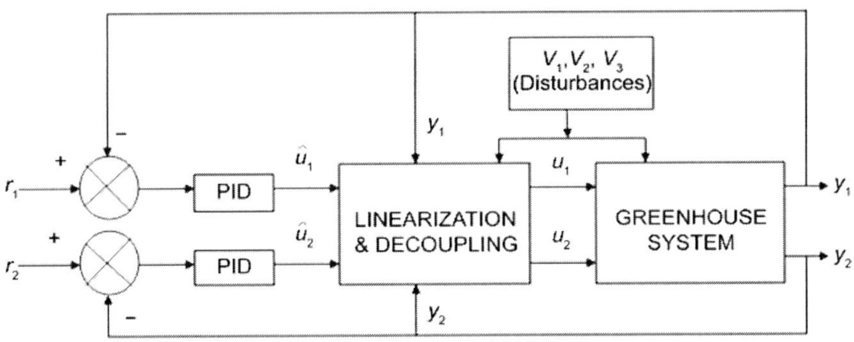

Figure 2. Closed loop control strategy for greenhouse system.

4.1. Servo Operation

The GHS output responses namely the inside temperature and inside humidity obtained using three different control schemes for the linearized and decoupled of the greenhouse are shown in Figure 3(a) and Figure 3(b) respectively.

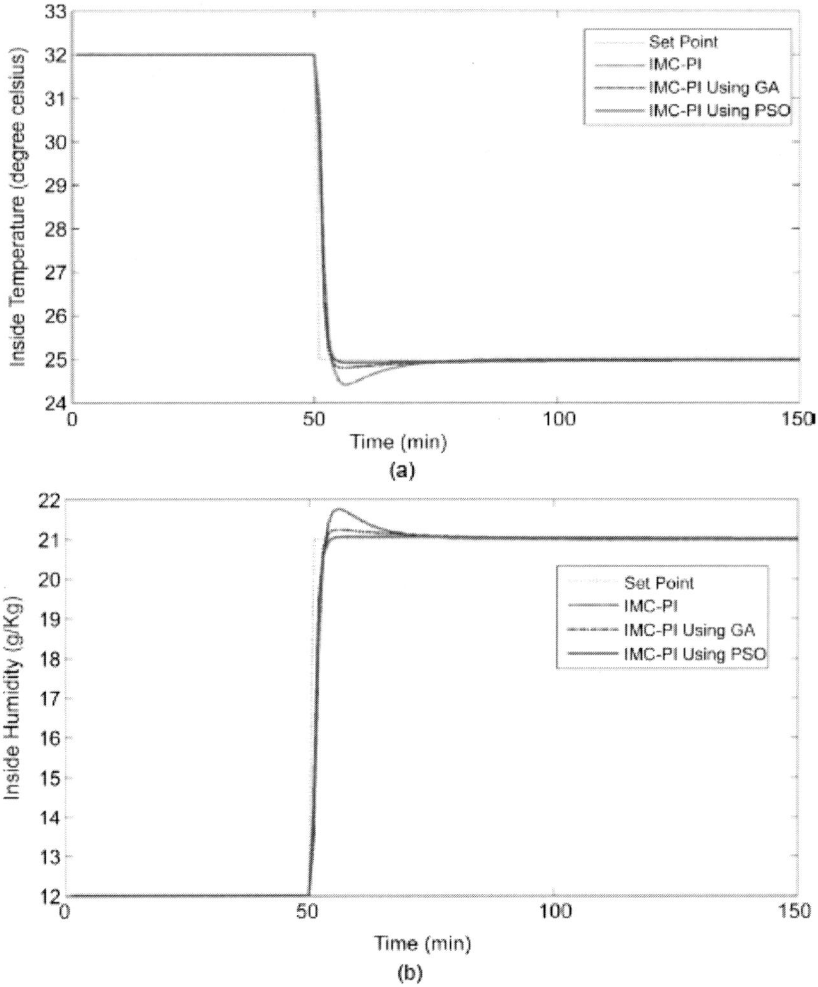

Figure 3. Comparison of closed loop performance of here types of IMC-PI controllers for servo operations. (a) Variation in inside temperature; (b) Variation in inside humidity.

From Figure 3(a) and Figure 3(b), it is observed that the performance of the IMC- PI controller tuned using PSO is better. This is reflected in the value of λ that is computed using the PSO-based optimization algorithm.

Control Signals

The corresponding variation in control signals namely the ventilation rate and water capacity of the fog system, for the linearized and decoupled model of the GHS are shown in Figure 4(a) and Figure 4(b), respectively. It is observed that the variation in the manipulated variable is smooth.

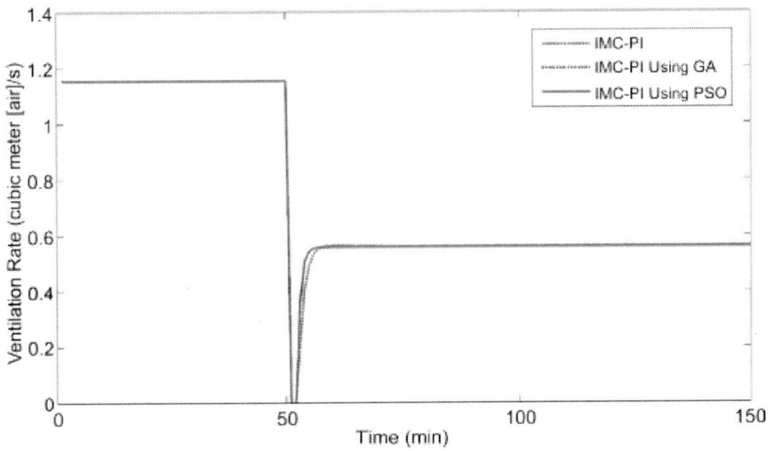

(a)

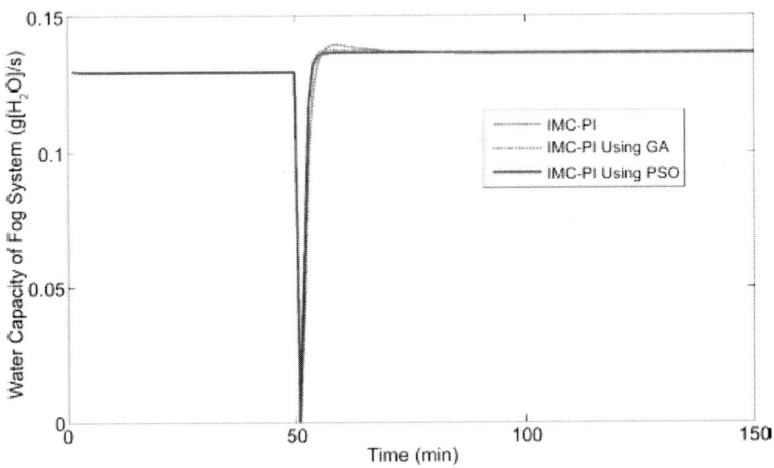

(b)

Figure 4. Corresponding variations in control signal for servo operations. (a) Variation of ventilation rate with respect to time; (b) Variation of water capacity of fog system with respect to time.

It is observed that the variation in the manipulated variable is smooth.

4.2. Regulatory Operation

4.2.1. For Disturbance in Temperature Loop

Regulatory responses for the three types of IMC-PI controllers are obtained by providing a load change of 5% at time t = 100 minutes on the temperature loop, and the variations observed in temperature and humidity are shown in Figure 5(a) and Figure 5(b), respectively.

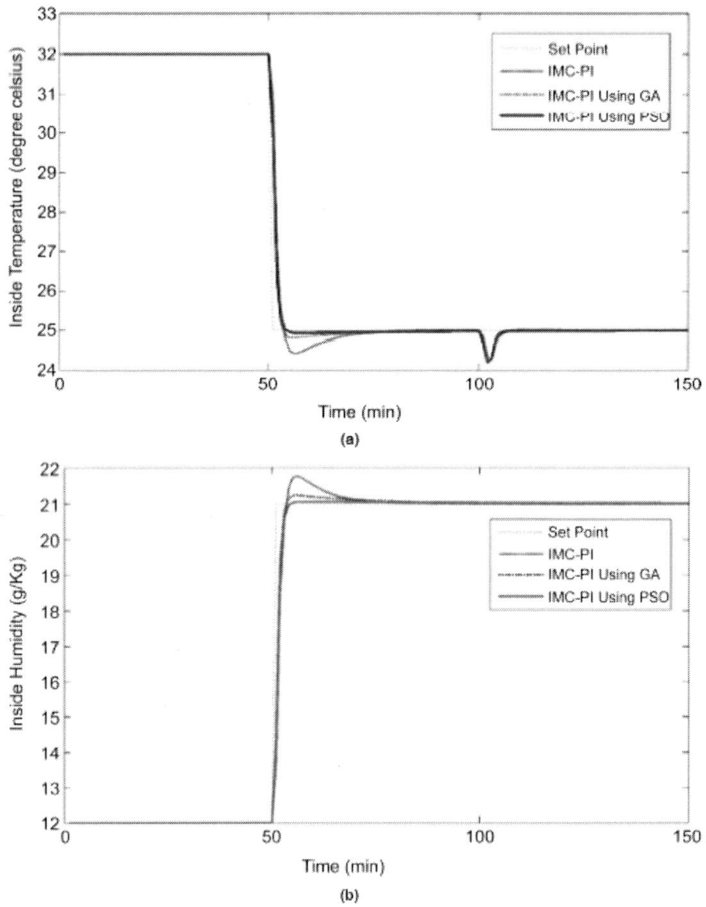

Figure 5. Comparison of closed loop performances of 3 types of IMC-PI controller's regulatory operation (for disturbances in temperature loop). (a) Variation in temperature for disturbance in temperature loop; (b) Variation in humidity for disturbance in temperature loop.

It is observed that the performance of the IMC-PI controller tuned using PSO is better. This is reflected in the value of λ that is computed using the PSO-based optimization algorithm.

4.2.2. Control Signals (for Disturbance in Temperature Loop)

The corresponding variations in control signals, namely, the ventilation rate and water capacity of the fog system, for the linearized and decoupled model of GHS for a disturbance in the temperature loop at time t = 100 minutes are shown in Figure 6(a) and Figure 6(b), respectively.

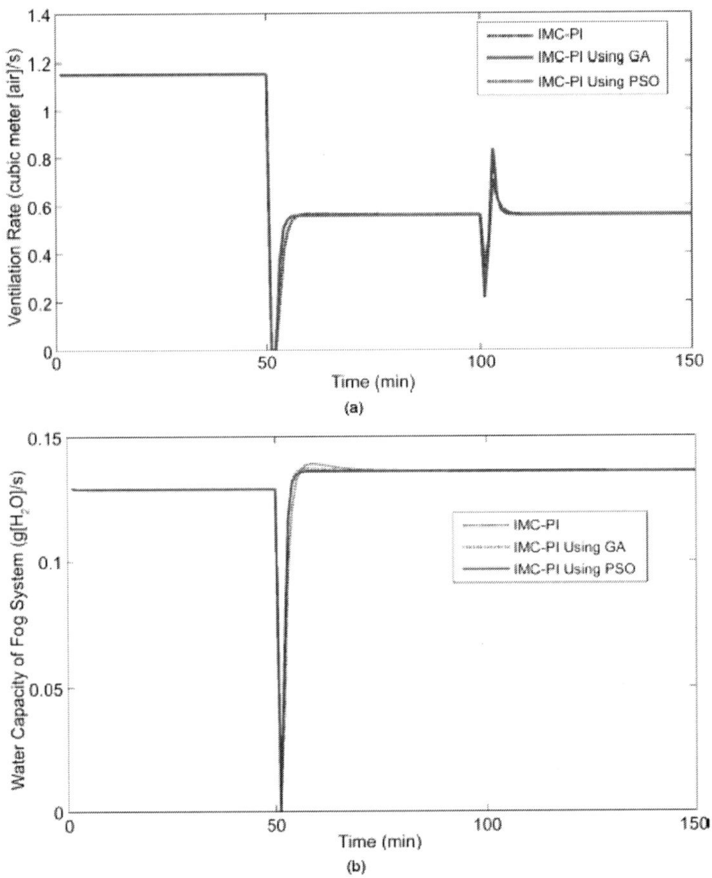

Figure 6. Corresponding variation in control signal for regulatory operation (for disturbance in temperature loop). (a) Variation for ventilation rate for disturbance in temperature loop; (b) Variation of water capacity of fog system for disturbance in temperature loop.

It is observed that the variation in the manipulated variable is smooth.

4.2.3. For Disturbance in Humidity Loop

Regulatory responses for the three types of IMC-PI controllers are obtained by providing a load change of 5% at time t = 100 minutes on the humidity loop and the variations are shown in Figure 7(a) and Figure 7(b) respectively.

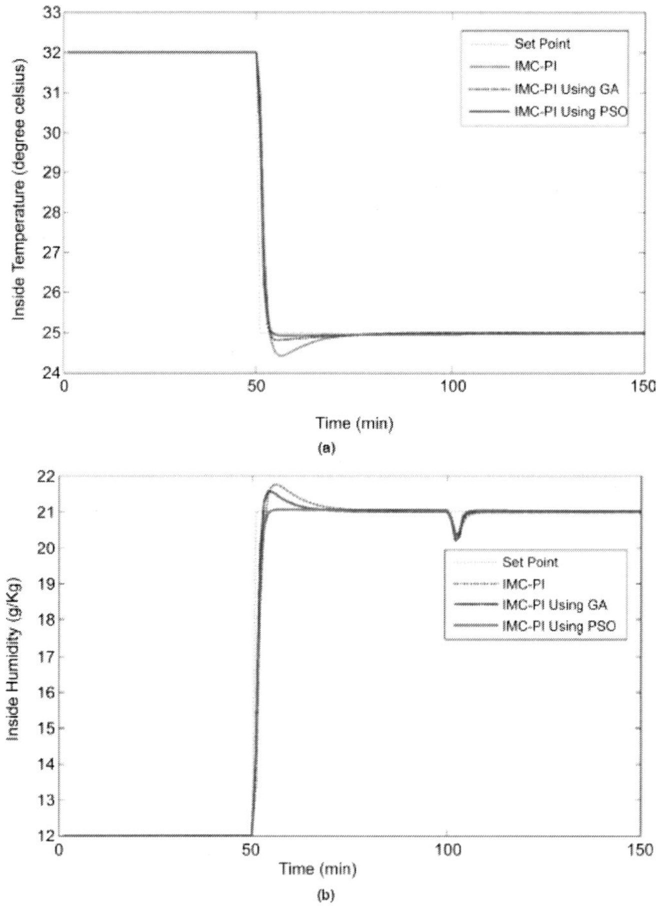

Figure 7. Comparison of closed loop performances of 3 types of IMC-PI controller's regulatory operation (for disturbances in humidity loop). (a) Variation in temperature for disturbance in humidity loop; (b) Variation in humidity for disturbance in humidity loop.

It is observed that the IMC-PI controller tuned using PSO has the minimum overshoot.

4.2.4. Control Signals (for Disturbance in Humidity Loop)

The corresponding variations in the control signals, namely, the ventilation rate and water capacity of the fog system, for the linearized and decoupled model of GHS for a disturbance in the humidity loop at time t = 100 minutes is shown in Figure 8(a) and Figure 8(b), respectively.

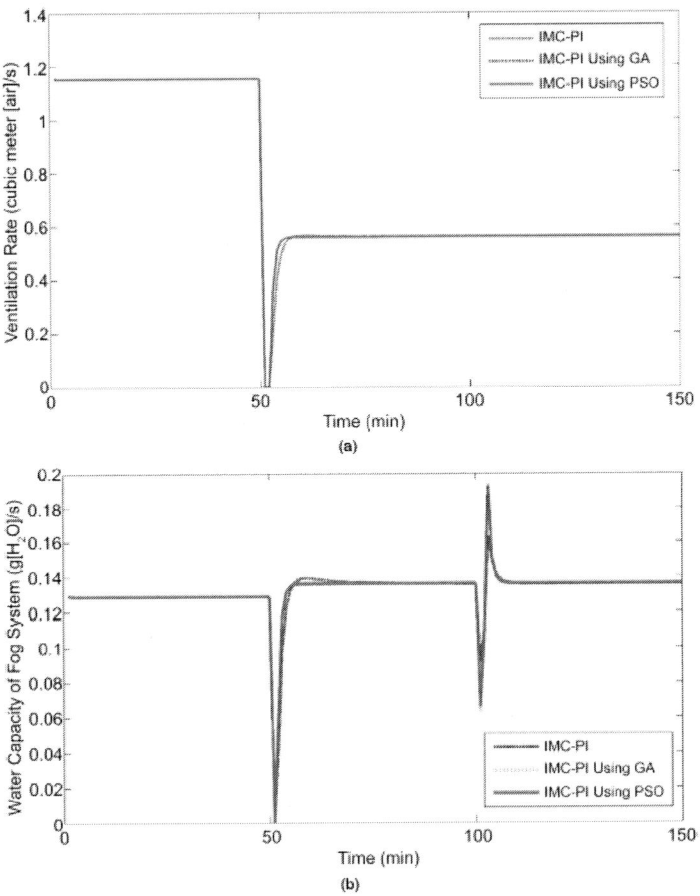

Figure 8. Corresponding variation in control signal for regulatory operation (for disturbance in humidity loop). (a) Variation of ventilation rate for disturbance in humidity loop; (b) Variation of water capacity of fog system for disturbance in humidity loop.

It is observed that the variation in the manipulated variable is smooth.

5. QUANTITATIVE COMPARISON

The performance measures for IMC-PI controller, IMC-PI tuned using Genetic Algorithm (GA) and IMC-PI tuned using Particle Swarm Optimization (PSO) for linearized and decoupled model of GHS are carried out using the Integral Absolute Error (IAE) and the Integral Square Error (ISE) as given in Equations (14) and (15) respectively.

$$IAE = \int_0^{\infty} |e(t)|\,\mathrm{d}t \tag{14}$$

$$ISE = \int_0^{\infty} |e(t)|^2\,\mathrm{d}t \tag{15}$$

The performance measures obtained for servo and regulatory operation (for disturbance in temperature loop and for disturbance in humidity loop) are tabulated in Tables 6-11 respectively. From tables it is observed that the error is minimum for the IMC-PI tuned using PSO when compared to IMC-PI using GA and conventional PI.

Table 6. Comparison of performance measures for servo response for Temperature.

Performance Criterion	Temperature		
	IMC-PI	IMC-PI Using GA	IMC-PI Using PSO
IAE	12.87	8.692	8.484
ISE	28.71	20.3	15.19
Overshoot (%)	2.29	0.756	0.54
Settling time (min)	40	38	35

Table 7. Comparison of performance measures for servo response for Humidity.

Performance Criterion	Humidity		
	IMC-PI	IMC-PI Using GA	IMC-PI Using PSO
IAE	15.42	11.16	10.23
ISE	44.83	32.56	30.85
Overshoot (%)	3.66	2.726	2.24
Settling time (min)	30	29	27

Table 8. Comparison of performance measures for regulatory response for Temperature (for disturbance in temperature loop).

Performance Criterion	Temperature		
	IMC-PI	IMC-PI Using GA	IMC-PI Using PSO
IAE	14.91	10	9.754
ISE	29.76	21.14	20.29
Overshoot (%)	2.9	3.2	2.8
Settling time (min)	29	25	15

Table 9. Comparison of performance measures for regulatory response for Humidity (for disturbance in temperature loop).

Performance Criterion	Humidity		
	IMC-PI	IMC-PI Using GA	IMC-PI Using PSO
IAE	15.42	10.46	9.63
ISE	44.83	32.81	29.48
Overshoot (%)	3.66	2.726	2.34
Settling time (min)	32	28	24

Table 10. Comparison of performance measures for regulatory response for Temperature (for disturbance in humidity loop).

Performance Criterion	Temperature		
	IMC-PI	IMC-PI Using GA	IMC-PI Using PSO
IAE	12.86	8.484	7.692
ISE	28.71	20.46	19.3
Overshoot (%)	2.29	0.75	0.52
Settling time (min)	26	24.6	23.2

Table 11. Comparison of performance measures for regulatory response for Humidity (for disturbance in humidity loop).

Performance Criterion	Humidity		
	IMC-PI	IMC-PI Using GA	IMC-PI Using PSO
IAE	17.46	11.36	10.56
ISE	45.87	34.66	32.63
Overshoot (%)	3.67	2.73	2.24
Settling time (min)	24.5	23.2	20.6

From these closed loop simulation studies, it is seen that IMC-PI controller tuned using Particle Swarm Optimization (PSO) gives better performance than IMC-PI controller using Genetic Algorithm (GA) and IMC-PI controller.

6. CONCLUSION

In this work, a model based on feedback-feed-forward linearization and decoupling for the GHS has been developed in order to avoid the interactions among the process variables. The IMC based PI controller is designed for the GHS. The controller settings are also tuned using Genetic Algorithm (GA) and Particle Swarm Optimization (PSO) to achieve minimum Integral Square Error (ISE). The results implicate that PSO based IMC-PI controller provides better performance compared to conventional PI and IMC-PI tuned using Genetic Algorithm (GA). Such a control scheme when used for GHS would result in better yield in the production of crops such as tomato, lettuce and broccoli.

CITE THIS PAPER

Manonmani, A., Thyagarajan, T., Sutha, S. and Gayathri, V. (2016) Design of Soft Computing Based Optimal PI Controller for Greenhouse System. Circuits and Systems, 7, 3431-3447.

REFERENCES

1. Albright, L.D., Gates, R.S., Arvanitis, K.G. and Drysdale, A. (2001) Environmental Control for Plants on Earth and in Space. IEEE Control System Magazine, 21, 28-47.
2. Pasgianos, G.D., Arvanitis, K.G., Polycarpou, P. and Sigrimis, N. (2003) A Nonlinear Feedback Technique for Greenhouse Environmental Control. Computers and Electronics in Agriculture, 40, 153-177.
3. Fourati, F. and Chtourou, M. (2007) A Greenhouse Control with Feed-Forward and Recurrent Neural Networks. Simulation Modelling Practice and Theory, 15, 1016-1028.
4. He, F. and Ma, C. (2010) Modeling Greenhouse Air Humidity by Means of Artificial Neural Network and Principal Component Analysis. Computers and Electronics in Agriculture, 71S, S19-S23.
5. Trejo-Perea, M., Herrera-Ruiz, G., Rios-Moreno, J., Miranda, R.C. and Rivas-Arazia, E. (2009) Greenhouse Energy Consumption Prediction Using Neural Networks Models. International Journal of Agriculture and Biology, 11, 1-6.
6. Hu, H.G., Xu, L.H., Wei, R.H. and Zhu, B.K. (2011) Multi-Objective Optimization for Greenhouse Environment Using Evolutionary Algorithms. Sensors, 11, 5792-5807.
7. Cao, Y.J. and Wu, Q.H. (1999) Teaching Genetic Algorithm Using Matlab. International Journal of Electrical Engineering, 36, 139-153.
8. Liu, W.K. and Juang, J.G. (2004) Application of Genetic Algorithm Based on System Performance Index on PID Controller. Proceedings of the Ninth Conference on Artificial Intelligence and Application.
9. Hasni, A. and Taibi, R. (2011) Optimization of Greenhouse Climate Model Parameters Using Particle Swarm Optimization and Genetic Algorithms. Energy Procedia, 6, 371-380.

CHAPTER 12

Dynamic Modeling of Mineral Contents in Greenhouse Tomato Crop

Antonio Juárez-Maldonado[1], Adalberto Benavides-Mendoza[1], Karim de-Alba-Romenus[1*], América B. Morales-Díaz[2]

[1]*Departamento de Horticultura, Universidad Autónoma Agraria Antonio Narro, Saltillo, México;*

[2]*Robótica y Manufactura Avanzada, Centro de Investigación y Estudios Avanzados (CINVESTAV), Saltillo, México*

ABSTRACT

Tomato is one the most important vegetables worldwide and mineral nutrition in tomato crops is considered as the second most important factor in crop management after water availability. Mathematical modeling techniques allow us to design strategies for nutrition management. In order to generate the necessary information to validate and calibrate a dynamic growth model, two tomato crop cycles were developed. Several mineral analyses were performed during crop development to determine the behavior of N, P, K, Ca, Mg and S in different organs of the plant. Regression models were generated to mimic the behavior of minerals in tomato plants and they were included in the model in order to simulate their dynamic behavior. The results of this experiments showed that the growth model adequately simulates leaf and fruit weight (EF > 0.95 and Index > 0.95). As for harvested fruits and harvested leaves, the simulation was less efficient (EF < 0.90 and Index < 0.90). Simulation of minerals was suitable for N, P, K and S as both, the EF and the Index, had higher values than 0.95. In the case of Ca and Mg, simulations showed indices below 0.90. These models can be used for planning crop management and to design more appropriate fertilization strategies.

Keywords: Licopersicon Esculentum; Mathematical Modeling; Simulation; Crop Growth; Nutrition Management

1. INTRODUCTION

Crop production in greenhouses has a great importance as it gives us an advantage over open-pit production because it provides a barrier between the external environment and the culture. Greenhouses create near optimal microclimate conditions for growing crops, protecting them from adverse conditions [1] and controlling factors such as temperature, radiation, CO_2 concentration, relative humidity, etc. In Mexico, the use of greenhouses for vegetable production has increased rapidly. Most recent information shows 12,000 hectares of greenhouses, without including 8000 hectares of mesh shade and macro tunnels [2].

Tomato (Licoperson esculentum Mill) is the most important vegetable crop in the world, it is used in both fresh and processed presentations [3-6]. Regarding the area under cultivation, tomato is the second most important crop after potato, but it ranks first as a processed crop [5]. In recent years, global production has increased by about 10% mainly because it is a significant source of vitamins and minerals [6]. In Mexico, tomato represents a 70% of crops grown in protected conditions, followed by pepper (16%) and cucumber (10%) [2]. Added to this, Mexico is the major international exporter, shipping the product to the United States, Canada and El Salvador. In 2011, a total of 1,872,000 tons were exported [7].

In order to promote the maximum productive potential of tomato, it is important to generate and apply crop management practices aimed at maximizing the inputs provided [8]. Nutrition is considered as the most important factor after water availability [8]. Although there are successful techniques such as fertigation, problems regarding fertilizer application are still encountered [9] since adequate fertilization plans appropriate to the real needs of nutrient intake along crop production are rarely found [8]. Therefore, fertilization practices should be defined based on the growth characteristics of the crop [10], for which dry matter accumulation to quantify the nutrient demand, is proposed [9].

Considering the current global scenario which emphasizes the need for friendly agricultural practices for environmentally sustainable food production [5], we must take into account issues such as the impact of excessive use of mineral fertilizers [3,11] as well as the increase in the cost of these and their availability in the future [4,5,11].

Technological advances provide novel techniques such as simulation of greenhouse crops. A crop growth simulation model is the application of systematic analysis and computer technology which integrates different disciplines such as crop physiology, ecology, meteorology and agriculture [12]. Thus, mathematical models allow us to improve the current knowledge of a system [13]. In agronomy, particularly in horticulture, these models have applications such as crop yield prediction and crop management, support systems for decision-making, greenhouse climate control, and root environment among others [13]. Crop models provide quantitative information from which decisions such as time of cultivation, irrigation, fertilization, crop protection, climate control measures, etc., can be taken at field level [14]. Protected environment models are necessary if one wishes to perform some optimization in production [14]. Although several studies involving the modeling of greenhouse tomato growth can be found [15,16], these models do not consider nutritional relations for the efficient management of mineral nutrition and some are based on variations of mineral concentration in nutrient solution or on drained solution [17].

The purpose of this study was to incorporate the behavior of major nutrients to a dynamic model of tomato growth to estimate crop mineral requirements. This will permit a more efficient management of fertilizer application to reduce production costs and the environmental impact from overuse.

2. MATERIALS AND METHODS

2.1. Development of Tomato Crop

An indeterminate growth habit tomato (Lycopersicon esculentum Mill), "Cayman" Enza Zaden® ball type hybrid fruit, was cultivated under a hood type greenhouse with polycarbonate cover and automatic temperature control. This took place during the years 2011 and 2012, from July 3rd to October 30th and from May 6th to September 23rd, respectively, in the northern region of Mexico. The culture was established in 19 L plastic pots with a density of 3 plants per m^2 and a soilless system using a mixture of perlite and peat moss substrate in a 1:1 ratio. A microtube irrigation system was used with high flow drippers for each pot. Automatic timers were also installed in order to irrigate at 4 different times per day (8:00, 12:00, 16:00

and 20:00 hr). The amount of water applied varied according to each phenological stage using 2.2 L per plant per day on high consumption stages. The crop nutrition was carried out with the application of Steiner solution [18], using different concentrations according to the phenological stage (50% at transplant, 75% from flowering and 100% after fruit set). One stalk plants were managed and they were limited in growth by removing the apical bud at 13 weeks after transplantation. The experimental setup was completely random, considering each plant as an experimental unit.

Considering the amount of water applied to plants along with the concentration of the Steiner solution for each phenological stage, the complete application of higher concentrations of minerals (N, P, K, Ca, Mg and S) was determined during the period time each culture was developed.

Climatic variables were measured inside the greenhouse during the development of both crops using a photosynthetically active radiation (PAR) sensor (LI-190) and an air temperature sensor (1400-101) connected to a data logger LI-1400 LI-COR Inc. In addition, concentration of CO_2 in the air was measured using a CO_2 meter© K-33 ELG sensor. Measurements were performed every 15 min and stored automatically in data loggers for a later download to a laptop.

In order to determine crop growth, destructive sampling was carried out weekly for four tomato plants. They were separated into leaves, stems and fruits and their fresh weights were obtained. After drying in oven at 80°C for 4 days, dry weight was obtained for the different parts of the plant. Furthermore, the total pruning on each plant and total fruit harvested were quantified in fresh and dry weight.

Minerals with higher concentration (N, P, K, Ca, Mg and S) were determined for different organs of the tomato plant (leaf, stem, fruit and root) at different stages of plant development.

2.2. Description of Tomato Growth Model

The dynamic model used was proposed by [19], and this was developed to indeterminate growth habit tomato under greenhouse conditions. This model starts from the flowering stage of the crop and has six state variables: mass balance for assimilation buffer (B), fruit dry weight (W_F), leaf dry weight (W_L), development stage (D_p), dry weight of harvested fruits (W_{HF}) and dry weight of harvested leaves (W_{HL}). The corresponding equations are described below:

$$\frac{dB}{dt} = P_g - b\left(fg_F W_F + vg_L \frac{W_L}{z} \right) - bR \qquad (1)$$

$$\frac{dW_F}{dt} = bg_F W_F - (1-b) r_F W_F - h_F W_F \qquad (2)$$

$$\frac{dW_L}{dt} = bg_L W_L - (1-b) r_L W_L - h_L W_L \qquad (3)$$

$$\frac{dD_p}{dt} = d_1 + d_2 \ln\left(\frac{T_g}{d_3} \right) - \ln\left(\frac{T_g}{d_3} \right) - d_4 t - h \qquad (4)$$

$$\frac{dW_{H_F}}{dt} = h_F W_F \qquad (5)$$

$$\frac{dW_{H_L}}{dt} = h_L W_L \qquad (6)$$

Climate variables measured inside the greenhouse where used as input variables to the model (temperature, photosynthetically active radiation (PAR) and CO_2). The outputs of the model are leaf dry weight ($g \cdot m^{-2}$), dry weight of fruit ($g \cdot m^{-2}$), harvested leaf dry weight ($g \cdot m^{-2}$) and harvested fruit dry weight ($g \cdot m^{-2}$). The complete description of the model is presented in [19] (Figure 1).

2.3. Calibration and Validation of the Tomato Growth Model

Model calibration was achieved by modifying the nominal values of a subset of parameters. These parameters, resulting from a sensitivity analysis, significantly affect model performance. The way this adjustment takes place is by modifying the nominal values of the model parameters and run the simulation model taking the input variables measured inside the greenhouse in the first growing season to find the values that give the best fit between simulated vs. observed outputs. Parameters were selected according to the sensitivity analysis performed by [19] plus some others that influence tomato varieties. Different values of these parameteres were tested until they met those providing more appropriate results when comparing the outputs of the model with real data of the first growing season (2011) of tomato.

To validate the model, climate data were taken from the second crop cycle as model inputs. The outputs of the same were compared with the actual data obtained from the second cycle (2012). To study the efficiency of the simulation of both the calibration and validation procedures, the "EF" and "Index" indexes given by [20] were used. These indices get values from 0 to 1, being 1 a perfect efficiency.

2.4. Modeling Mineral Content in Tomato

Nutrient demand mainly depends on the growth of the different plant organs, so that the maximum requirements are those when the plant has no limitation on the availability of these nutrients [21]. To generate behavioral models of the various minerals evaluated in the tomato plant, a simple regression technique is used. Time from flowering as well as actual mineral content in different organs of the plant for the first growing season (2011), were used. Six regression models corresponding to the six minerals (N, P, K, Ca, Mg and S) of higher concentration in each organ of the tomato plant were generated. The whole process of generating models was processed in Regress© (v. 2.21) for Excel.

The models generated were incorporated into the growth dynamic model to dynamically simulate the behavior of tomato minerals, because nutrients simulated concentration is equal to the plant demand if there is no limitation in the availability of nutrients [21]. This was done as follows:

$$OMC = \frac{ODW * \text{regression model}}{100} \tag{7}$$

where OMC is the mineral content per organ ($g \cdot m^{-2}$) and ODW is the organ dried weight ($g \cdot m^{-2}$).

Since tomato growth model does not include dry weights of stem and root as outputs, nor the total dry weight of leaves and fruits generated by the crop, the following equations were used:

$$W_S = 0.3391 * TW_L - 0.5377 \tag{8}$$

$$W_R = TW_L * 0.6 \tag{9}$$

where W_S is the stem dry weight ($g \cdot m^{-2}$) and W_R is the root dry weight ($g \cdot m^{-2}$). Both equations were obtained from linear regression.

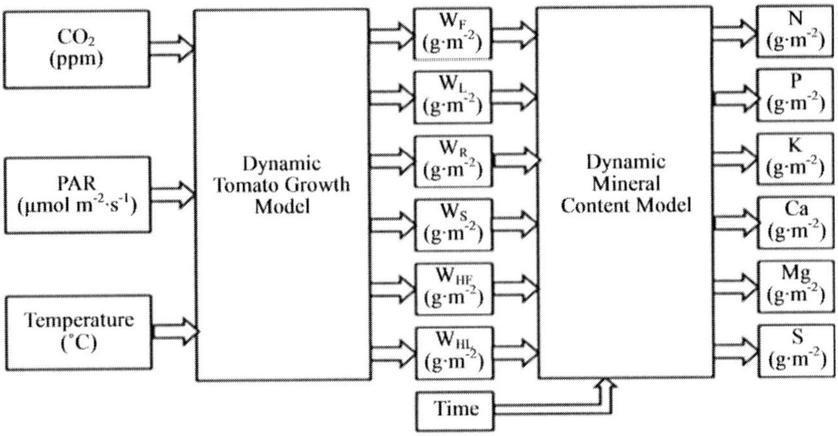

Figure 1. Block diagram of the tomato growth and mineral content models.

$$TW_L = W_L + W_{HL} \quad (10)$$

$$TW_F = W_F + W_{HF} \quad (11)$$

where TW_L is the total generated leaves dry weight ($g \cdot m^{-2}$) y TW_F is the total generated fruit dry weight ($g \cdot m^{-2}$).

In order to assess the total mineral quantity for the plant, only mineral contents from the different organs were added.

$$TMC = MCW_S + MCW_R + MCTW_L + MCTW_F \quad (12)$$

where TMC is the total mineral content ($g \cdot m^{-2}$), MCW_S is the mineral content in the stem ($g \cdot m^{-2}$), MCW_R is the mineral content in the root ($g \cdot m^{-2}$), $MCTW_L$ is the mineral content in the leaves ($g \cdot m^{-2}$), y $MCTW_F$ is the mineral content in the fruit ($g \cdot m^{-2}$).

TMC was considered only for the evaluation of the simulations of the six minerals (N, P, K, Ca, Mg and S) (Figure 1). Analogously to the modeling of growth of tomato plants, first crop cycle (2011) was used for the calibration of the mineral models used in the second cycle (2012). The whole programming process and dynamic simulation was performed in Matlab© R2011a.

3. RESULTS AND DISCUSSION

3.1. Modeling Growth of Tomato Plant

Following the calibration process, 11 parameters were modified out of a total of 27 available for the tomato growth model. Table 1 shows the evaluation for the different model outputs. It was observed that at 77 DAF only harvested leaves EF was less than 0.9 (0.8541), the rest was higher. In general at 84 and 91 DAF rates are lower than at 77 DAF, this is because growth was limited in the plants at 70 DAF, so real growth declined while the simulated one continued to grow (Figure 2).Table 2 shows the parameters of the model with their nominal values [19] and the values generated by the calibration process. Six of these parameters were slightly modified, and the rest (yL, mF, mL, f1 and d4) were greater in magnitude (Table 2). It is worth noticing that these parameters differ from those used by [19]. These differences are due to two main factors: the climate of the region where the experiments were carried out and the plant material used, since, each newly generated hybrid generated its own peculiarities regarding its technical management and absorption of nutrients [8].

Figure 2 shows the comparison of the actual data of the growing cycle of 2011 with the simulation result data obtained from the calibration. It is noted that the adjustment of the experimental observations concerning simulation is very good, because, in conducting the assessment with the indices "EF" and "Index" both showed a value greater than 0.95. Furthermore, WHL output gets values above zero from the start of the simulation (tomato flowering) unlike [19], which began tomato leaf pruning at the same time as tomato fruit harvesting (Figure 2). These differences are due to the use of different tomato hybrids, since as mentioned, each material has its own characteristics [8].

For the validation process of tomato growth model, unlike the work of [19], indices to assess the quality of the simulation model were used. In Table 3, the results of that assessment is shown. It was observed that the two indices used (EF and Index) are greater than 0.94 in the assessments made at 77 and 84 days after flowering (DAF) for WF, WL y WHF outputs.

Figure 3 shows the graphical comparison between the four outputs of the model and data from the tomato crop cycle 2012. It can be noticed the high quality of simulation at 84 DAF for W_F, W_L and W_{HF} values. The output variable W_{HL}, was underestimated by the model, this was because the management of pruning in the second growing season changed due to certain stress factors, forcing some extra mowing.

Table 1. List of indices used to evaluate the efficiency of the dynamic model simulation for different outlets in the calibration process.

	Fruits in Plant		Leaves in Plant		Harvested Fruits		Harvested Leaves	
DAF*	*EF*	*Index*	*EF*	*Index*	*EF*	*Index*	*EF*	*Index*
77	0.9888	0.9971	0.9507	0.9873	0.9978	0.9995	0.8541	0.9671
84	0.9893	0.9972	0.9259	0.9816	0.6358	0.9412	0.8855	0.9714
91	0.9812	0.9950	0.8828	0.9717	0.4647	0.9233	0.9037	0.9775

*Days considered for evaluation. DAF Days after Transplanting. The "*EF*" and "*Index*" are the indexes given by [20]. These indices get values from 0 to 1, being 1 a perfect efficiency.

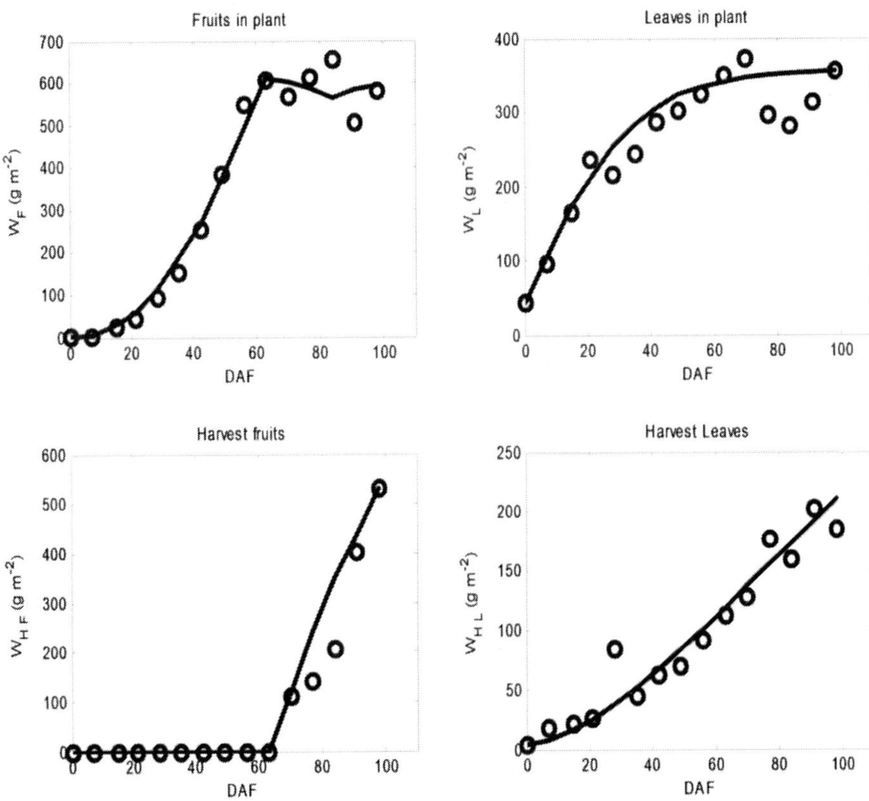

Figure 2. Calibration plots of model outputs. Real (o) vs. simulated (-) data comparison corresponding to cycle 2011.

Table 2. Parameter description, nominal values [19] and calibrated values.

Parmeter	Nominal value	Calibrated value	Units	Description
b_1	2.7		$m^2 \cdot g^{-1}$	buffer switching function coefficient
d_1	2.13e−07		s^{-1}	plant development rate parameter
d_2	2.47e−07		s^{-1}	plant development rate parameter
d_3	20		°C	plant development rate parameter
d_4	7.50e−11	3.46e−13	-	plant development rate parameter
F	1.2		-	fruit assimilate requirement quotient
f_1	8.10e−07	8.10e−06	s^{-1}	fruit growth rate coefficient
f_2	4.63e−06		s^{-1}	fruit growth rate coefficient
M	2.511		-	LAI-correction function parameter
m_F	1.16e−07	9.57e−08	s^{-1}	fruit maintenance respiration coefficient
m_L	2.89e−07	2.89e−09	s^{-1}	vegetative maintenance respiration coefficient
p_3	577		$W \cdot m^{-2}$	net-photosynthesis parameter
p_4	221		$g \cdot s^{-1} \cdot m^{-2}$	net-photosynthesis parameter
P_m	2.25e−03	3.55e−03	$g \cdot s^{-1} \cdot m^{-2}$	maximum photosynthesis
Q_G	1		-	fruit groeth rate temperature Q_{10}-value
Q_R	2		-	maintenance respiration Q_{10}-value
T	86.400		s	time
T_G	20		°C	growth rate temperature effect reference temperature
T_R	25		°C	maintenance respiration reference temperature
V	1.23	2.23	-	vegetative assimilate requirement quotient
v_1	1.3774	0.45	-	vegetative fruit growth ratio parameter
v_2	−0.168	−0.068	$°C^{-1}$	vegetative fruit growth ratio parameter
v_3	19	20	°C	vegetative fruit growth ratio parameter
w_R	32.23		$g \cdot m^{-2}$	LAI-correction function parameter
y_F	0.5983	1.9183	-	fruit harvest coefficient parameter
y_L	0.5983	8.70e−08	-	leaf harvest coefficient parameter
Z	0.6081		-	leaf fraction of vegetative dry weight

Table 3. Indices used to evaluate the efficiency of the dynamic model simulation for different outputs in the validation process. *Days considered for evaluation.

	Fruits in plant		Leaves in plant		Harvested Fruits		Harvested Leaves	
DAF*	*EF*	*Index*	*EF*	*Index*	*EF*	*Index*	*EF*	*Index*
77	0.9724	0.9929	0.9516	0.9895	0.9917	0.9979	0.6489	0.9031
84	0.9647	0.9904	0.9467	0.9886	0.9593	0.9877	0.6320	0.8771
91	0.9676	0.9912	0.8907	0.9772	0.9679	0.9906	0.6350	0.8779

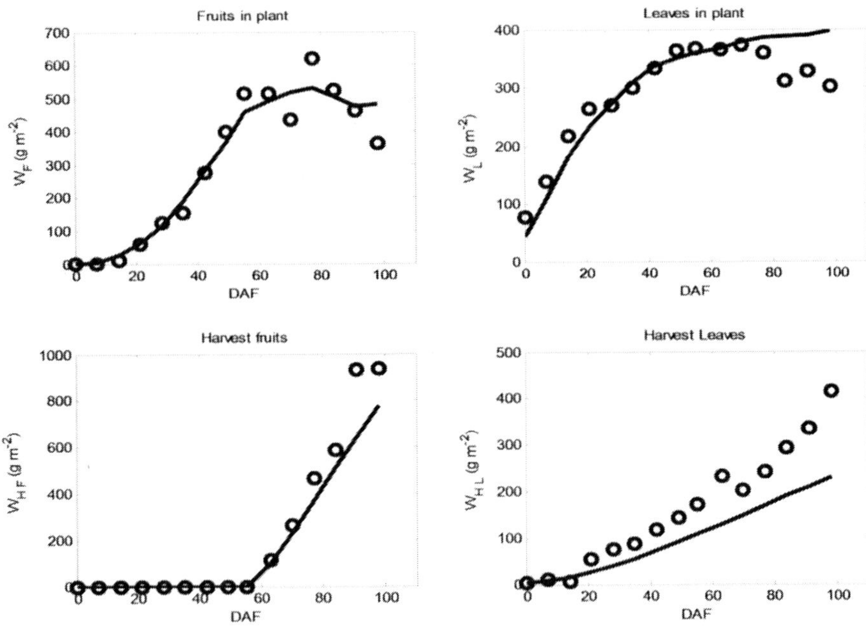

Figure 3. Model output plots. Cycle 2012 Real Data (o) vs. simulated output (-) are compared.

When evaluated at 91 DAF, best results were obtained for W_{HF} (EF = 0.9676, Index = 0.9912) and W_F (EF = 0.9679, Index = 0.9906). On the other hand, EF < 0.90 was obtained for both W_L and W_{HL} outputs (Table 3). These results should be explained as a result of the limited growth of tomato plants to 70 DAF, because in this case, the number of leaves on the plant as well as the number of leaves taken from the plant has decreased since then (Figure 3). Since the model is designed for indeterminate growth habit tomato, the simulation obtained by the same continues to grow while the actual values obtained from culture decrease (Figures 2 and 3). In the case of fruits, this is not the case, because by the time when growth was limited, there were still small fruits and flowers, so they continued to grow over time and harvesting continued (Figures 2 and3).

3.2. Mineral Modeling in Tomato Plants

The necessary models to mimic the behavior over time of minerals in tomato plants were generated from regression techniques. Table 4 shows the 24 regression models generated for the six minerals with higher concentration (N, P, K, Ca, Mg, S) and for the four tomato plant organs (leaf,

Table 4. Regression equations used to estimate minerals content in tomato plant.

	Mineral	Regression Model
Leaf	N	$y = 0.0027x^2 - 0.0959x + 2.7678$
	P	$y = -0.015x + 0.5504$
	K	$y = -0.0053x^2 + 0.0854x + 2.0369$
	Ca	$y = -0.0073x^2 + 0.1394x + 2.3713$
	Mg	$y = 0.0071x + 0.7181$
	S	$y = -0.0018x^2 + 0.0795x + 0.7042$
Stem	N	$y = 0.0016x^3 - 0.0251x^2 + 0.0094x + 1.82$
	P	$y = 0.0003x + 0.4964$
	K	$y = 0.0016x^4 - 0.0396x^3 + 0.2968x^2 - 0.5476x + 2.0415$
	Ca	$y = -0.0023x^2 + 0.0447x + 0.9849$
	Mg	$y = 0.0005x^3 - 0.003x^2 - 0.0273x + 0.4775$
	S	$y = 0.0169x + 0.303$
Fruit	N	$y = -0.0119x^2 + 0.1305x + 1.6867$
	P	$y = -0.0042x + 0.3828$
	K	$y = 0.0023x + 1.9929$
	Ca	$y = 0.004x + 0.0878$
	Mg	$y = 0.0057x + 0.2273$
	S	$y = 0.0029x + 0.1566$
Root	N	$y = 0.0101x^2 - 0.2082x + 1.9307$
	P	$y = 0.0025x^2 - 0.044x + 0.391$
	K	$y = 0.0122x^2 - 0.2538x + 1.6305$
	Ca	$y = -0.0081x^2 + 0.1291x + 0.9659$
	Mg	$y = -0.0069x + 0.3373$
	S	$y = -0.0045x + 0.3488$

In the regression equation "y" represents the mineral content expressed in % of dry weight based on the organ. "x" represents the time, expressed in weeks, from flowering.

Table 5. Number of times minerals were applied during cultivation cycles.

	N	P	K	Ca	Mg	S
2011	2.70	5.13	4.23	3.97	2.29	7.17
2012	1.95	3.39	2.88	1.86	3.89	4.25

stem, fruit and root). It was noted that 11 of the generated models are linear, 10 are quadratic, 2 are cubic, and just one of fourth order (Table 4). These results agree with [21], since the mineral concentration showed ontogenetic changes. In the particular case of the fruit, only nitrogen showed changes over time, and the remaining minerals (P, K, Ca, Mg and S) remained virtually unchanged (Table 4).

After entering the regression models generated for minerals in the dynamic model of tomato growth, a comparison between the simulated data and actual data of the growing season for 2011 was performed. Efficiency results of simulation are shown in Table 5. It is observed that for all minerals both indexes were higher than 0.95, a normally considered good efficiency. Although [21] performed nutrient modeling on peppers, there is no index to evaluate the efficiency of the model simulation.

Additionally, data corresponding to minerals applied to the crop by Steiner solution were included (Figure 4). It can be seen in the graphs presented in Figure 4 the good fit between the model simulated data and actual data of the six minerals in question.

To validate the modeling of minerals (N, P, K, Ca, Mg and S) in tomato plants, simulated values were compared against those obtained from the crop cycle conducted in 2012 (Figure 5). Efficiency results of simulation are shown in Table 5. In this case it was found that the simulations corresponding to N, P, K and S showed very good fit because both indices used were higher than 0.95. This is borne out graphically in Figure 5. Regarding Ca and Mg, EF values of 0.75 and 0.89, respectively, were obtained. Model underestimated the Ca content as measured value was higher in plants by approximately 61% at 98 days after transplantation (DAT). On the other hand, the Mg content was overestimated by the model by 62% in the same time period (Figure 5). These results can be explained due to errors in the simulation of growth and its partitioning into the different plant organs [21], since as shown in Table 3, simulation efficiency for harvested leaves was very low (EF = 0.64).

Although mineral simulation in tomato has already been reported, this was based on variations in the concentration of minerals in the nutrient solution or drain using a mass balance equation based on the concept of concentration ion extraction [17]. In contrast, in this paper, actual mineral concentrations in the different organs of the tomato plant are considered.

Considering the presented results, some adjustments can be made based to the management of fertilization of greenhouse tomato plants by means of daily quantification of nutrient demand [9]. This helps us prevent shortfalls or excesses of nutrients in the tomato plants, which in turn will reduce production costs without affecting production either in quantity or in quality. In addition, by allowing the simulation of minerals to be based on tomato growth, which in turn depends on the climate in which it operates, we can design management practices that increase productivity while minimizing environmental impact caused by the agriculture activity [10].

Regarding the actual application of minerals to tomato plants during the time the cultures were grown, Figures 4 and 5 show the comparison of these data with the simulated and measured data in the culture. Additionally, Table 5 shows the relationship between the amount of minerals applied and measured in tomato plants during both cycles of cultivation. It can be noticed that S is most excessively applied mineral in both tomato crop cycles, namely 7.17 times in 2011 and 4.25 times in 2012. P was the second most applied mineral with 5.13 times in 2011 and 3.39 times in 2012. The N and Ca were those applied in smaller excess in the 2012 cycle, with 1.95 and 1.86 times respectively. It should be considered that while the S is essential for the growth and development of plants, it has been documented that higher applications of S, lower Cu concentration in the tomato plant, and soil salinity increases [22].

These results show that we must consider several important issues for the production: the increase in the cost of mineral fertilizers and their availability in the future [4,5,11] as well as the impact on the environment on overuse of fertilizers [3,11]. In this sense, the modeling techniques proposed in this work can support in managing the nutrition of tomato plants. Extraction curves allow nutrient demand according to the phenological stage of the crop, so that can be used to establish fertilization programs in order to maximize its efficiency [8]. This is consistent with [12], since the use of these models has a significant impact on the economic and ecological benefits.

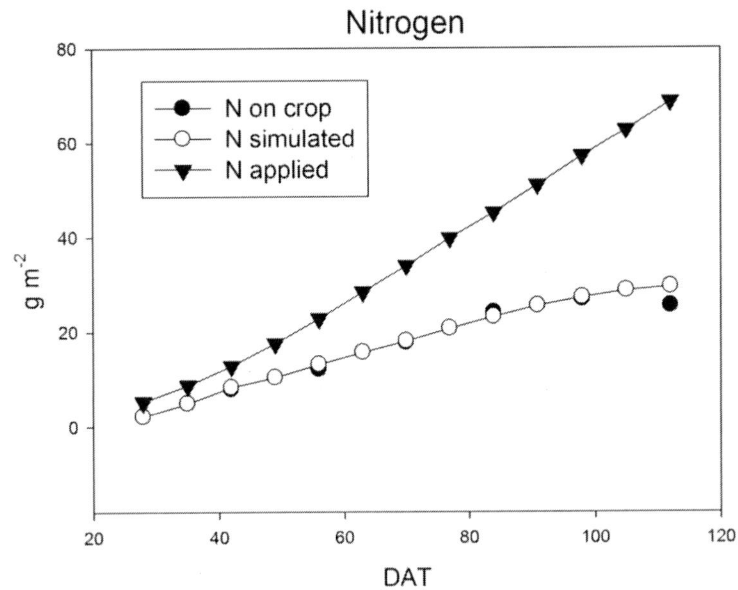
Nitrogen
N on crop
N simulated
N applied
g m-2
DAT
80
60
40
20
0
20
40
60
80
100
120

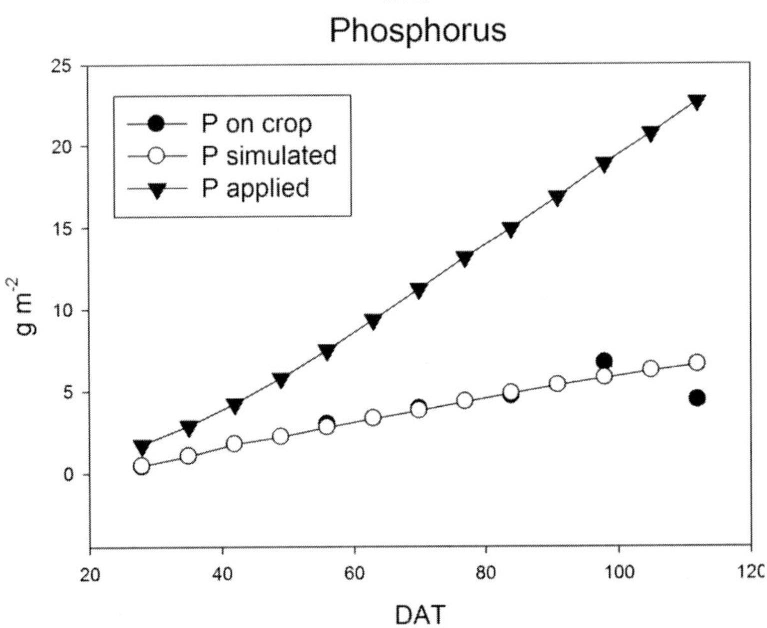
Phosphorus
P on crop
P simulated
P applied
g m-2
DAT
25
20
15
10
5
0
20
40
60
80
100
120

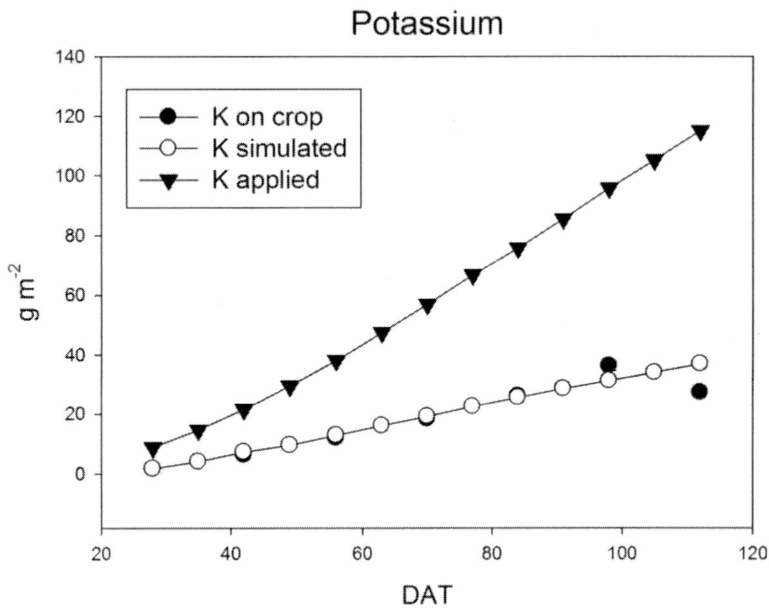
Potassium
K on crop
K simulated
K applied
g m-2
DAT
0
20
40
60
80
100
120
140
20
40
60
80
100
120

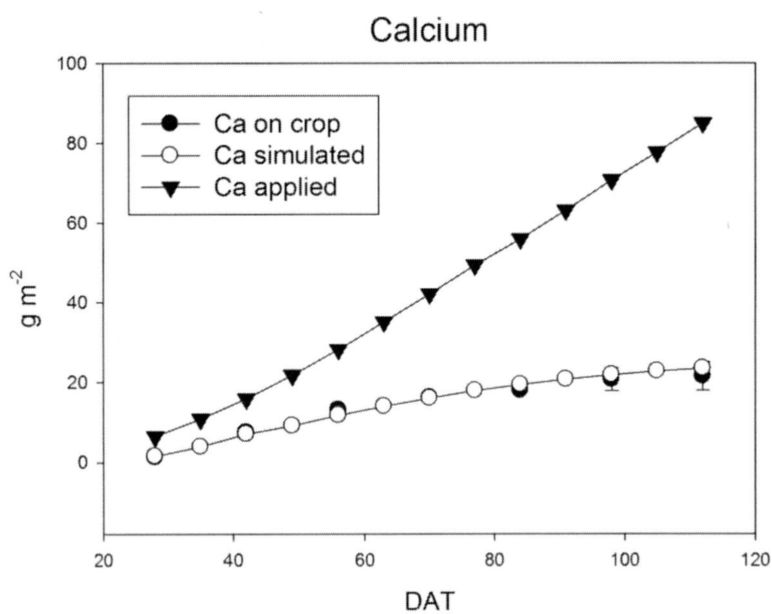
Calcium
Ca on crop
Ca simulated
Ca applied
g m-2
DAT
0
20
40
60
80
100
20
40
60
80
100
120

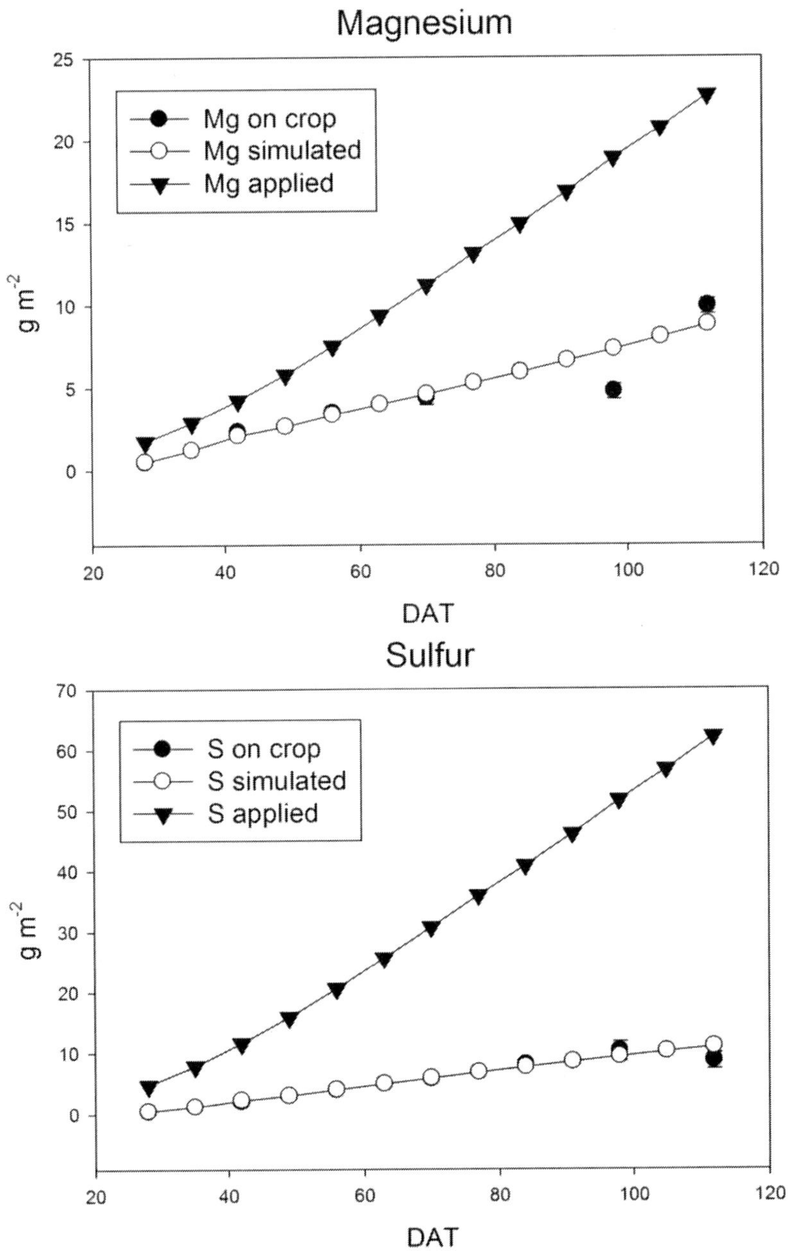

Figure 4. Calibration plots showing mineral content in tomato plant. Real quantity of minerals applied using Steiner solution is included.

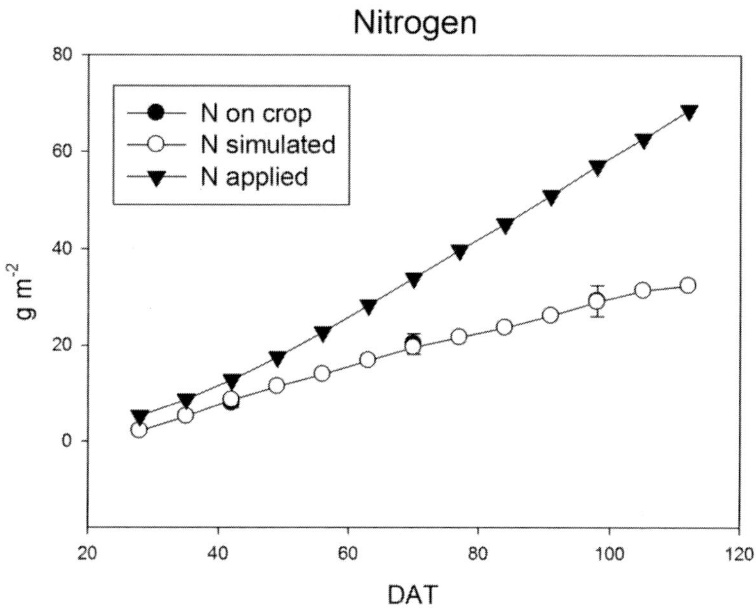
Nitrogen
N on crop
N simulated
N applied
g m-2
80
60
40
20
0
20
40
60
80
100
120
DAT

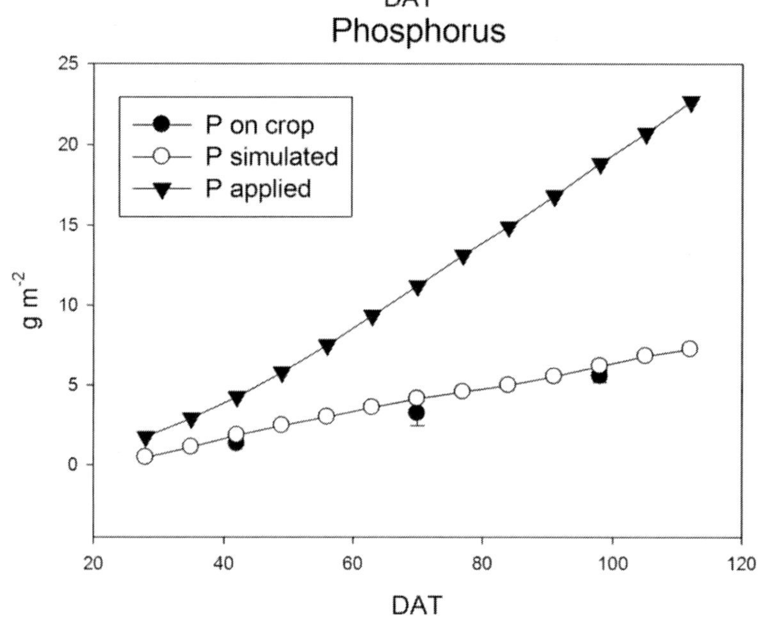
Phosphorus
P on crop
P simulated
P applied
g m-2
25
20
15
10
5
0
20
40
60
80
100
120
DAT

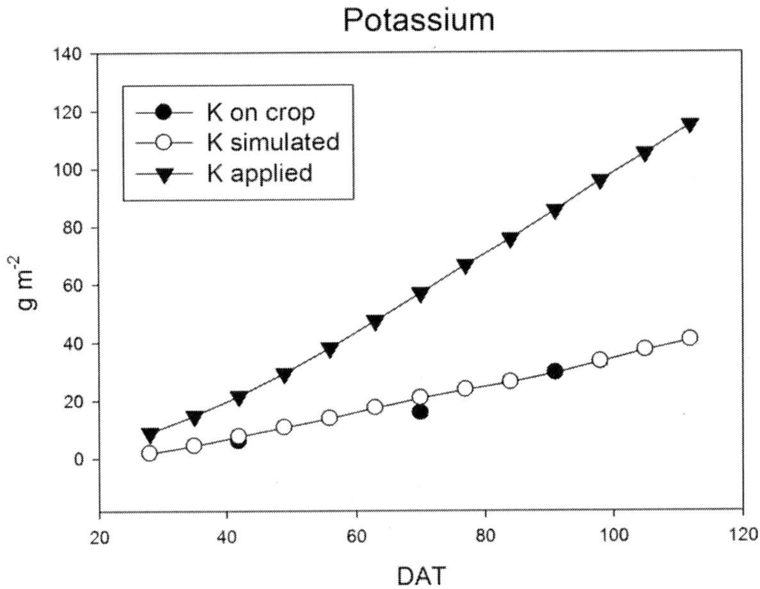
Potassium
K on crop
K simulated
K applied
g m-2
DAT
140
120
100
80
60
40
20
0
20
40
60
80
100
120

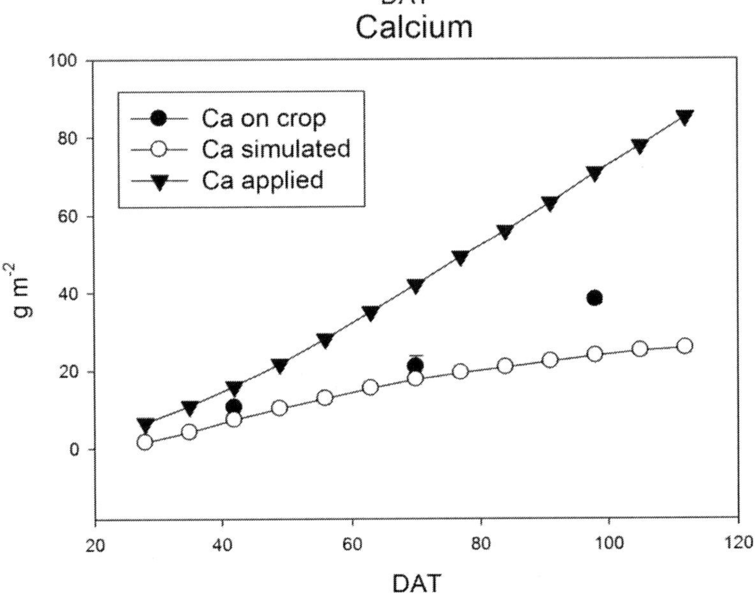
Calcium
Ca on crop
Ca simulated
Ca applied
g m-2
DAT
100
80
60
40
20
0
20
40
60
80
100
120

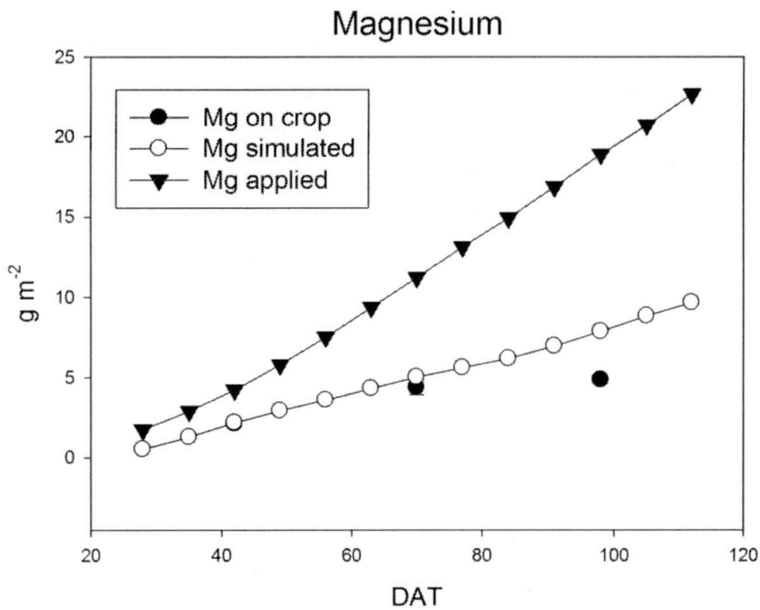

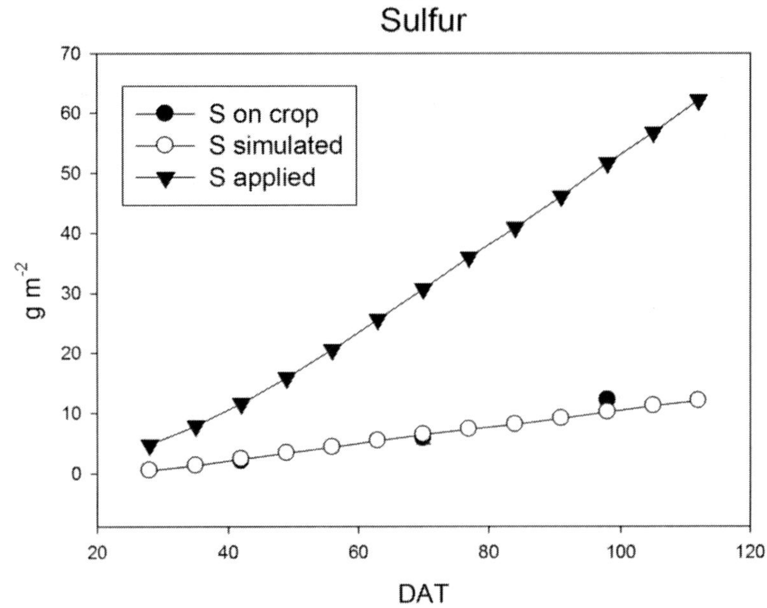

Figure 5. Validation plots showing mineral content in tomato plant. Real quantity of minerals applied using Steiner solution is included.

4. CONCLUSSIONS

In this work, it was calibrated and validated a dynamic model of tomato growth under greenhouse conditions which can be used to design appropriate strategies for the management of this crop.

Several regression models generated from measurements of the concentrations of N, P, K, Ca, Mg and S in the various organs of the tomato plants and they were incorporated into the dynamic model of tomato growth. This model allows us to simulate the nutritional requirements for tomato plants under greenhouse conditions from weather conditions existing within it. This model can be used as support for the management of mineral fertilization of tomato.

Antonio Juárez-Maldonado, Adalberto Benavides-Mendoza, Karim de-Alba-Romenus, América B. Morales-Díaz The application of these modeling and simulation techniques can minimize the use of fertilizers for greenhouse tomato crop, allowing us to estimate very precisely the nutritional requirements. This will generate, as a result, a decrease in production costs and reduce environmental impact by excessive application of fertilizers in the cultivation of tomato, all without affecting the production and quality.

REFERENCES

1. Martínez-Ruiz, A., López-Cruz, I.L., Ruiz-García, A. and Ramírez-Arias, A. (2012) Calibración y validación de un modelo de transpiración para gestión de riegos de jitomate (Solanum lycopersicum L.) en invernadero. Revista Mexicana de Ciencias Agrícolas, 4, 757-766.
2. SAGARPA (2013) Agricultura protegida 2012.http://www.sagarpa.gob.mx/agricultura/Paginas/Agricultura-Protegida2012.aspx
3. Sepat, N.K., Kumar, A., Yadav, J. and Srivastava, R.B. (2012) Effect of integrated nutrient management on growth, yield and quality of tomato in trans Himalayan. Annals of Plant and Soil Research, 14, 120-123.
4. Gad, N. and Hassan, N.M.K. (2013) Role of cobalt and organic fertilizers amendments on tomato production in the newly reclaimed soil. World Applied Sciences Journal, 22, 1527-1533.
5. Mehdizadeh, M., Darbandi, E.I., Naseri-Rad, H. and Tobeh, H. (2013) Growth and yield of tomato (Lycopersicon esculentum Mill.) as influenced by different organic fertilizers. International Journal of Agronomy and Plant Production, 4, 734-738.
6. Shalaby, T.A. and El-Banna, A. (2013) Molecular and horticultural characteristics of in vitro induced tomato mutants. Journal of Agricultural Science, 5, 155-163.
7. MÉXICOPRODUCE (2013) Productos: Jitomate.http://www. mexicoproduce.mx/productos.html#jitomate
8. Quesada-Roldán, G. and Ber tsh-Hernández, F. (2013) Obtaining of the absorption curve for the fb-17 tomato hybrid. Terra Latinoamericana, 31, 1-7. [Citation Time(s):6]

9. Bugarín-Montoya, R., Galvis-Spinola, A., Sánchez-García, P. and García-Paredes, D. (2002) Daily accumulation of aboveground dry matter and potassium in tomato. Terra Latinoamericana, 20, 401-409. [Citation Time(s):3]
10. Enriquez-Reyes, S.A., Alcántar-González, G., CastellanosRamos, J.Z., Suárez, E.A., González-Eguiarte, D. and Lazcano-Ferrat I. (2003) NUMAC-N Tomato: Mineral nutrition fit at growth. The nitrogen nutrition in tomato greenhouse production. 1. Model description and parameters adjust. Terra Latinoamericana, 21, 167-175.
11. Basheer, M. and Agrawal, O.P. (2013) Effect of vermicompost on the growth and productivity of tomato plant (Solanum lycopersicum) under field conditions. International Journal of Recent Scientific Research, 4, 247-249.
12. Zhang, J. and Wang, S. (2011) Simulation of the canopy photosynthesis model of greenhouse tomato. Procedia Engineering, 16, 632-639.
13. Marcelis, L.F.M., Elings, A., Bakker, M.J., Brajeul, E., Dieleman, J.A., de Visser, P.H.B. and Heuvelink, E. (2006) Modelling dry matter production and partitioning in sweet pepper. Acta Horticulturae, 718, 121-128.
14. Gary, C. (1999) Modeling greenhouse crops: State of the art and perspectives. Acta Horticulturae, 495, 317-322.
15. De Gelder, A., Heuvelink, E. and Opdam, J.J.G. (2005) Tomato yield in a closed greenhouse and comparison with simulated yields in closed and conventional greenhouses. Acta Horticulturae, 691, 549-552.
16. Marcelis, L.F.M., Elings, A., de Visser, P.H.B. and Heuvelink, E. (2009) Simulating growth and development of tomato crop. Acta Horticulturae, 821, 101-110.
17. Massa, D., Incrocci, L., Maggini, R., Bibbiani, C., Carmassi, G., Malorgio, F. and Pardossi, A. (2011) Simulation of crop water and mineral relations in greenhouse soilless culture. Environmental Modelling and Software, 26, 711-722.
18. Steiner, A.A. (1961) A universal method for preparing nutrient solutions of a certain desired composition. Plant Soil, 15, 134-154.
19. Tap, R.F. (2000) Economics-based optimal control of greenhouse tomato crop production. Ph.D. Dissertation, Wageningen Agricultural University, Wageningen. [Citation Time(s):8]
20. Wallach, D. (2006) Evaluating crop models. In: Wallach, D., Makowski, D. and Jones, J.W., Eds., Working with Dynamic Crop Models. Evaluation, Analysis, Parameterization and Applications, Elsevier, Amsterdam, 11-54.
21. Marcelis, L.F.M., Brajeul, E., Elings, A., Garate, A. and Heuvelink, E. (2005) Modelling nutrient uptake of sweet pepper. Acta Horticulturae, 691, 285-292. [Citation Time(s):5]
22. Orman, S. (2012) Effects of elemental sulphur and farmyard manure applications to calcareous saline clay loam soil on growth and some nutrient concentrations of tomato plants. Journal of Food, Agriculture & Environment, 10, 720-725.

CHAPTER 13

Above and Below Ground Interactions in Monoculture and Intercropping of Onion and Lettuce in Greenhouse Conditions

Jessica L. de Haan, Liette Vasseur*

Department of Biological Sciences, Brock University, St. Catharines, Canada

ABSTRACT

Intercropping has been seen as an advantageous strategy in sustainable agriculture. Plants how- ever interact with one another both above and below ground with members of the same species (intraspecific) or members of a different species (interspecific) for nutrients, water and light. It is therefore essential to understand these interactions when intercropped. The objective was to examine the above and below ground interactions between onion and lettuce in monocrop and intercrop systems. We examined the various possible interactions (no competition, above ground, below ground, or full) using a full factorial randomized design under greenhouse conditions. Onion yield was highest in intraspecific above ground competition and lowest in below ground and full interspecific competition with lettuce. Dry weight of onions in above ground competition with lettuce was significantly greater than that of the control group. Fresh weight of lettuce leaves were highest in below ground and full interspecific competition treatments. The hectare model and yield results suggest that there is strong below ground competitive effect between onion and lettuce in intercrop. Asymetric interspecific facilitation was found: facilitation by onion led to in- creased lettuce yield but a negative effect of lettuce on onion yield was observed. Knowledge of competitive interactions between component crops can have several applications in sustainable agricultural as it helps to match the most efficient species under specific conditions.

Keywords: Intraspecific and Interspecific Competition, Onion, Lettuce, Sustainable Agriculture, Facilitation

1. INTRODUCTION

Intercropping is an agricultural technique based on the management of plant interactions to maximize crop yield and involves cultivating two or more crops in the same field within the same growing season [1] [2] . Intercrop- ping can provide increased yields in an environmentally sustainable manner through resource complementarity, since niche overlap and competition between species is assumed to be reduced, permitting crops to capture a greater range and quantity of resources than in monoculture [3] [4] .

Competitive interactions between two plants, brought about by a shared requirement for soil nutrients, water or light, can have negative effects on either species by decreasing growth rate, survival, reproduction or yield of one or both plants [5] [6] . In intercropping, the net effect of competition includes not only interspecific competi- tion but also intraspecific competition [3] . In addition, competition can be above ground when plants modify their leaf areas in function of light interception, or below ground when plants alter the length, density, and spa- tial distribution of their roots in response to the presence of a neighbour [1] [6] . Lettuce (Lactuca sativa), for example, is expected to be more competitive in intercrop because it is a fast emerging, tall and broad-leaf plant that is able to obtain a larger share of available light than slower growing and smaller plants like onion [7] . In- deed, onion crops (Allium cepa) are known to be a weak competitor because of their initial slow growth rate af- ter planting (juvenile growth), development of shallow fibrous roots and a small above ground canopy, which does not cover the soil [7] .

In order to study competition in intercrop treatments and fully explore the yield potential of intercrops, we must examine and distinguish between the different types of interactions [3] . The cumulative and individual ef- fects of above and below ground competition are not predictable; they vary with availability of resources and the species used [1] . Only by physically separating above ground competition from below ground competition can we distinguish between competition for light and competition for nutrients and water.

Research into cereal-legume intercrops is extensive and common in tropical and subtropical regions such as Africa and Asia. There is however limited research dedicated to vegetable or non-legume intercrops, intercrop competition and the specific interactions occurring between intercropped species, especially in temperate re- gions [2] [8] [9] . The objective of this experiment was to examine the interactions between onion and lettuce in monocultures and intercrops to determine whether the interactions were above or below ground, interspecific or intraspecific in nature, and if the interactions were positive (facilitative) or negative (competitive). It is hypothe- sized that the biomass for lettuce and onion intercrops will be greater than that of the monocultures. From this, we expect that intraspecific competition between lettuce plants will occur mainly above ground while interspe- cific competition between lettuce and onion will occur mainly below ground. Intraspecific competition between onions will be limited to minimal below ground interactions, while the biomass of onion plants in above ground competition with lettuce is expected to be negatively affected.

2. MATERIALS AND METHOD

2.1. Experimental Design

We used a complete random factorial design in a greenhouse setup where both intraspecific and interspecific competitions were examined using 11 treatments with 10 replicates: one single plant of onion 1) or lettuce 2) acted as the control groups; two plants of onion 3) or of lettuce 4) or one onion and one lettuce 5) planted in the same pot with full (above and below ground) interactions; two plants of onion 6) or of lettuce 7) or one onion and one lettuce 8) planted in a pot where the above ground portions were separated with greenhouse plastic (no above ground interaction); and two plants of onion 9) or of lettuce 10) or one onion and one lettuce 11) planted in two different pots but placed side by side (no below ground interaction). Seeds were sown directly into pots: onion at a depth of 2 cm and lettuce at a depth of 1 cm (according to the seed company). Seeds were spaced ap- proximately 10 cm from each other in either the same pot for below ground and full competition, or in two pots for above ground competition treatments (one per pot) (note that the volume of soil was the same for all plants, whether in a larger pots for two plants or smaller pots for single plants). Plants were watered when the soil was dry to the touch. Fertilizer (20-20-20) was mixed into the water once every two weeks. Temperature varied be- tween 15°C and 25°C with photoperiod of 14 h day/10 h night.

Lettuce was harvested after 55 days (on May 1, 2013). Fresh leaves were weighed and the samples were placed into a drying cabinet at 45°C for two weeks, at which point dry weight was recorded for each sample. The onion plants were left to grow for another 33 days and were harvested on June 3 - 4, 2013. Fresh bulb weight was immediately recorded, samples were placed in the drying cabinet, and after two weeks the dry onion samples were removed from the oven and weighed.

2.2. Data Analyses

Fresh and dry weight data were transformed using a natural logarithmic transformation to satisfy normality and compared between treatment groups with an analysis of variance followed by a multiple range Dunnett T3 post- hoc test.

A model was created to estimate the yield of lettuce and onion in intercrop and monoculture in a simulated area of 1 hectare (100 m × 100 m). The hectare model uses average yield from both the onion and lettuce, controls, intraspecific full competition and interspecific full competition treatments. In the onion monoculture model, rows were spaced 40 cm apart, and onion plants were spaced 10 cm apart within rows, in accordance with onion- growing fact sheet [10] . In the lettuce monoculture model, lettuce rows were spaced 30 cm apart and plants were spaced 25 cm apart within rows. In the intercrop model, onion and lettuce were alternated every 12 cm with 30 cm between rows. Values were compared using a standard normal distribution (z-test).

The Relative Interaction Index (RII, [11]) was used to determine the types of interactions occur-ring between species in each treatment group. The fresh weight of onion bulbs and lettuce leaves were compared to their respective control groups using the following Equation (1):

$$RII=\frac{Mw - Mo}{Mw + Mo}$$

where Mw is the fresh mass of plant matter (lettuce leaves or onion bulbs) in competition with neighbours, and Mo is the fresh mass of isolated plants (control). RII has values ranging from −1 to 1, is symmetrical around zero and is negative for competition and positive for facilitation [11] . Given a 95% confidence interval, a treatment was considered competitive if the RII value was less than −0.10, facilitative if the RII value was greater than 0.10, and neutral if the RII value was between −0.10 and 0.10 [11] . All statistical tests were performed using SPSS version 17.0.

3. RESULTS

3.1. Biomass and Yield Analyses-Onions

Onion bulbs exposed to intraspecific above ground competition had the highest fresh weight. In interspecific competition, fresh bulb weight was the lowest in below ground and full competition with lettuce (Figure 1). Onions in interspecific below ground and full competition had significantly lower dry weight of bulbs than the onion control group, while onions growing in intraspecific above ground competition had the largest dry weight of bulbs (Figure 2). Additionally, onions in above ground competition with lettuce had significantly greater dry weight of bulbs than those in the control group.

When individual onions were grown in pots, i.e. with no competition, and extrapolated for yield in a field of 1 hectare, the results predicted the highest yield with a mean of 247 kg/ha (Figure 3). However, the mean value decreased to 185 kg/ha (Figure 3) as soon as two onions grown in the same pot (simulating intraspecific mono- culture) were considered. While this value considered the same spacing, it is clear that there is an interaction between individuals. The lowest yield is predicted when onions are planted in interspecific competition with lettuce (63 kg/ha, Figure 3). This value was statistically different than the monoculture yields ($p < 0.001$).

3.2. Biomass and Yield Analyses-Lettuce

Lettuce yield was greatest in below ground and full interspecific competition with onion compared to intraspe- cific or no competition (Figure 4 and Figure 5). The hectare model for lettuce showed no significant difference between lettuce in individual pots (control) and in monocultures ($p = 0.464$, 6850 kg/ha, 6893 kg/ha) (Figure 3). The yield of lettuce in interspecific competition with onion in the hectare model was both the highest lettuce yield observed, and statistically different than the yield of the monoculture models ($p = 0.012$, 9821 kg/ha) (Figure 3).

3.3. Competition vs Facilitation

The effect of lettuce on onion was competitive in the below ground and full competition treatments, but was neutral in the above ground competition

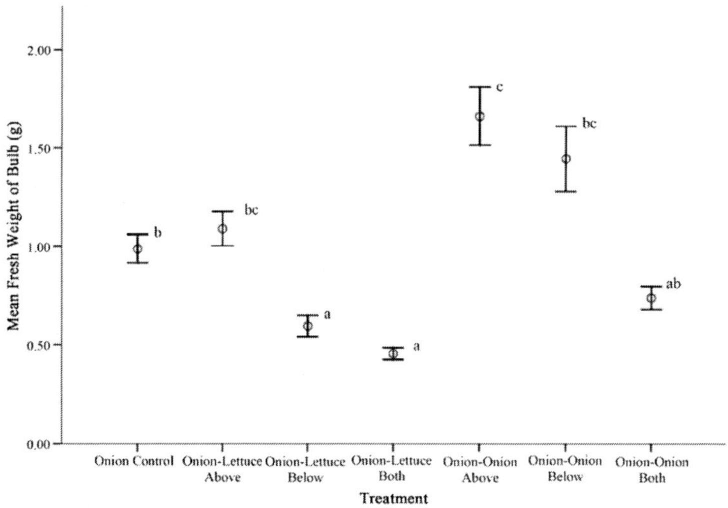

Figure 1. Average fresh weight of onion bulbs grown in control, intraspecific or interspecific competition with lettuce. An analysis of variance with a Dunnett T3 Post-hoc test was used to find significance between treatments. Letters next to the standard error bars note significance df = 6, MS = 2.62, F = 17.04, $p < 0.001$).

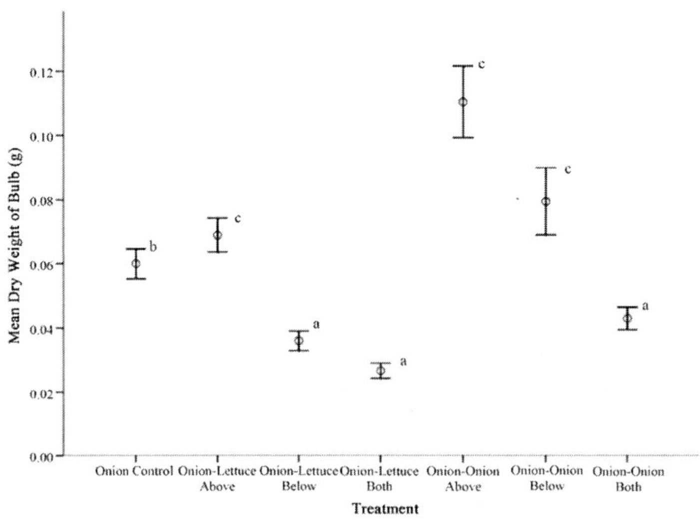

Figure 2. Comparison of average dry weight of onion bulbs grown in control, or competition with lettuce or onion. Significance between treatments is represented using letters next to the standard error bars, and was found using an analysis of variance with a Dunnett T3 Post-hoc test (df = 6, MS = 2.82, F = 13.94, $p < 0.001$).

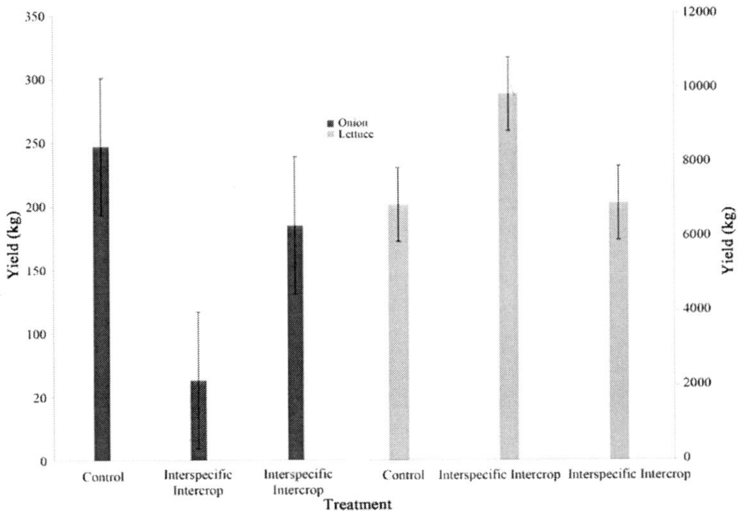

Figure 3. Predicted yield comparison of onion bulbs (kg/ha) for onion grown in monoculture assuming no competition, onion grown in monoculture with intraspecific competition and onions grown in an intercrop under interspecific competition with lettuce. Error bars are +/−1 standard error.

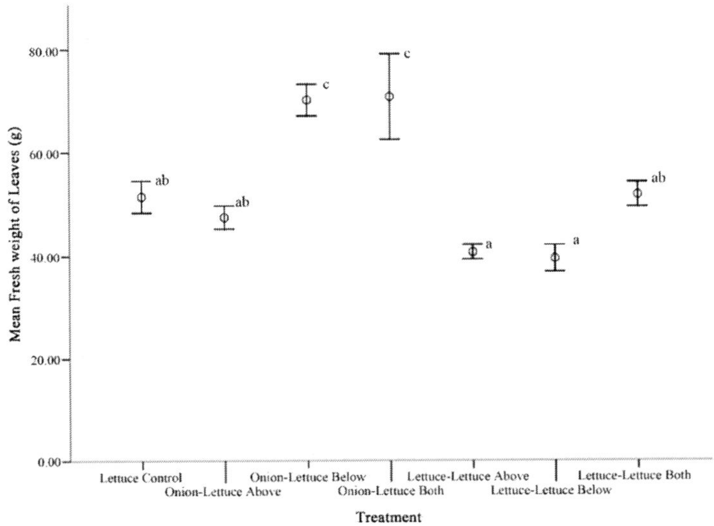

Figure 4. Comparison of average fresh weight of lettuce leaves growing in control and intraspecific and interspecific competition with onion. Significance was determined using analysis of variance with a Dunnett T3 post-hoc test, and is represented using letters adjacent to the standard error bars (df = 6, MS = 0.69, F = 10.50, $p < 0.001$).

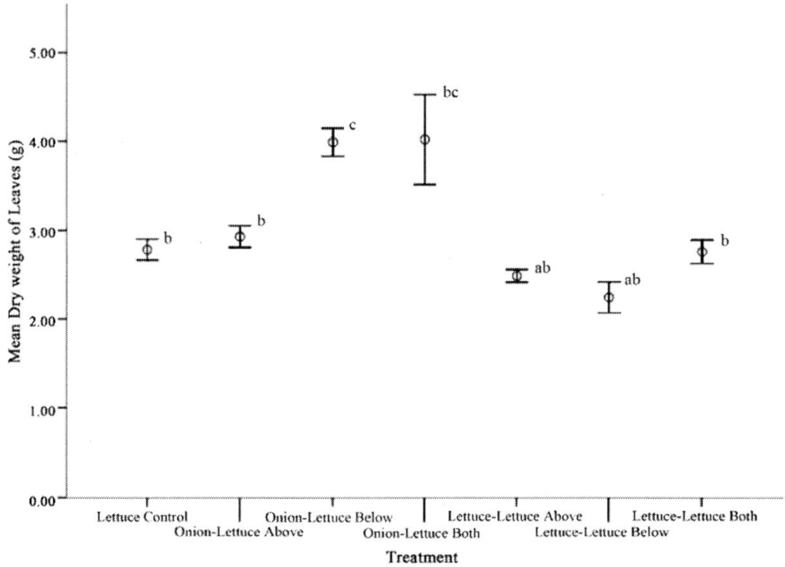

Figure 5. Average dry weight of lettuce leaves for lettuce grown in control, intraspecific or interspecific competition with onion. Letters beside error bars signify significance between treatments found using analysis of variance with a Dunnett T3 post-hoc test (df = 6, MS = 0.63, F = 8.67, $p < 0.001$).

treatment. Onion monoculture was facilitative in above ground interacttion as well as below ground interaction, but when paired in full competition, the effects were slightly competi- tive (Table 1(a)).

When lettuce was paired with onion above ground, there was a neutral effect, while a facilitative effect was observed when lettuce was paired with onion in below ground and full interaction treatments. The effects of let- tuce in intraspecific interaction above ground and below ground were competitive, while there was a neutral ef- fect for lettuce paired with lettuce in full interaction (Table 1(b)).

4. DISCUSSION

Competitive interactions among plants of the same or different species can have either a positive or a negative effect on growth, reproduction, and survival [1] [5] . In this study, we hypothesized that the biomass for lettuce and onion intercrops would be greater than that of the monocultures.

Indeed, for lettuce, we found that the bio- mass in intercrops with onion was generally greater than in monoculture. However, we also found that the yield of onion bulbs in all cases of intercrop with lettuce was equal to or lower than the biomass of monocultures. These results are especially visible in the predicted yield (hectare) model. It is known that onions are usually not very competitive against weeds or crops that are planted and growing at relatively the same time because of its slow growing habit [12] . Even once the lettuce removed, the recovery of onion in growth (measure in terms of plant height) was not significant (data not shown).

This unbalanced benefit of both crops in intercropping has been previously observed, showing the complexity of the system and its interactions [2] . Similar results were observed in a leek/celery inter-crop mixture where to- tal yield was equal to the yield in the two species in monoculture, but celery competition reduced leek yield and quality [13] . In a cabbage/radish intercrop, yield of cabbage was significantly less in intercrop compared to mo- noculture [14] . This is known as asymmetric interspecific facilitation where the intercrop combination increases the yield of one species, but causes a decrease in the other [15] .

The results suggested that the interactions were not necessarily similar for above and below ground. In interspecific competition, onions faired best in above ground interaction with lettuce. We expected that lettuce plants would shadow onions, but it appears that below ground competition with lettuce limited onion growth more than above ground competition. The combined effect of full competition is therefore dominated by the below ground portion. A similar reduction of onion bulb yield due to intercropping with vegetables was noted by [16] .

Our comparison between individual onion and onion in full or below ground competition also suggest that onions required greater spacing than what was stated by the seed company due to mainly below ground effect. Our results were similar to Peach et al. [17] who reported that intraspecific onion yield was improved by elimi- nating either above or below ground competition, but full competition greatly reduced yield. Aerts et al. [18] al- so found similar results.

It was expected that broader and more horizontal leaves of lettuce would lead to increased shading, as well as respiration and maintenance costs, thus likely reducing yield in intraspecific above ground competition [7] [19]. However, this was not necessarily the case; suggesting that light competition, at least at the soil surface, was not as much as a factor as expected [5] .

Table 1. Mean fresh weight of plant tissue (g), Relative Interaction Index (RII) values and type of interaction defined by RII for (a) onions when in intraspecific or interspecific competition with lettuce and (b) lettuce when in intraspecific or interspecific competition with onion.

(a)

Treatment	Fresh Bulb Weight (g)	RII	Interaction
Onion Control	0.99		
Onion-Onion Above	1.66	0.25	Facilitative
Onion-Lettuce Above	1.09	0.05	Neutral
Onion-Onion Below	1.45	0.19	Facilitative
Onion-Lettuce Below	0.59	−0.25	Competitive
Onion-Onion Both	0.74	−0.14	Slightly Competitive
Onion-Lettuce Both	0.45	−0.38	Competitive

(b)

Treatment	Fresh Leaf Weight (g)	RII	Interaction
Lettuce Control	51.47		
Lettuce-Lettuce Above	40.68	−0.12	Facilitative
Onion-Lettuce Above	47.44	−0.04	Neutral
Lettuce-Lettuce Below	39.43	−0.13	Facilitative
Lettuce-Onion Below	70.13	0.15	Competitive
Lettuce-Lettuce Both	51.75	0.00	Slightly Competitive
Lettuce-Onion Both	70.71	0.16	Competitive

Intercropping has been shown to have yield advantage presumably resulting from both above and below ground facilitation between intercropped species [15] . Contrary to our initial hypothesis, the RII of onion showed a facilitative effect in intraspecific above and below ground competition. In intercropping, onion played a neutral to facilitative role in the presence of lettuce. RII results for lettuce showed that in monoculture, neutral or slight- ly competitive effects would be found. The extensive root system of lettuce most likely exploited a large soil volume, contributing to its high interspecific competitive ability below ground, thus limiting soil moisture and nutrient availability for onions [20] .

In the face of interactions or changes in environmental cues, plants can respond by emitting different biogenic volatile organic compounds or by modifying their internal hormonal balance [21] . Allelochemicals have been found to influence interactions between plants of different species thus reducing or improving their productivity potential [2] . While this aspect was not considered in this study, we can assume that some chemical interactions might have influenced plant responses and performance. The facilitative outcome observed in lettuce in inters- pecific below ground competition with onions may be due to the excretion of certain allelopathic chemicals such as growth promoting compounds [22] [23] . It may be interesting to consider these chemicals (most likely alle- lochemicals) in future studies.

Results from this study have promising applications in crop yield optimization and sustainable agriculture. While intercropping has been practiced in some countries for centuries, several questions remain unanswered regarding optimization of plant combinations, optimal density, spatial arrangements, or selective breeding ne- cessary to produce maximum yields in full scale agricultural systems [2] [3] . Field testing, as mentioned in [2] will also be required to examine the influence of weeds and pests in such systems and determine whether alle- lochemicals from other species like weeds can affect the outcomes.

ACKNOWLEDGEMENTS

Jessica L. de Haan, Liette Vasseur We thank V. De Luca for his involvement in the planning and execution of this study and the technical assis- tance of C. Carpenter-Cleland, A. Shidu, B. Amout. K. Dundas and A. Gheorghe. Funding to LV provided by Brock University.

REFERENCES

1. Mariotti, M., Masoni, A., Ercoli, L. and Arduini, I. (2009) Above and below Ground Competition between Barley, Wheat, Lupin and Vetch in a Cereal and Legume Intercropping System. Grass and Forage Science, 64, 401-412.
2. Lithourgidis, A.S., Dordas, C.A., Damalas, C.A. and Vlachostergios, D.N. (2011) Annual Intercrops: An Alternative Pathway for Sustainable Agriculture. Australian Journal of Crop Science, 5, 396-410. [Citation Time(s):6]
3. Park, S.E., Benjamin, L.R. and Watkinson, A.R. (2002) Comparing Biological Productivity in Cropping Systems: A Competition Approach. Journal of Applied Ecology, 39, 416-426.
4. Morsy, S.M., Drgham, E.A. and Mohamed, G.M. (2009) Effect of Garlic and Onion Extracts or Their Intercropping on Suppressing Damping-Off and Powdery Mildew Diseases and Growth Characteristics of Cucumber. Egyptian Journal of Phytopathology, 37, 35-46.
5. Tilman, D. (1988) Plant Strategies and the Dynamics and Structure of Plant Communities. Princeton University Press, New Jersey.
6. Jose, S., Williams, R. and Zamora, D. (2006) Below Ground Ecological Interactions in Mixed-Species Forest Plantations. Forest Ecology and Management, 233, 231-239.
7. Baumann, D.T., Bastiaans, L. and Kropff, M.J. (2001) Competition and Crop Performance in a Leek-Celery Intercrop- ping System. Crop Science, 41, 764-774.
8. Andersen, M.K., Hauggaard-Nielsen, H., Ambus, P. and Jensen, E.S. (2004) Biomass Production, Symbiotic Nitrogen Fixation and Inorganic N Use in Dual and Tri-Component Annual Intercrops. Plant and Soil, 266, 273-287.

9. Andersen, M.K., Hauggaard-Nielsen, H., Weiner, J. and Jensen, E.S. (2007) Competitive Dynamics in Two- and Three- Component Intercrops. Journal of Applied Ecology, 44, 545-551.
10. Jaime, M., Roberts, L. and McDonald, M.R. (2013) Growing Onion Transplants in Plug Trays. Ontario Ministry of Agriculture and Food. http://www.omafra.gov.on.ca/english/crops/facts/01-019.htm
11. Armas, C., Ordiales, R. and Pugnaire, F.I. (2004) Measuring Plant Interactions: A New Comparative Index. Ecology, 85, 2682-2686.
12. Qasem, J.R. (2006) Response of Onion (Allium cepa L.) Plants to Fertilizers, Weed Competition Duration, and Planting Times in the Central Jordan Valley. Weed Biology and Management, 6, 212-220.
13. Baumann, D.T., Bastiaans, L., Goudriaan, J., van Laar, H.H. and Kropfl, M.J. (2002) Analysing Crop Yield and Plant Quality in an Intercropping System Using an Eco-Physiological Model for Interplant Competition. Agricultural Systems, 73, 173-203.
14. Guvenc, I. and Yildirim, E. (2006) Increased Productivity with Intercropping Systems in Cabbage Production. Journal of Sustainable Agriculture, 28, 29-44.
15. Li, L., Sun, J., Zhang, F., Guo, T., Bao, X., Smith, A.F. and Smith, S.E. (2006) Root Distribution and Interactions between Intercropped Species. Oecologia, 147, 280-290.
16. Narla, R.D., Muthomi, J.W., Gachu, S.M., Nderitu, J.H. and Olubayo, F.M. (2011) Effects of Intercropping Bulb Onion and Vegetables on Purple Blotch and Downy Mildew. Journal of Biological Sciences, 11, 52-27.
17. Peach, L., Benjamin, L.R. and Mead, A. (2000) Effects on the Growth of Carrots (Daucus carota L.), Cabbage (Brassica oleracea var. capitata L.) and Onion (Allium cepa L.) of Restricting the Ability of the Plants to Intercept Resources. Journal of Experimental Botany, 51, 605-615.
18. Aerts, R., Boot, R.G.A. and van der Aart, P.J.M. (1991) The Relation between Above- and Below Ground Biomass Allocation Patterns and Competitive Ability. Oecologia, 87, 551-559.
19. Tei, F., Scaife, A. and Aikman, D.P. (1996) Growth of Lettuce, Onion, and Red Beet. 1. Growth Analysis, Light Interception, and Radiation Use Efficiency. Annals of Botany, 78, 633-643.
20. Armas, C. and Pugnaire, F.I. (2011) Plant Neighbour Identity Matters to Below Ground Interactions under Controlled Conditions. PLoS One, 6, 1-7.
21. Kegge, W. and Pierik, R. (2009) Biogenic Volatile Organic Compounds and Plant Competition. Trends in Plant Sci- ence, 15, 126-132.
22. Amin, A.A., Rahad, El-Sh.M. and El-Abagy, H.M.H. (2007) Physiological Effect of Indole-3-Butyric Acid on Growth, Yield and Chemical Constituents of Onion Plants. Journal of Applied Sciences Research, 3, 1554-1563.
23. Shrestha, H. (2013) A Plant Monograph of Onion (Allium cepa). http://acepa.net76.net /Allium%20cepa.pdf

CHAPTER 14

Impact of Production System on Quality Indices Distribution in Butterhead Lettuce: A Comparative Study among Open Field and Greenhouse

Goñi María G[1,2,3]*, Moreira María del R[1,3], Agüero María V[1,3] and Roura Sara I[1,3]

[1]*Consejo Nacional de Ciencia y Tecnología (CONICET), Rivadavia 1917 (C1033AAJ). Ciudad Autónoma de Buenos Aires, Argentina*

[2]*Agencia Nacional de Promoción Científica y Tecnológica (ANCyT). Godoy Cruz 2370 (C1425FQD). Ciudad Autónoma de Buenos Aires, Argentina*

[3]*Grupo de Investigación en Ingeniería de Alimentos, Facultad de Ingeniería, UNMdP, Juan B. Justo 4302, Mar del Plata (7600), Argentina*

ABSTRACT

Butterhead lettuce quality was followed through changes in physical, physiological, microbiological, nutritional and sensory indices during postharvest storage. Production was in greenhouses, traditional and mulch, and in the field. Greenhouse lettuce heads had higher nutritional and sensory quality at harvest and lower enzymatic browning. Increased exposure to environmental conditions for open field grown lettuce heads may initiate defense mechanisms that could affect texture, color and appearance. During refrigerated postharvest storage, greenhouse lettuce heads had better leaf color and texture and were less susceptible to enzymatic browning. There was no difference in postharvest shelf life for greenhouse and field grown plants. Production in greenhouses produced important marketing benefit, along greater weight per plant and leaf number, by increasing producer profitability.

Keywords: Ascorbic acid; Enzymatic browning; *Lactuca sativa*, Overall visual quality; Postharvest; Preharvest; Shelf life

INTRODUCTION

Lettuce (*Lactuca sativa* var. Lores, variety Longifolia, Longifolia Group) is often associated with health benefits due to the presence of antioxidant components [1-3]. Postharvest quality of lettuce is influenced by time of harvest or maturity at harvest and postharvest storage temperature and relative humidity; growing season environmental conditions that include light, average temperature, minimum temperature, and night-day temperature difference, and irrigation, fertilization, and cultivation method [4,5].

Butterhead lettuce is basically grown under two production systems: open field and protected culture using greenhouses. The need to provide fresh quality products during prolonged periods of the year, along with a more efficient use of water had led to the adoption of greenhouses [6]. Greenhouse production is widespread worldwide, and lettuce is globally one of the vegetables produced using this technology. It is common in lettuce production to either use or not use soil covering plastic mulch in the greenhouse system.

The plastic cover used on the greenhouses changes the internal radiation balance with respect to the external environmental reducing light input by at least 30% [7], which may affect the lettuce quality. The soil cover changes the immediate environment around the plant, especially affecting the radiation balance around the lettuce head. Plastic mulch, affects the water balance and the temperature regime of the plant microenvironment [8].

The use of mulching improves quality and productivity [9]. Under greenhouse conditions with or without there were differences in Butterhead lettuce initial counts, evolution and tolerance to refrigeration temperatures of some microbial populations [10]. Goñi et al. [11] reported for freshly harvested Butterhead lettuce there were differences in water status indices between mulched and unmulched lettuce plants. The impact of the production system on postharvest quality retention during storage was not established.

Placing the plant in dispirit production systems constitutes a biological model to elucidate effects of conditions during production on quality indices. Postharvest changes in quality indices of Butterhead lettuce, as affected by production systems, were followed during storage.

MATERIALS AND METHODS

Plant Material and Sample Preparation

The experiments took place in Sierra de los Padres, Mar del Plata, Argentina (South 37°57' West 57°42'). Butterhead lettuce plants, var. Lores, were utilized for all studies. Three production systems were used; two were mulched (black plastic sheeting) and unmulched plants in greenhouses and, in the third system, plants were grown under field conditions. The mulch used in the greenhouses was polyethylene (Cristal Agroman, Beniplast, Argentina, 150 μm thickness) and the material used for mulching was black low-density polyethylene (Agroman Mulching, Beniplast, Argentina, 100 μm thick).

In all production systems, lettuce plants were subjected to the recommended irrigation regime to their specific requirements. The fields were on closely located to assure that the geographical and weather effects were similar; lettuce were grown in the same season (spring/summer). Harvest was at optimal maturity after head formation was completed.

After harvest, heads were immediately pre-cooled and transported to the laboratory. Five heads were analyzed to determine quality indices values at harvest. Other lettuce heads were placed individually in polyethylene bags (with an O2 permeability 600 $cm^3m^{-2}d^{-1}$, CO_2 permeability 4000 $cm^3m^{-2}d^{-1}$, and water vapor permeability 4 $g.m^{-2}d^{-1}$), heat sealed and stored in a refrigerated chamber at 0-2°C and 97-99% relative humidity. Under this temperature, respiration rate, ethylene production, and other physiological changes were minimized.

Vegetative Growth Parameters

Yield was evaluated by: lettuce head weight, number of leaves per head, and leaf area. Each plant was weighed and all leaves detached and counted. Leaf area was determined with a method of Goñi et al. [11]. The technique consisted of tracing the leaf shape onto clear paper, cutting it and measuring it using an Area Measurement System (Delta T-Devices Ltd., Cambridge, England).

Quality Indices

Quality indices measured were from the external (outer leaves), middle (mid leaves) and internal (inner leaves) zones. For each head, zones were visually delimited, according to organoleptic criteria established by Agüero et al. [8]. Each zone had a mean of approximately 6 to 9 leaves.

Physiological Quality

Water status was determined as reported by Viacava et al. [12] with the indices: Water Content (WC), Relative Water Content (RWC), Free Water (FW), Bound Water (BW) and the ratio FW and Total Water (FW/TW).

Nutritional Quality

Reduced ascorbic acid content (AA) was determined by the titrimetric assay described by Roura [13] using 2, 6-dichloroindophenol as titrant solution.

Microbiological Studies

Total microbial counts were evaluated with the methodology of Ponce et al. [10]. Ground lettuce from each head zone was macerated (Stomacher 400 Circulator Homogenizer, London, UK) in a buffer solution (PO_4K_3, pH=7.2). Total microbial counts were on plate count agar after incubation for 48 hr at 30°C [14]. Total microbial counts were performed in triplicate.

Physicochemical Quality

Color determination was done using a tristimulus colorimeter SP60 (Lovibond, Solstice Park, Amesbury, United Kingdom). The instrument was calibrated with a standard white plate (Y=93.2, X=0.3133, Z=0.3192). Leaf color was measured by L*, a* and b* chromaticity coordinates of the CIELab scale [15]. Measurements were performed in ten different positions on leaves. Using the L*, a* and b* values, a color index (CI) was determined [16].

Chlorophyll content of each zone was determined following the methodology of Moreira using a cold solution 18:1 propanone:ammonium hydroxide (0.1 N) as extraction solution [17]. The homogenate was filtered through sintered glass and water removed from the filtrate with anhydrous sodium sulfate. The filtrate absorbance at 660.0 and 642.5 nm was measured with a UV 1601 PC UV-visible spectrophotometer (Shimadzu Corporation, Kyoto, Japan).

Sensory Quality

Sensory analysis was performed as described by Agüero et al. [8]. At each storage time, each individual head was subjected to a sensory panel to evaluate the overall visual quality (OVQ) of each plant zone. The panel consisted on 9 judges, aged 30 to 55 years old, with previous sensory evaluation experience in leafy vegetables. Samples were presented in random order to judges who made independent evaluations. The OVQ was evaluated on leaf color (shade and uniformity), brightness, texture, and presence or absence of defects. A 9-point scoring scale was used, in which 9 was excellent quality and 1 was very poor quality; the acceptance limit was 5 (poor quality). Judges were asked to evaluate the presence of enzymatic browning in the butt stem (EBB) and leaves (EBL). A 5-point scale was used for these indices, in which 1 was absence of enzymatic browning and 5 was excessive enzymatic browning; 3 was the acceptance limit.

Statistical Analysis

Data were analyzed with SAS, ver. 9.0 [18]. PROC GLM (general linear model procedure) was used for the analysis of variance (ANOVA) and PROC UNIVARIATE was used to validate the ANOVA. Factors used as sources of variation were: PREHARV (production system: open field, unmulched or mulched in greenhouses), DAY (storage time, day of sampling: 0, 2, 4 and 10), ZONE (zones of lettuce head: external, middle and internal), and the corresponding interactions were used for postharvest changes analysis. Only PREHARV and ZONE were used as factors for quality indices at harvest. Results reported are LSMEAN values (estimators of means by the least squares method). When significant effects were found, the Tukey–Kramer multiple comparison test was performed [19].

RESULTS AND DISCUSSION

Vegetative Growth Parameters

Values of vegetative growth parameters to compare yield varied (**TABLE 1**). There was an effect of the factor PREHARV, with lettuce heads from open field being lighter than those from greenhouses (mulched and unmulch, without significant differences between them). Produce obtained using greenhouses systems resulted in the heaviest heads and this could be due to the more controlled environmental conditions in the greenhouse.

TABLE 1: Mean values at harvest of vegetable growth parameters of lettuce heads produced under different systems.

		Production system		
		Traditional Greenhouse	Mulch greenhouse	Open field
Lettuce head weight (g)*		$373.8^{A} \pm 96.7$	$388.7^{A} \pm 70.8$	$352.0^{B} \pm 59.1$
Total leaf number*		$20^{A} \pm 1$	$23^{A} \pm 1$	$34^{B} \pm 2$
Leaf area‡ (cm^2)	External	$306.6^{Aa} \pm 47.7$	$323.0^{Aa} \pm 41.7$	$238.8^{Ba} \pm 55.0$
	Middle	$240.9^{Ab} \pm 95.7$	$254.8^{Ab} \pm 76.5$	$188.0^{Bb} \pm 39.1$
	Internal	$42.7^{Ac} \pm 12.3$	$116.5^{Bc} \pm 14.6$	$80.2^{Cc} \pm 16.8$

* Zones were not defined; therefore only between production systems comparisons were performed. Different uppercase letters indicate significant differences between production systems ($p<0.05$)
‡Lsmean values are reported, different lowercase letters indicate significant difference between zones ($p<0.05$) and uppercase letters indicate significant differences between production systems ($p<0.05$)

Total leaf number was reduced in lettuce plants from greenhouses compared to open field. Plants from field culture had the higher leaf number in the external, older and green mature zone with respect to greenhouse lettuce.

Inversely, leaf area corresponding to leaves from external and middle zones in greenhouse lettuce heads, was significantly higher compared to leaves from the open field. Internal leaves presented a similar profile, being higher for greenhouse lettuce in respect to those from the open field. Leaf area is an important variable considered in most physiological and agronomic studies as it is correlated with light interception, photosynthetic efficiency, evapotranspiration and plant growth [20]. The single biggest driving force for biomass generation is light interception, which could be improved when leaf area is increased. Light intensity is often limited in a greenhouse and can be half that found in the field. Additionally, the plastic cover used in greenhouses changes the internal radiation balance compared to external conditions [6]. Differences found in leaf area among lettuce heads from open field or greenhouses could be associated with radiation-related parameters. Leaf emergence and leaf expansion rates are influenced by temperature with increasing temperature an indirect way to improve light capture [21]. Greater leaf area in greenhouse lettuce heads may be an adaptation to enhance light capture to cover photosynthesis demands.

Quality Indices Distribution In Freshly Harvested Lettuce

RWC provides information about leaf water content by measuring the amount of water in the leaf tissue in relation to fully turgid tissue [22]. An interaction occurred between ZONE and PREHARV. For each zone, no differences were detected in RWC between greenhouse lettuce, with mean values of 83.48, 85.15 and 85.77% for outer-, midand inner-leaves, respectively (**TABLE 2**) while RWC in open field lettuce was nearly 8% lower. Inner leaves presented higher RWC values than outer leaves for open field and unmulched greenhouse lettuce, but these increments were absent in mulch greenhouse heads that had similar RWC in all three zones. Goñi et al. [11] had similar results by doing a characterization of Butterhead lettuce heads produced in mulched and unmulched greenhouses. At harvest, open field lettuce had the lowest RWC values in the three zones. High air relative humidity, coupled with lower wind velocity inside the greenhouse, tends to reduce water vapor exchange between the canopy and the atmosphere compared to the open field. An 8% drop in RWC could be regarded as mild water stress following Hsiao´s classification [23]. Under moderate stress there is additional turgor reduction, leading to partial closure of stomata aperture and a reduction in photosynthetic activity. The lower RWC values for open field lettuce heads may indicate a moderate level of water stress compared with greenhouse produced lettuce heads. Inside greenhouses, lettuce plants develop in a more moderate environment. Differences in RWC between inner (younger) and outer leaves (older) could be attributed to metabolic differences instead of different exposure level to environmental conditions. Guerra et al. [24] found similar results for greenhouse celery (*Apium graveolens L.*), attributing differences to degree of tissue development.

Table 2: Quality indices distribution by both lettuce zone (external, middle and internal) and production system (open field, traditional greenhous and mulching greenhouse) at harvest.

Quality index*	Production System								
	Traditional Greenhouse			Mulch Greenhouse			Open field		
Lettuce zone	*External*	*Middle*	*Internal*	*External*	*Middle*	*Internal*	*External*	*Middle*	*Internal*
Relative Water Content (%)	84.50^{Aa}	85.43^{Aa}	85.00^{Aa}	82.47^{Aa}	84.87^{Aab}	86.53^{Ab}	75.80^{Ba}	76.02^{Ba}	78.97^{Bb}
Water Content (%)^{nsi}	95.82^{Aa}	95.53^{Aa}	95.07^{Aa}	93.93^{Ba}	93.13^{Ba}	93.07^{Ba}	91.83^{Ca}	91.57^{Ca}	92.20^{Ca}
Free Water Content (g/g)	9.57^{Aa}	10.10^{Aa}	10.20^{Aa}	16.19^{Ba}	12.72^{Ab}	10.18^{Ac}	3.80^{Ca}	6.18^{Bb}	6.17^{Bb}
Bound Water Content (g/g)	3.67^{Aa}	5.92^{Ab}	7.81^{Ac}	4.35^{Aa}	6.68^{Ab}	12.70^{Bc}	12.10^{Ba}	12.51^{Ba}	11.59^{Ba}
Free Water/Total Water Ratio	0.73^{Aa}	0.63^{Ab}	0.57^{Ab}	0.79^{Aa}	0.66^{Ab}	0.44^{Bc}	0.24^{Ba}	0.33^{Bb}	0.35^{Bb}
Acid Ascorbic Content (mg/100g FT)	29.25^{Aa}	26.93^{Aa}	16.93^{Ab}	9.01^{Ba}	17.23^{Bb}	14.14^{Ab}	16.79^{Ca}	20.72^{Bb}	18.00A^{ab}
Total Mesophyll Bacteria (log(CFU/g FT))	5.31^{Aa}	6.45^{Ab}	6.36^{Ab}	5.98^{Ba}	6.44^{Ab}	5.98^{Bc}	5.35^{Aa}	6.90^{Bb}	6.71^{Cb}
Color Index (CI)	-6.64^{Aa}	-6.17^{Ab}	-5.20^{Ac}	-6.66^{Aa}	-5.98^{Ab}	-3.32^{Bc}	-6.10^{Ba}	-5.58^{Ab}	-3.66^{Bc}
Chlorophyll content (mg/100gFT)	37.27^{Aa}	32.23^{Aa}	12.21^{Ab}	43.05^{Ba}	34.44^{Ab}	9.89^{Ac}	45.03^{Ba}	41.20^{Ba}	18.91^{Bb}
Overall Visual Quality Score	8.98^{Aa}	8.94^{Aa}	9.00^{Aa}	9.00^{Aa}	8.97^{Aa}	9.00^{Aa}	7.60^{Ba}	8.42^{Bb}	8.67^{Bc}
Enzymatic Browning in Leaves Score	1.00^{Aa}	1.00^{Aa}	1.00^{Aa}	1.00^{Aa}	1.00^{Aa}	1.00^{Aa}	2.06^{Ba}	1.23^{Bb}	1.50^{Bc}
Enzymatic Browning in Butt stem Score**	1.00^{A}			1.00^{A}			1.55^{B}		

*Lsmean values are reported, different lowercase letters indicate significant difference between zones ($p<0.05$) and uppercase letters indicate significant differences between production systems ($p<0.05$)

nsi No significant interaction

** Zones were not defined for Enzymatic Browning in Butt stem Score, therefore only between production system comparisons were performed

Only PREHARV affected WC (**TABLE 2**). Open field lettuce heads had the lowest WC while greenhouse produced lettuce heads had the highest values for mulched and unmulched greenhouses. Open field produced lettuce heads underwent more severe climatic conditions than greenhouse produced lettuce plants, which may be responsible for lower WC values.

Production system affected initial FW and BW (**TABLE 2**) distribution, with an interaction between PREHARV and ZONE. No significant differences were detected in FW values between zones in plants cultivated in unmulched greenhouses; for mulched lettuce FW was higher in outer than middle and inner leaves. For open field lettuce higher values were in inner and middle leaves. For BW, lettuce heads from mulched and unmulched greenhouses had higher values in younger (inner) leaves; in open field plants higher values were in older leaves (outer). External leaves in mulch system had the highest FW value compared to those in the unmulched greenhouse and open field. Middle and internal zones had similar trends, but without differences to unmulched greenhouse system. For external and middle zones, the highest BW values were in open field plants, while the highest internal value was in mulched greenhouse lettuce heads. Once again, it was clear that the preharvest environment affected physiological indices distribution. A more efficient water use regime in the mulched lettuce plants could increase FW along the whole plant, resulting in better water status compared to the other production systems. Goñi et al. [11] reported higher FW for lettuce plants cultivated using mulch than in plants cultivated in the open field. As BW is often associated with tolerance to abiotic stress [25] increased values for outer- and midleaves from open field lettuce may be a tissue response to exposure to a preharvest moderate water stress.

The FW/TW profile in the lettuce head was dependent on production system (**TABLE 2**). Plants from greenhouses had a decrease in values from outer to inner leaves while plants cultivated in open field had the highest values in inner leaves decreasing towards outer leaves. External and internal zones had higher values in traditional greenhouse than mulched greenhouse and open field when comparison was in the same zone. The middle zone had a higher value in mulched than in unmulched greenhouses and open field.

The external and middle leaves from unmulched greenhouse lettuce heads had the highest AA while mulched lettuce had the lowest AA (**TABLE 2**). For inner leaves, AA was not different between production systems. Goñi et al. [11] found differences in initial AA, associating this result to production system and leaf maturity stage. The AA concentrations can vary, depending on maturity, genetic variations, preharvest conditions, postharvest handling, processing, and preparation [26-28]. Wojciechowska et al. [28] found lower AA in lettuce plants and celery grown in a greenhouse with mulching compared to unmulched greenhouses. Black plastic mulch intercepts incident radiation, affecting quantity and quality of photosynthetic active radiation (PAR) received by the plant [29] decreasing amounts of reactive oxygen species (ROSs) formed in leaves [30,31]. This decrease in oxidative stress may diminish the physiological response, inhibiting the AA synthesis in mulched lettuce plants. High AA concentrations in chloroplasts help protect tissues against damaging oxygen-derived species often produced during photosynthesis [32]. Mulched lettuce plants were grown in a more protected environment than those from open field and the milder conditions might cause a reduction in synthesis of AA. Oh et al. [33] found that slight variations in temperature or small periods of radiation intensities higher than normal, were sufficient to trigger defense mechanisms in lettuce tissue against oxidative stress. The AA variability at harvest could be associated to production system and also to leaf position within plants. Unmulched greenhouse lettuce plants had higher AA in outer and middle leaves. The AA in vegetable tissue is synthesized in a metabolic path closely related to photosynthesis [34]. External and middle leaves functioned as photosynthetic leaves and were exporters of photoassimilates. The fact that AA in these leaves was significantly higher than AA of inner leaves was a predictable result.

Total microbial bacteria counts provide a useful estimate of total viable microbial population, and it is indicative of endogenous microflora and contamination undergone by the produce [35]. Bacteria found in fresh produce are the same than those normally found in field [36]; production system may have an effect on TMB of lettuce at harvest. The microbiological quality of lettuce heads was affected by the production system. The TMB counts were higher than 5 log (**TABLE 2**) in the lettuce head zones whatever production system was used. Ponce found initial values higher than 5 log in lettuce heads cultivated in greenhouses [10]. For unmulched greenhouse and open field, the external zone had lower TMB (mean of 5.33 log for both systems) than middle and inner zones (mean of 6.61 log for the two zones and two production systems). Fonseca [37] found similar results for Iceberg lettuce and they attributed this result to exposure of outer leaves to intermittent water stress which can result in higher accumulation of metabolites that diminish bacterial growth. For mulched lettuce heads, TMB was higher in the middle zone than in outer and inner zones, which were not different. Middle leaves of mulched lettuce heads may provide a safe environment to microorganisms, allowing native microflora to become established, resulting in higher counts. Open field and unmulched greenhouse lettuce heads were subject to environmental conditions more susceptible to change than mulched greenhouse lettuce heads; some defense mechanisms may be elicited as a result of those changes [37].

Production system also affected color parameters. Plants from unmulched greenhouse had a more uniform color distribution and a stronger green hue than open field or mulched greenhouse lettuce heads. The two latter have a significant color zonification within the plant (**TABLE 2**), with the lower CI values (greener) in outer leaves. Mid- and inner-leaves of unmulched greenhouse lettuce plants have darker green color than mulched and open field lettuce heads. The total chlorophyll content (TC) was affected by production method and plant zone. The initial TC of outer leaves was similar for open field and mulched lettuce plants (mean value of 44.0 mg/100 g FT) while it was slightly lower for unmulched greenhouse. Mid- and inner-leaves of open field lettuce plants had a higher TC than greenhouse lettuce. Agüero et al. [8] reported differences in TC between lettuce zones, and attributed the differences to amount of sun light exposure of outer leaves, which may have increased chlorophyll synthesis. Production system, affected TC with open field lettuce plants having the highest initial values of TC in all plant zones, probably associated with longer radiation exposure during growth. These results agree with Caldwell and Britz [38] working with lettuce.

Sensory quality was affected by production system and the head zone. Greenhouse lettuce had lower enzymatic browning scores in leaves and butt stem than open field lettuce (**TABLE 2**). Lettuce heads in unmulched and mulched systems produced optimal values in the plant zones. Lettuce heads produced in open field had slight enzymatic browning in leaves with higher scores (worst) in the external zone than in middle and internal ones. External leaves were more exposed to environmental conditions, and more susceptible to mechanical damage during harvest. Within the damaged leaf biochemical events that cause enzymatic browning could be triggered. Open field lettuce plants were exposed to UV-B solar radiation (280-320 nm) that leads to generation of ROSs. Outer leaves from lettuce cultivated in the open field may increase synthesis of phenolic compounds as a response to protect themselves from harmful radiation. As a consequence, an adjustment of antioxidant systems (polyphenols and other compounds) at cell and whole organism levels could take place [39,40]. This protective effect involves an increment susceptibility to enzymatic browning due to higher concentration of browning substrates, phenolic compounds. Similar enzymatic browning occurred in butt stem: optimal scores were obtained for the greenhouse systems while a slight degree of EBB was detected in open field lettuce plants.

Finally, overall visual quality of lettuce heads at harvest was optimal in all zones from lettuce plants obtained from greenhouses, without significant differences due to zone and mulch or unmulched systems. Open field lettuce plants had significantly lower OVQ scores in the three zones (**TABLE 2**). The reduction in OVQ score was especially noteworthy for external zone, with 11% reduction with respect to the other zones and 16% reduction with respect to the other production systems. Increased exposure to environmental conditions in open field plants when compared to greenhouse lettuce plants, may initiate defense mechanisms that affected quality. Changes in leaf morphology increases lignification of the cell wall or polyphenolic compounds synthesis might affect perception of texture, color and appearance of the plant causing lower OVQ scores [41]. These changes can be a plant response to cope with mild water stress and higher radiation exposure.

Postharvest Changes In Quality Indices

A significant triple interaction affected the evolution of RWC during postharvest storage (**FIGURE 1**). Mulched lettuce maintained almost constant RWC values throughout storage with slight differences among zones for external, middle and inner leaves, respectively. An increase in RWC occurred for unmulched greenhouse lettuce until day 10 of storage for all three zones; by day 14, RWC values were similar values to harvest. These results could be attributed to saturated relative humidity inside the package (97-98%) maintaining cell membrane permeability. Differences between zones could be associated with location of the leaf inside the plant (related to degree of exposure to environmental conditions) and physiological differences among leaf tissues associated with leaf age, outer leaves older than inner leaves.

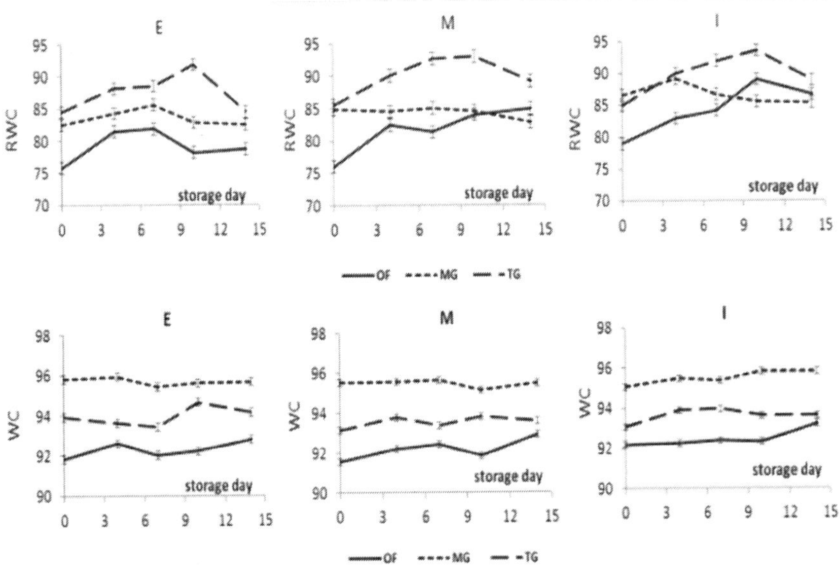

Figure 1: Relative Water Content (RWC) and Water Content (WC) evolution during refrigerated storage (0 - 2ºC) of Butterhead lettuce by zone (E: external, M: middle, I: inner) and production system (OF: open field, MG: mulch greenhouse, TG: traditional greenhouse).

The WC (**FIGURE 1**) was strongly affected by production system with mulched lettuce having higher WC, followed by unmulched greenhouse and lower values for open field lettuce. There were no differences due to DAY but were differences for PREHARV. The mean WC for the whole plant in open field lettuce plants was lower than WC of unmulched greenhouse lettuce plants and mulched lettuce. In the middle and internal zones, the PREHARV and DAY interaction was significant. Both zones had similar results for greenhouse lettuce plants, where WC remained similar to that at harvest, and mulched WC was always superior to traditional greenhouse WC. Open field lettuce had increased WC by day 14 compared to harvest (1.5% and 1.1% for middle and internal zones). Frequently, WC has been used to quantify water deficit in leaf tissue, so it is a useful indicator of plant water balance. Use of mulch reduces evaporation from the soil, resulting in a more uniform soil moisture regime. The whole plant water balance was better equilibrated in mulch compared with unmulched greenhouse, and was better than in open field.

The WC from the lettuce zones and heads from the production systems during refrigerated storage varied (**FIGURE 1**). Regardless of production system, no changes were detected in outer leaves, with WC values almost constant during storage. Differences detected at harvest among production systems were maintained during storage, with mean WC in external zone of open field lettuce plants lower than those for mulched and unmulched greenhouse heads. For middle and internal zones, no changes were detected in plants from greeenhouses, mulched and mulched during refrigerated storage, maintaining values similar to those at harvest. Open field lettuce increased WC values of middle and internal zones during storage (increments of 1.5% and 1.1%, respectively). Frequently, WC is been used to quantify water deficit in leaf tissue, so it is a useful indicator of plant water balance. In the greenhouse evaporation is reduced resulting in a more uniform soil moisture regime. The whole plant water balance was better equilibrated and there was better performance in these plants. The increase in WC in middle and internal zones of lettuce plants produced in open field could indicate higher respiration activity, but this hypothesis must be established with specific studies measuring, for example, carbon reserves in the cell.

During storage, the tendency of FW was to remain unchanged or decline slightly, always higher in greenhouse lettuce plants (**FIGURE 2**). A triple interaction was found and results are again presented by ZONE. The FW in external leaves of open field lettuce plants maintained its value. Although higher FW occurred in freshly harvested mulched lettuce respect to unmulched greenhouse lettuce, no differences were found by day 14, with a mean value of 12.44 $g.g^{-1}$. In the middle zone, FW of open field lettuce plants had a decrease by the end of storage (65% less) while the inner leaves FW remained relatively constant (5.62 $g.g^{-1}$). In both zones, unmulched greenhouse lettuce maintained similar FW during storage, with mean value of 10 $g \cdot g^{-1}$. The middle leaves from mulched lettuce had a 20% decreased in FW by day 14, while it remained constant for inner leaves. Agüero et al. [8] also reported similar results for lettuce.

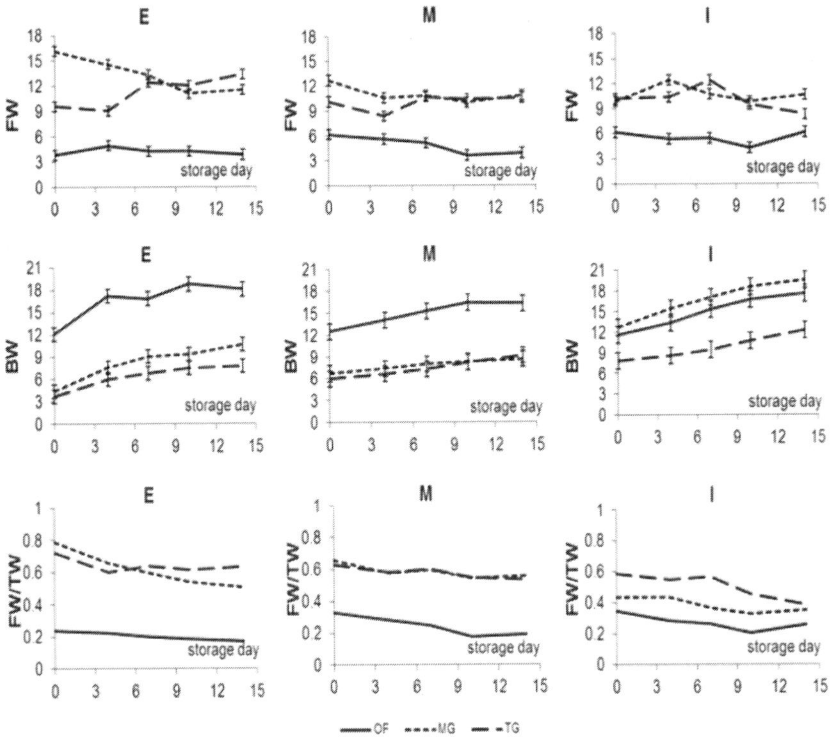

Figure 2: Free Water Content (FW), Bound Water Content (BW) and Relation Free Water /Total Water (FW/TW) evolution during refrigerated storage (0-2°C) of Butterhead lettuce by zone (E: external, M: middle, I: inner) and production system (OF: open field, MG: mulch greenhouse, TG: traditional greenhouse).

The BW increased for all production systems and zones (**FIGURE 2**). The increased BW during storage could be a tissue response to harvest stress. An increase in BW can be associated with solute accumulation, leading to better osmotic adjustments and tolerance to water stress [42]. Since high relative humidity inside packages impaired dehydration, the BW increase during storage cannot be associated with water stress but be associated with maintenance of structural integrity and turgidity [8]. Harvest implies the removal of nutrient and water and tissues respond by increasing BW in cells.

The FW/TW had similar results compared to BW (**FIGURE 2**), where the only significant interaction was PREHARV and ZONE. Open field lettuce plants had lower FW/TW than greenhouse lettuce plants, regardless of the zone. External and middle leaves had similar FW/ TW for mulched and unmulched greenhouse production while in the internal zone, unmulched greenhouse lettuce plants had higher FW/ TW than did mulched. Agüero reported that FW/TW play a key part in the shelf-life of each lettuce zone, being related to leaf texture which is an important quality parameter in lettuce [8,43].

The FW/TW during storage changed (**FIGURE 1**). Open field lettuce plants had lower FW/TW than greenhouse lettuce plants, regardless of zone. No differences were detected in FW/TW values of outer- and mid-leaves for lettuce heads produced in greenhouses. In the internal zone, unmulched greenhouse lettuce plants had higher FW/TW than mulch. Slight decreases were detected in FW/TW values during storage. Agüero reported that FW/TW had a key part in shelf-life of each zone, and was related to leaf texture an important quality parameter [8,43].

Nutritional quality of lettuce heads was affected by the production system used and by the placing of the leaves within the plant (zones) as presented in **FIGURE 3**. As early as day 4, a significant reduction in ascorbic acid content occurred in external and middle zones, especially in unmulched greenhouse plants. These reductions were detected in other production systems but with lower values for mulched and open field lettuce, without differences between production systems. For external and middle zones of heads cultivated in open field and mulched greenhouse, a linear model was fitted to the data. The AA degradation rate (slope) was higher in open field lettuce plants than in mulched plants. For traditional greenhouse lettuce plants, no linear model could be fitted to AA data from these two zones because they exhibited faster degradation with greater changes in the first days of storage. In the internal zone, no differences occurred in AA behavior in lettuce heads cultivated under the production systems, with a downward trend during postharvest storage. In these leaves, AA reduction followed a linear tendency. At day 14, AA content was similar for all lettuce zones and production systems with a mean value of 5.85 ± 0.67 mg/100 g FT. Percent of reduction in the present work were expected given the liability of AA [44] and allowed reaffirming the need for development of new pre- and post-harvest technologies to maintain the nutritional value of freshly harvested vegetables.

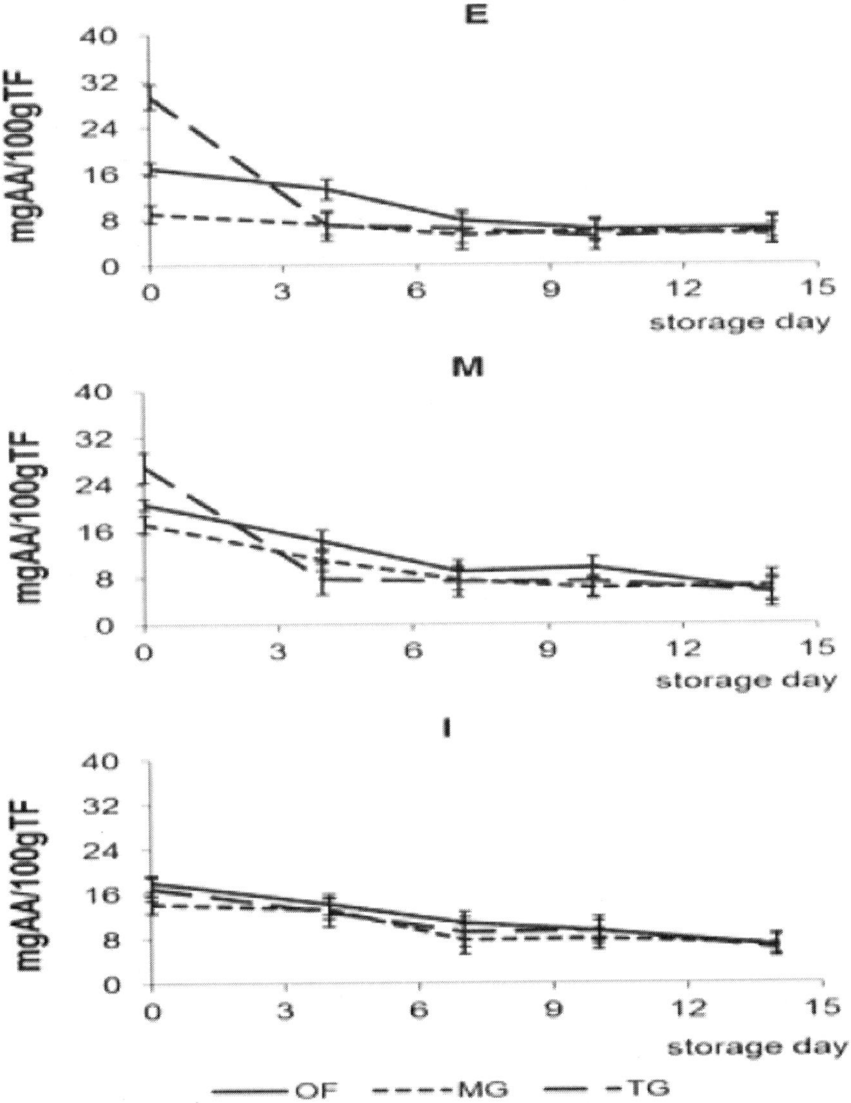

Figure 3: Reduced Ascorbic Acid (AA) evolution during refrigerated storage (0-2ºC) of Butterhead lettuce by zone (E: external, M: middle, I: inner) and production system (OF: open field, MG: mulch greenhouse, TG: traditional greenhouse).

Regardless of production system and zone, increases in TMB occurred during storage (**FIGURE 4**). Preharvest conditions seem to have an effect on microbial populations during refrigerated storage. Although no differences were found between growth rate of TMB from open field and traditional greenhouse lettuce plants, growth rate in mulched lettuce plants was lower. Ponce and Agüero found differences in the dynamic of microbial populations due to production system and attributed it to environmental conditions during primary production [8,10]. They proposed that growth of mulch lettuce in a controlled environment allowed bacteria to proliferate in a more benign environment and could enhance their susceptibility to cold storage. The TMB during refrigerated storage were different between lettuce zones. At day 14, the differences in TMB were not significant among zones (9.3 log) but had increments of 63, 40 and 51% for external, middle and internal zones, respectively.

Changes in CI during postharvest storage (**FIGURE 5**) indicated effects on color retention, according to production system. Unmulched greenhouse lettuce maintained its color in external and middle zones throughout storage reaching, at day 14, similar values to those at harvest. Open field and mulched lettuce plants exhibited alternating increases and decreases during storage. For internal leaves, unmulched greenhouse lettuce plants had greener leaves compared to mulched and open field plants, which were similar. During storage, increased yellowing of these leaves occurred with a linear function. For inner leaves, unmulched greenhouse lettuce, with the greenest value at harvest, had a yellowing response with a linear function, while inner leaves from the other production systems maintained this index almost constant until the end of storage.

Changes occurred in TC during storage (**FIGURE 5**). Significant reductions occurred in all situations. As at the beginning of storage, open field lettuce plants had the highest TC values, compared to greenhouse plants. No significant differences among greenhouse systems occurred. The external zone, in all production systems, had higher TC values than did middle and internal zones. The TC losses during refrigerated storage were fit linear functions in all lettuce zones, and production systems, with a degradation rate of -0.5 mg TC/100 g FT. These results indicate sensitivity of chlorophyll pigment to postharvest storage. The TC reduction by day 14 (26% for open field and 22% for both greenhouse lettuce) might be responsible for the CI increased, even though other factors may also affect overall color perception, such as enzymatic browning dehydration [45-47].

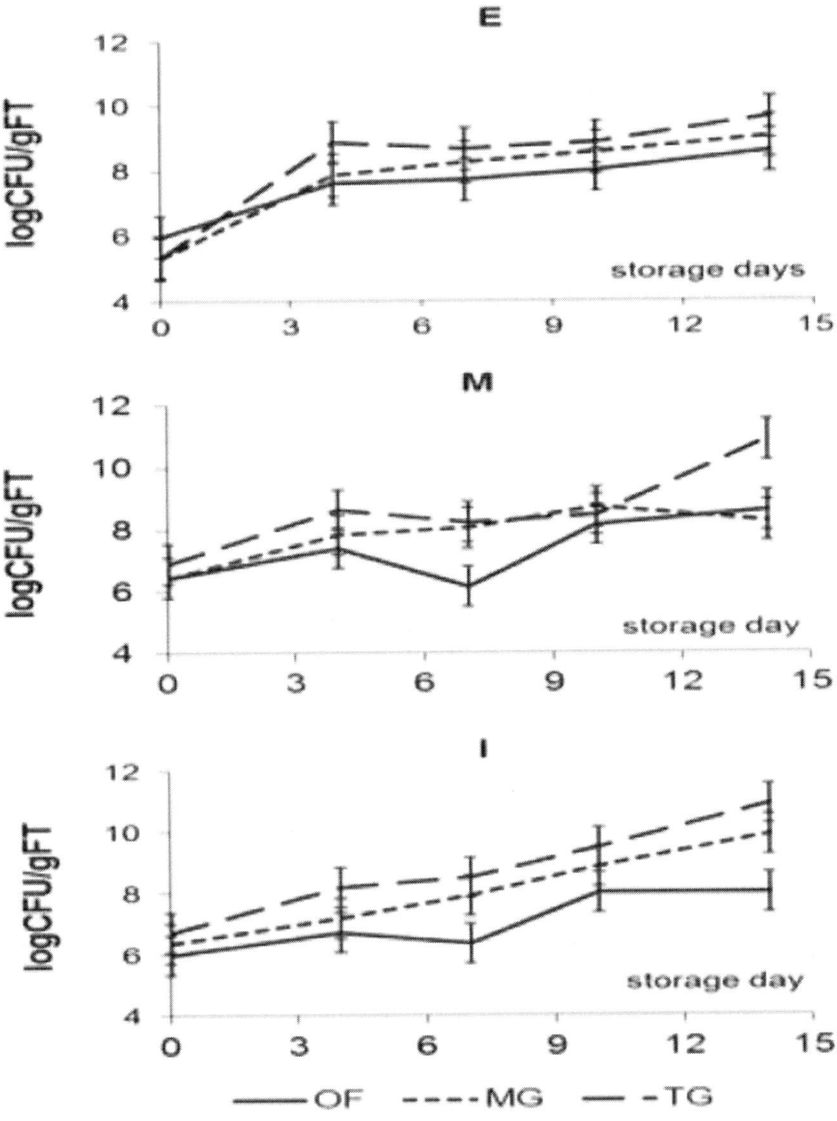

FIGURE 4: Mesophylic Total Count (log CFU / g FT) evolution during refrigerated storage (0-2°C) of Butterhead lettuce by zone (E: external, M: middle, I: inner) and production system (OF: open field, MG: mulch greenhouse, TG: traditional greenhouse).

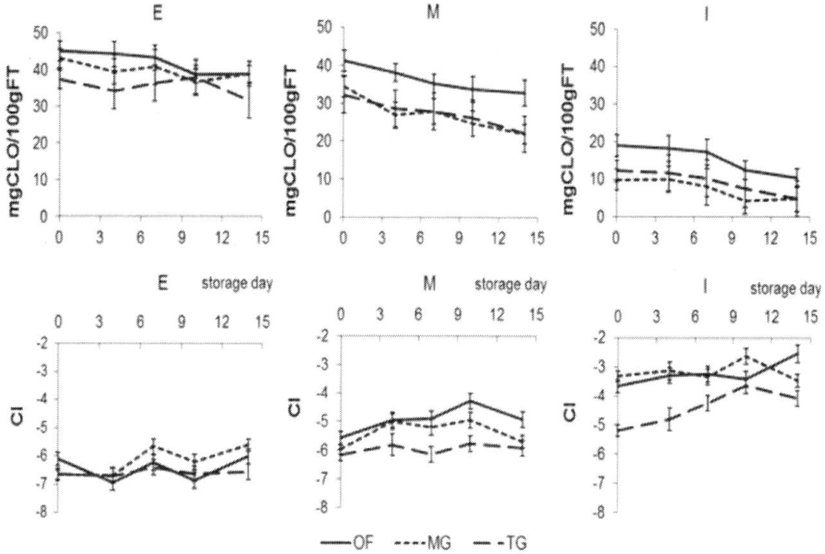

FIGURE 5: Chlorophyll content (CLO) and Color Index (CI) evolution, during refrigerated storage (0-2°C) of Butterhead lettuce by zone (E: external, M: middle, I: inner) and production system (OF: open field, MG: mulch greenhouse, TG: traditional greenhouse).

Enzymatic browning is a cause of quality loss during postharvest storage of lettuce, and often occurring at the end of shelf-life [32]. Enzymatic browning in the butt stem (EBB) (**FIGURE 6**) increases during storage in plants in all production systems. The EBB score for open field lettuce heads reached a value by day 4, indicating rejection of the product, while traditional greenhouse lettuce was rejected at day 10. Mulched lettuce heads did not surpass the acceptance limit in the present study. Enzymatic browning in leaves (EBL) (**FIGURE 7**) had a similar trend to EBB, with a better performance in greenhouse lettuce zones (a slight increase only in external and middle zones towards the end of storage reaching values below the acceptance limit), but with increases in zones from plants cultivated in open filed. The acceptance limit for EBL in these plants was surpassed by day 7 in the external zone, and by day 14 in middle and internal zones.

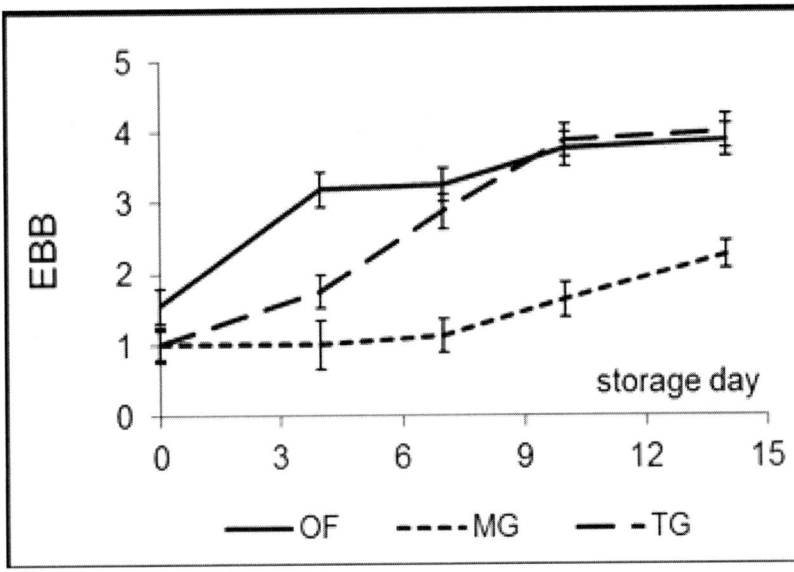

FIGURE 6: Enzymatic browning in butt stem (EBB) evolution during refrigerated storage (0-2°C) of Butterhead lettuce by production system (OF: open field, MG: mulch greenhouse, TG: traditional greenhouse).

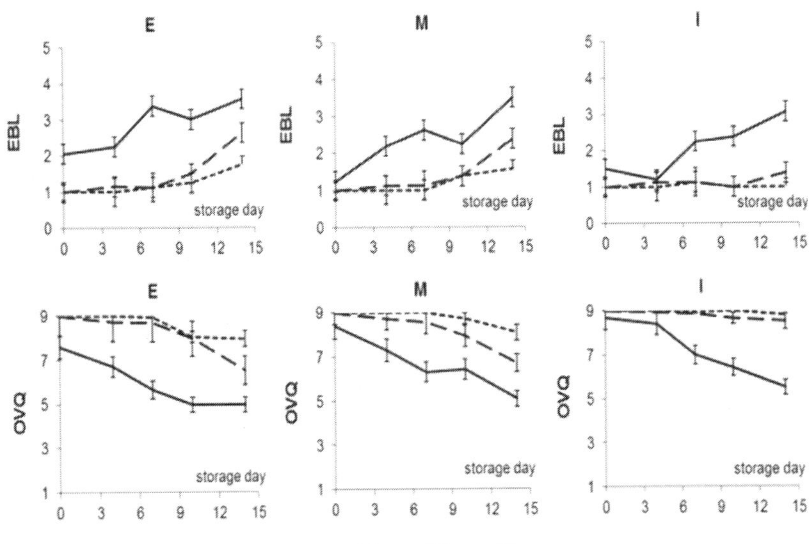

FIGURE 7: Enzymatic browning in leaves (EBL) and Overall Visual Quality (OVQ) evolution during refrigerated storage (0-2°C) of Butterhead lettuce by zone (E: external, M: middle, I: inner) and production system (OF: open field, MG: mulch greenhouse, TG: traditional greenhouse).

The OVQ varied (**FIGURE 7**). Lettuce heads cultivated in the greenhouse had high OVQ values until day 8 of storage. After this values in external and middle zones decreased, were higher in unmulched than in mulched greenhouse, but OVQ values were always higher than 5 (acceptability limit). The internal zone of these plants did not change during storage. The three zones had decreases from the first days of storage, reaching the acceptability limit value towards the end of the period. The external zone reached this limit before (day 10) than did middle and internal zones. Open field lettuce plants lost near 40% of quality, by day 14. At the end of storage, the unmulched greenhouse lost almost 20% of quality, as a function OVQ; mulched lettuce plants only lost 8%. These results, combined with those for enzymatic browning, represent positive commercial advantages since shelf-life of mulched lettuce was longer than open field or unmulched greenhouse.

At harvest, greenhouse plants were heavier than open field lettuce, with fewer, but larger, leaves. This could be considered an advantage by consumers because less preparation would be required. Greenhouse lettuce plants had better water status at harvest. Open field lettuce had higher BW, probably as a physiological response to mild water stress during development. Nutritional quality at harvest was affected by the production system, with middle leaves of unmulched greenhouse lettuce having higher AA at harvest. No differences were found for microbiological indices, with microbial counts higher than 5 log in all situations. Open field lettuce plants had lower OVQ scores and more enzymatic browning than greenhouse lettuce plants.

During postharvest storage use of greenhouses produces heads with higher initial quality, and better tolerance to refrigerated storage, with no differences between plants from unmulched or mulched greenhouses. The most noticeable changes during postharvest storage were to sensory quality. Open field lettuce plants had lower OVQ scores and more enzymatic browning than greenhouse lettuce plants, indicating lower quality. Open field lettuce plants reached the limit of acceptability by day 7 for the external head zone and by day 14 for middle and internal zones. Greenhouse lettuce plants did not reach the acceptance limit during the14 days of storage.

The use of greenhouses for production of Butterhead lettuce may be beneficial due to increased yields and may represent improvement of initial quality and better retention of the quality during refrigerated storage. Shelf-life could be increased, another advantage for producers because longer shelf-life may lead to the possibility of reaching other markets.

ACKNOWLEDGEMENTS

Work supported by Consejo Nacional de Investigaciones Científicas y Técnicas (CONICET), Agencia Nacional de Promoción Científica y Tecnológica (ANCyT) and Universidad Nacional de Mar del Plata (UNMDP). Paper is a partial fulfillment of requirements for the degree of Magister Scientiae in Vegetable Production, College of Agronomy, UNMDP of María G. Goñi.

REFERENCES

1. Negi PS, Roy SK (2000) Effect of blanching and drying methods on ß -carotene, ascorbic acid and chlorophyll retention of leafy vegetables. LWT - Food Science and Technology 33: 295-298.
2. Romani A, Pinelli P, Galardi C, Sani G, Cimato A, Heimler D (2002) Polyphenols in greenhouse and open-air-grown lettuce. Food Chem 79: 337-342.
3. Nicolle C, Cardinault N, Gueux E, Jaffrelo L, Rock E, et al. (2004) Health effect of vegetable-based diet: lettuce consumption improves cholesterol metabolism and antioxidant status in the rat. Clin Nutr 23: 605-614.
4. Kader AA (2002) Quality parameters of fresh-cut fruit and vegetable products. Lamikanra, O. (Edn) CRC Press.
5. Wien HC (1997) Lettuce. Wien, H.C. edn. In: The Physiology of Vegetables Crops. CAB International. New York. pp 479- 509.
6. Casanova PM, Messing I, Joel A, Cañete MA (2009) Methods to estimate lettuce evapotranspiration in greenhouse conditions in the central zone of Chile. Chilean J Agric Res 69: 60-70.
7. Radin B, Reisser CJ, Matzenauer R, Bergamaschi H (2004) Crescimento de cultivares de alface conduzidas em estufa e a campo. Horticultura Brasileira 22: 178-181.
8. Agüero MV, Barg MV, Yommi A, Camelo A, Roura SI (2008) Postharvest changes in water status and chlorophyll content of lettuce (Lactuca sativa L.) and their relationship with overall visual quality. J Food Sci 73: 176-185.
9. Fontanetti verdial M, Santos de lima M, Morgor AF, Goto R (2001) Production of Iceberg lettuce using mulches. Scientia Agricola 58: 737-740.
10. Ponce AG, Agüero MV, Roura SI, Del Valle CE, Moreira MR (2008) Dynamics of Indigenous microbial population of Butterhead lettuce grown in mulch and on bare soil. J Food Sci 73: M257-M263.
11. Goñi MG, Agüero MV, Moreira MR, Ponce A, Roura SI (2010) Ring characterization of quality indices in butterhead lettuce cultivated under mulch and bare soil. J Food Quality 33: 439-460.
12. Viacava GE, Roura SI, Agüero MV (2011) Improvement of water status methodology for leafy vegetables reduces consumption of time, skilled labor, and laboratory resources. Food Analytical Methods 4: 307-312.
13. Roura SI, Moreira MR, Ponce A, del Valle CE (2003) Dip treatments for fresh Romaine lettuce. Italian Journal of Food Science 15: 405-415.
14. ICMSF (1983) Métodos recomendados para el análisis microbiológico en alimentos. En: Microorganismos de los alimentos I. Técnicas de análisis microbiológicos, (2nd edn). Zaragoza, España.
15. Westland J(2001) Qué es el espacio de color CIE L*a*b*?.
16. Vignoni LA, Cásari RM, Forte M, Mirábile ML (2006) Determinación del Índice de Color en Ajo Picado. Información Tecnológica 17: 63-67.

17. Moreira MR, Roura S, Del valle C (2003) Quality of Swiss chard produced by conventional and organic methods. LWT - Food Science and Technology36: 135-141.
18. SAS Inc. (2002) SAS software, Version 9.0 of the SAS System for Windows. SAS Institute Inc., Cary, NC, USA.
19. Kuehl R (2001) Diseño de Experimentos 2da ed. Thompson Learning Intl. 666 p.
20. Blanco FF, Folegatti MV (2005) Estimation of leaf area for greenhouse cucumber by linear measurements under salinity and grafting. Scientia Agricola 62: 305-309.
21. Frantz JM (2010) The greenhouse environment: heat and light. USDA.
22. Eitel JUH, Gessler PE, Smith AMS, Robberecht R (2006) Suitability of existing and novel spectral indices to remotely detect water stress in Populus spp. Forest Ecology and Management 229: 170-182.
23. Hsiao TC, Acevedo E (1974) Plant response to water deficits, water-use efficiency and drought resistance. Agricultural Meteorology 14: 59-84.
24. Guerra N, Carrozzi L, Goñi MG, Roura S, Yommi A (2010) Quality characterization of celery (Apium graceolens L.) by zones of plant and two harvest dates. J Food Sci 75: S327-S332.
25. Rascio A, Russo M, Platani C, Di fonzo N (1998) Drought intensity effects on genotypic differences in tissue affinity for strongly bound water. Plant Sci 132: 121-126.
26. Roura SI, Moreira MR, Crapiste G, Del valle CE (2001) Biochemical characterization of two pepper varieties in green and red maturation stages. Italian Journal of Food Science 4: 391-397.
27. Toledo MEA, Ueda Y, Imahori Y, Ayaki M (2003) L-ascorbic acid metabolismo in spinach (Spinacia oleracea L.) during postharvest storage in Light and dark. Postharvest Biology and Technology 28: 47-57.
28. Wojciechowska R, Siwek P, Libik A (2007) Effect of mulching with various films omn the yield quality of butterhead lettuce and celery stalks with special reference to nitrate metabolism. Folia Horticulturae 19: 37-44.
29. Liakatas A, Clark JA, Monteith JL (1986) Measurements of the heat balance under plastic mulches. I: Radiation balance and soil heat flux. Agricultural and forest meteorology 36: 227-239.
30. Arrigoni O, De Tullio MC (2002) Ascorbic acid: much more than just an antioxidant. Biochim Biophys Acta 1569: 1-9.
31. Bartoli CG, Tambussi EA, Diego F, Foyer CH (2009) Control of ascorbic acid synthesis and accumulation and glutathione by the incident light red/far red ratio in Phaseolus vulgaris leaves. FEBS Letters 583: 118-122.
32. Franck C, Lammertyn J, Tri Ho Q, Verboven P, Verlinden B, et al. (2007) Browning disorders in pear fruit. Postharvest Biology and Technology 43: 1-13.
33. Oh MM, Carey EE, Rajashekar CB (2009) Environmental stress induce health-promoting phytochemicals in lettuce. Plant Physiol Biochem 47: 578-583.
34. Smirnoff N (1996) The function and metabolism of ascorbic acid in plants. Annals of Botany 78: 661-669.
35. Ponce AG, Roura SI, Del valle CE, Fritz R (2002) Characterization of native microbial population of Swiss Chard (Beta vulgaris, type cicla). LWT - Food Science and Technology35: 331-337.
36. Zagory D (1999) Effects of post-processing handling and packaging on microbial populations. Postharvest Biology and Technology 15: 313-321.
37. Fonseca JM (2006) Postharvest quality and microbial population of head lettuce as affected by moisture at harvest. J Food Sci 71: M45-M49.
38. Caldwell CR, Britz S (2006) Effect of supplemental ultraviolet radiation on the carotenoid and chlorophyll composition of greenhouse-grown leaf lettuce (Lactuca sativa L.) cultivars. Journal of Food Composition and Analysis 19: 637-644.
39. Winkel-shirley B (2002) Biosynthesis of flavonoids and effects of stress. Current Opinion in Plant Biology 5: 218-223.

40. Carletti P, Masi A, Wonisch A, Grill D, Tausz M, et al. (2003) Changes in antioxidant and pigment pool dimensions in UV-B irradiated maize seedlings. Environmental and Experimental Botany 50: 149-157.
41. Ingram J, Bartels D (1996) The molecular basis of dehydration tolerance in plants. Annual review of plant biology47: 377-403.
42. Singh V, Pallaghy CK, Singh D (2006) Phosphorouos nutrition and tolerance of cotton to water stress II. Water relations, free and bound water and leaf expansion rate. Field Crops Research 96: 199-206.
43. Burdon J, Clark C (2001) Effect of postharvest water loss on "Hayward" kiwifruit water status. Postharvest Biology and Technology 22: 215-225.
44. Rodrigues CMA, Della lucia CM, Azeredo RMC, Cota AM, Santana AMC, Pinheiro-santana HM (2010) Control of vitamin C losses in vegetables prepared at a food service. Food Control 21: 264-271.
45. Tomás-Barberán FA, Gil MI, Castañer M, Artes F, Salveit ME (1997) Effect of selected browning inhibitors on phenolic metabolism in stem tissue of harvested lettuce. J Agric Food Chem 45: 583-589.
46. Tano K, Oulé MK, Doyon G, Lfncki RW, Arul J (2007) Comparative evaluation of the effect of storage temperature fluctuation on modified atmosphere packages of selected fruit and vegetables. Postharvest Biology and Technology 46: 212-221.
47. Vega-gálvez A, Lemus-Mondaca R, Bilbao-sáinz C, Fito P, Andrés A (2008) Effect of air drying temperature on the quality of rehydrated dried red bell pepper (var. Lamuyo). J Food Eng 85: 42-50.

CHAPTER 15

REDD Roses for a Green Lady – Target Setting for Deforestation in the Brazilian Amazon

Valny Giacomelli Sobrinho[1]

[1] Department of Economics, Federal University of Santa Maria (UFSM), Santa Maria, Rio Grande do Sul, Brazil

1. INTRODUCTION

In the lyrics written by Sid Tepper and Roy C. Bennett of a popular 1948 song, a broken-hearted guy, who had the day before argued with his girlfriend, rushed to the florist to buy some "red roses for a blue lady". His hope was that those pretty flowers could chase her blues away. In short, he wished some red flowers could compensate for the damage he had caused to his lover's heart.

Likewise, climate policy has recently[1] - come up with the REDD mechanism (Reduced Emissions from Deforestation and forest Degradation) to protect natural standing forests before they are damaged by deforestation or degradation. Within the REDD framework, carbon credits can be earned for deforestation avoidance, rather than as in forestry CDM (Clean Development Mechanism), for afforestation or reforestation.

Since deforestation gives off nearly one-fifth (1.6 Gt) of global carbon emissions [2], avoiding it is claimed to be the most effective and cheapest way of control. However, until recently deforestation avoidance had been kept out of international climate accords, mainly because of the fear that credits supplied in return could flood carbon markets, thereby lowering the price of carbon credits [2]. But, with the closing, in 2012, of the first commitment period of the Kyoto Protocol (2008-2012), avoiding deforestation has been taken up as a post-2012 proposal for an international agreement that might bring in developing (non-Annex I) countries also to comply with emission targets.

Although the forest conservation feature of REDD means a step forward regarding carbon offsets from either energy or forestry-CDM [3], the credits arising from avoiding deforestation sound like roses offered as rewards for *not* causing damages to any "green lady" (forest). However, public incentive systems seldom pay people for not doing something — in the case of REDD, for not deforesting or not degrading. Hence, since a reward means a payment due for not doing something, it might encourage people to do what is already forbidden. In addition, it could make room for compensation seekers, whose anti-social and imprudent behaviour would otherwise have been inhibited by moral or legal censure [4].

Anyway, such a reward largely rests upon the creation of a market for REDD credits, in which Payments for Ecosystem Services (PES) are made to landowners and users, to compensate them for keeping forests — which are worth more alive than dead. PES schemes are defined as voluntary, conditional transactions — the payment is only made if the service is delivered — with at least one seller, one buyer, and a well-defined environmental service [2]. However, none of these conditions are yet in place in most REDD countries: the commodity is hard to quantify, the sellers are not well defined, big buyers do not exist and the rules of the game are not well established [5].

Besides these practical hindrances, REDD architecture is not without ideological critiques. Polanyi's [6] followers have charged not only REDD with having a privatising and marketing nature, but also PES with allowing elites and large-scale commercial actors to profit from what is called "forest capitalism". The gist of the criticism is that the largest share of deforestation is carried out by larger and wealthier players than by the poor, who REDD notwithstanding claims to be the greatest beneficiaries of the payments for avoiding deforestation [5]. Thus, of course, the bulk of these reward payments is expected to go to the players whose monetary foregone benefits are the greatest, when deforestation is halted.

Another pitfall of REDD and PES is that they rely too much on financial support. Although forests are assigned a high value in public debate, the relatively low returns accruing to forest management are due to the many and diverse non-financial benefits that forests can provide [7]. Both global REDD regimes and national REDD strategies call for the provision of financial resources — in the form of compensation or incentive payments

— to alter the political economy of commercial access to forest resources [2]. Yet, REDD finance hinges on political will, institutional settings and long bureaucratic chains to work properly [2], [5].

At the national level, REDD policies demand the removal of support and subsidies for companies or activities that increase the pressure towards forest conversion — agricultural expansion, destructive logging, settlement schemes, plantation and road development in forest areas [2]. At the international level, a large proportion of finance will need to come from developed countries. Therefore, the scale of such finance will depend on the political will within these countries [5]. Nevertheless, developed countries are not likely to transfer payments to fragile states, where long-term efforts would be required to create or reform institutions, strengthen governance processes and build capacity to deal with the new models of forest management underlying REDD policies [2].

In any event, disturbances of this sort rest not only upon the promise to serve different actors and interests, but also upon that to bridge the environment and development agendas. While this proposal sounds so appealing and distinct from past efforts in the forestry sector, it not only has turned REDD into a successful idea, but also made it move from single (carbon) to multiple objectives. Such a move, though, is now threatening and overshadowing the main characteristic of REDD, which comes down to large-scale funding and performance-based support. So far, the vast majority of both developed and developing countries lack concrete strategies on how to implement REDD. Therefore, REDD finance remains unresolved, because the cost of reducing emissions from deforestation also depends on the strategy chosen [5].

Arguably, new strategies must emphasise carbon-stock protection [2]. Many studies have indicated that, after wood production, carbon sequestration is the most valuable output from forests [7]. However, a conservation market (REDD credits) for forest protection should draw on a form of compensation for producing something additional (new carbon stored), rather than on a reward for "not doing something" (not deforesting or not degrading) [4]. To begin with, this is supposed to help the political economy of REDD build the argument for its public support [4], thus allaying the mistrust that lurks over it of rewarding the wealthiest forest users.

A further step towards that shift lies in emphasising a *stock maintenance* rather than an *emissions avoidance* approach. This turns carbon conservation strategies from output (performance) into input-driven ones. After all, forest sequestration of carbon emissions is primarily a matter of forestland availability rather than of emissions avoided. If forest stocks are maintained or even increased, they do not emit carbon, which is kept there. Although this might sound like a "two-sides-of-the-same-coin" problem, the stock maintenance approach highlights the positive side of conservation — the stewardship of carbon stocks — whereas the emissions avoidance approach stresses its negative side — the discouragement or closure of activities causing emissions to rise in spite of delivering economic benefits. Quite often, the latter is seen as an unproductive strategy while the former might well evoke the production of some useful commodity.

The commodity at stake is carbon storage rather than money. Yet, REDD finance claims that economic and monetary incentives can, through price signals, alter the decisions of individual land users and compensate them for foregone benefits from not converting or degrading the forest [2]. Since carbon storage is increasingly needed, the demand for carbon credits is expected to go up, thereby generating finance for forest conservation [5]. At present, however, there is deep uncertainty as to whether and how a future international climate agreement would value carbon sequestration provided by forests. In addition, when standing forests compete with high-value agricultural and mining commodities, no one can ensure anymore whether and how REDD funding — particularly output, results-based finance — would be available in the future. So far, from current REDD finance, it is widely recognised that more money alone cannot solve the deforestation problem and that the expectations of more money can even make it worse [8].

As argued elsewhere [3], [9], [10], since the ascent of money in modern societies, a community's wealth now has two components: real goods, accumulated through real investments, and fiat or paper "goods", made of nothing. Unlike the former, the latter kind of wealth is not subject to the natural law of decrease, the entropy law. Whereas the accumulation of real goods, which hold use value, does meet physical limits, there is no limit at all to the accumulation of "virtual wealth" [10], used for exchange only [9]. So, in this world, where the substance is exchanged for the shadow, what determines the value of money is the amount of wealth people prefer to do without, and that is the same as the amount of credit they retain as money [10]. In monetary economies, the greater this "virtual wealth" is, the higher the price of the real goods it can afford; conversely, the smaller the former, the lower the latter[2] - . It is puzzling how the protection of environmental goods and services should be commanded by such a virtual wealth rationale [3], [9].

In this regard, a non-monetary Bio-Economic model for carbon Sequestration by Forests (BESF) [3] is applied to a deforestation scenario taking place in the Brazilian "Legal Amazon" region (*Amazônia Legal*, Figure 1), to show how forest stocks can be prevented from being degraded. After all, if these stocks wither away, they will no longer be able to store carbon emissions. The model assumes that the growth of natural stocks follows an upper limited path. Therefore, if emissions from economic growth (k) have to be taken in by standing natural forests, their stocks could not fall below a certain limit. Those forest storage and tree growth constraints cap emissions.

Building upon fishery and forest bio-economics, in the BESF model emitters play the catchers, whereas natural forests supply the catch — namely, the environmental service of emissions storage. But unlike in fishery and forest models, the "catch", in this case, is an input rather than an output. As emissions increase, so does the demand for their storage, and the supply of forest stocks goes down. Thus, forest stocks turn out to be priced biophysically rather than monetarily.

Such a biophysically set price is called the *bio-economic exchange rate* (ε). Found by dynamic optimisation methods, it works as a *shadow price* measuring the shortage of the environmental input (carbon removal) on demand. At any given time, the more (less) this service is demanded, the less (more) of it is carried over into the future. However, the supply of current removal stocks varies across the carbon sinks. Such spatial differences, measured by the *bio-diversity ratio* λ, are called "exports" (Z) of carbon removal to elsewhere and correspond to *conservation loans* (REDD credits) between carbon sinks. On the other hand, the provision of such an environmental service in the future is "imported" (borrowed) from the savings of removal stocks that happened in the past. Therefore, these transfers of carbon removal over time are called "imports" (M) and are referred to as *compensation loans* (CDM offsets) between the sinks.

Whereas Z means that the ecological burden of removing carbon emissions is "exported" to elsewhere, M implies carrying the ecological burden within an economy's boundaries over time [3]. The former translates into an ecological credit ($0 < \varepsilon < 1$) and the latter into an ecological debt ($\varepsilon > 1$).

When conservation is low (high), compensation is supposed to decrease (increase), unless ε depreciates (appreciates). Since ε exchanges future carbon removal stocks (M) for current ones (Z), its rise (depreciation) means that Z is relatively deteriorating, whereas its fall (appreciation) means that Z is relatively increasing. Otherwise, for a given ε, Z and M vary positively with it (Figure 2.b).

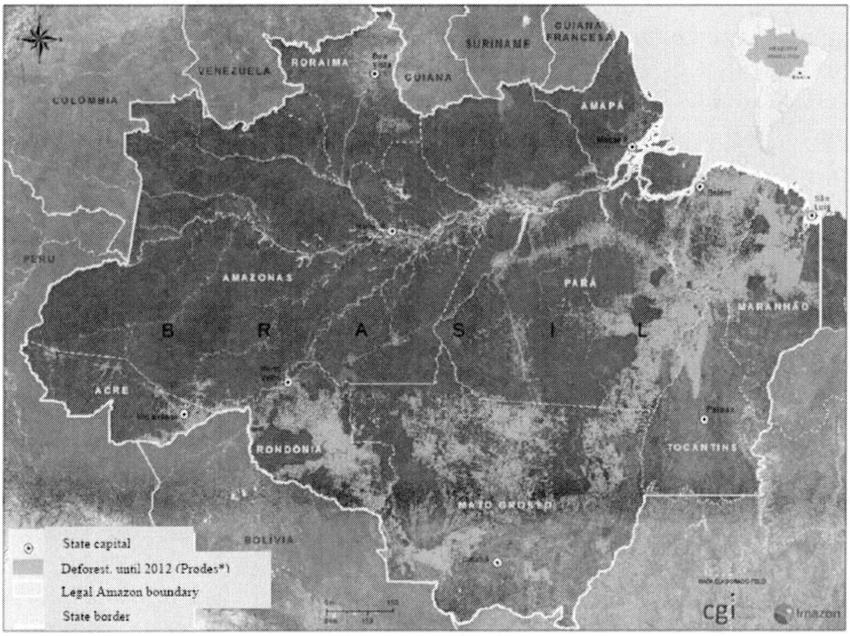

(*) Prodes Project — Satellite-borne monitoring of the Brazilian Amazon Rainforest [23]; Source: [12]

FIGURE 1. Deforestation accumulated until 2012 in the Brazilian Legal Amazon

Just like bond markets are grounded in existing loan supply (savings) and demand (investment), removal loans for either carbon conservation or compensation are backed up by the biological growth of actual stocks (removal supply) of forest sinks set aside for curbing emissions from economic growth (removal demand). Forest-wide, Z grows with λ, whose growth, in turn, causes ε to fall (appreciate). On the other hand, the faster (slower) the speed k of economic activity, the greater (smaller) the demand for M and the depreciation (rise) of ε should be. Hence, at this forestland level of aggregation, ε is a negative function of the spatial distribution of carbon sinks (λ) and a positive function of emissions given off by the growth of the economy over time (k) (Figure 2.a).

The objective of probing these relationships is to demonstrate that the macro-scale determination of ε (Figure 2.a) can help find: a) the optimal supply of conservation and compensation in the loan market for carbon removal (Figure 2.b); b) biophysically attainable deforestation targets at the micro-scale (Figure 2.c), according to the rates at which emissions from economic growth ($\ln k$) are given off and the ratio of forest to deforested land— the bio-diversity ratio — varies (ln λ).This analysis will be carried out for three scenarios:

i. BAU (Business-As-Usual), in which ln $k >$ ln λ;

ii. REDD, in which ln $k <$ ln λ;

iii. CDM, in which ln $k =$ ln λ.

To start with, the methodology sections will describe the BESF model, its geometry, basic assumptions, parameters, functions and variables. After that, empirical data on deforestation in Brazilian Amazonia will be used to account for the model equilibrium points — both at the micro and macro-scale. Finally, the aforementioned three scenarios will be assessed to determine, for either of them, how much forestland would have to be used for conservation (REDD) and for offsetting carbon emissions (CDM). According to this allocation, carbon sequestration provided by forests could be "paid" at its real, biophysical value, rather than according to the virtual monetary benefits such an environmental service is supposed to deliver.

2. RESEARCH QUESTION AND ANALYTICAL FRAMEWORK

The creation of a market for carbon is based on the assumption that monetary payments for carbon storage might make economic agents opt for forest conservation rather than forest conversion [4]. Although such payments might sound like a working solution, the heavier any forest-related decision falls back on them, the lighter it is supposed to lean on its biophysical and environmental footings. Thus, how should a market fit into turning this logic upside-down? Put differently, how could ecosystem services, such as carbon sequestration, be priced biophysically rather than monetarily? What does that mean? How would it work?

The answers to whether prices can be equated with value or considered only indirect means of measuring values [13] rest upon the proposition that economic value should not be reduced to an ultimate one dimensional gauge, as held by the labour, utility and energy theories of value[3] -. Economic value should be thought of as not bearing a single substance out of which it should be drawn or within which it should be found. In a biophysical sense, value is limited to the degree to which an item contributes to an objective or condition in an ecosystem [16]. For instance, the biophysical value of a tree species could come from its contribution to controlling soil erosion in steeply sloped areas [16]. Elsewhere, the same tree species might be worth for, say, sequestering carbon. So, although the tree species has not changed, the substance or content from which its value is drawn has.

Furthermore, just like man cannot reach too deep into the material microcosm [13], he is likely unable to search too deeply for the one and only source of economic value. According to the Heisenberg principle of indeterminacy in quantum physics, and given the high complexity of microstructures, the probability of, say, building a living cell from scratch is extremely small [13]. By the same token, the probability of drawing economic value out of a single underlying source is very low.

Within the economy's biophysical realms, useful goods and services should be valued by their usefulness[4] - rather than by their exchange properties. What is meant by "usefulness" builds on Daly's [18] notion of *ultimate efficiency*, defined as the ratio of service to throughput. The services yielded by the stocks of artefacts are the ultimate benefit of economic activity. The throughput is the inevitable cost of maintaining the stocks of people and artefacts. Thus, in the final analysis, the stock of physical wealth is an accumulated flow of throughput, which is a cost to be minimised [18].

Likewise, if carbon removal is the service to be used, then it must be valued by the biophysical cost of delivering carbon removal stocks. Yet if not technically estimating price or value, a method that estimates costs should at least be a fairly good approximation to price and value, when markets are in equilibrium [16]. As known, whenever a commodity has a much higher value than its cost of production, its profits will be higher. The commodity will then be produced with increasing marginal costs until cost just equals price and profits are zero. The opposite happens when the commodity cost is much higher than its value. Since the commodity profits are lower, it will not be produced. The marginal costs then decrease until cost and price are equal.

As real markets are seldom in equilibrium, cost and price are expected to diverge, so that commodities could become overpriced or underpriced. The *shadow price* $\varepsilon = M \div Z$, calculated by the BESF model, is the biophysical sign of that distortion. It informs how far or close the growth rates of the economy's emissions ($\ln k$) and of the forest stocks to remove them ($\ln \lambda$) are from each other (Figure 2.a). As explained before, the greater λ is, the larger Z will be, which, *ceteris paribus*, makes ε go down (appreciate); on the other hand, the greater k is, the larger M will be, which, *ceteris paribus*, makes ε go up (depreciate). The optimal balance between k and λ defines not only ε^*, in Figure 2.a, but also the optimal level of carbon conservation (Z^*) and compensation (M^*), in Figure 2.b, as well as the optimal growth ($G(Xt)$) of and demand (ht) for carbon removal stocks (Xt), in Figure 2.c. Hence, any bio-economic distortion in carbon prices is communicated by the bio-economic exchange rate ε.

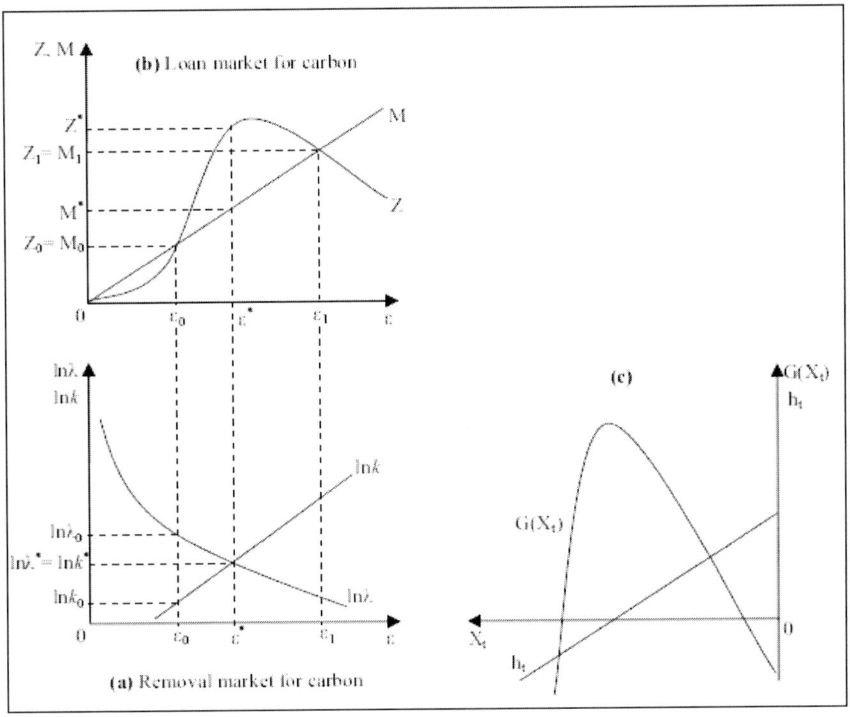

Figure 2. Loan and removal markets for carbon

For instance, given the hypothetical functional forms shown in Figure 2, ε^*, found in Figure 2.a, requires carbon conservation (Z^*) to be greater than carbon compensation (M^*), in Figure 2.b. Similarly, as ε^* was the outcome of a specific value taken on by λ and k, this bio-diversity ratio (λ^*) and economic growth rate (k^*) will then be applied to yield the curves depicted by Figure 2.c. It is these curves that set optimal deforestation targets and rates if forestland is split into deforested (u) and conserved (v) patches.

3. MODEL ASSUMPTIONS

Figure 2.c draws on bio-economic models, such as forestry and fishery models, which are concerned with the age and size of their biomass stocks (trees and fish) [3][19]. As the growth of a forest depends on the age of its trees, forestry models consider the time at which the maximum biological growth — or Maximum Sustainable Yield (MSY) — will be reached. In contrast, since fish growth depends on the size instead of on the age of fish stocks, fishery models involve the stock size at which the Maximum Economic Yield (MEY) will be attained [3].

Despite their different biological and economic emphasis, both such models are concerned with the output (trees and fish) provided by the corresponding natural source (forests and oceans). Neither of them, though, cares about the *input* — environmental service — supplied by the natural *sink* upon which the corresponding resource stock grows. Therefore, in the BESF model, the biomass stock is an *input* rendering an environmental service (carbon removal), instead of an *output* yielding biological (MSY) and economic (MEY) gains [3].

In this case, the stock harvested ($\hat{}ht$) stands for the *throughput* of emissions removal — or the environmental cost of storing in the forest biomass the emissions given off by the production and consumption of the economy's output. Provided that emissions from economic growth have to be removed by natural standing forests, the supply of removal forest stocks ($\hat{}G(Xt)$) need to meet an upper boundary (Kh) somewhere, since tree growth typically has an upper bound [3].

4. MODEL PARAMETERS AND VARIABLES

Such an upper bound (Kh) depends both on *space* (λ) and *time*-related (k) variables (Table 1 and Table 2). On one hand, Kh is a function of λt (Eq. (3)) — the spatial distribution of forest biomass Xt(Eq. (1)), in GtC, between $j = 1,..., n$ conserved (**v**) and deforested (**u**) sinks at each time period $t = 1,..., m$, where $\mathbf{u} = (u_1, u_2,..., um) = (x_{11}, x_{21},..., xm_1)$; $\mathbf{v} = (v_1, v_2,..., vm) = (x_1n, x_2n,..., xmn)$; $\mathbf{u} < \mathbf{v}$; xtj = carbon-equivalent emissions by sources at time t to be stored (removed) by sink j. On the other hand, the distribution of forest biomass is highly affected by economic activity and the rate at which land-use changes take place. In other words, λt depends on kj (Eq. (12)) — which measures emissions from deforestation or carbon emissions of sink j given off by its economic growth over time.

As time goes by, the ratio of energy changes ($\ln k$) to land changes ($\ln \lambda$), measured by ε (Eq. (14)), indicates in each period how much emissions from economic growth can be removed by each hectare of forestland [20]. Whereas economic growth (k) demands increasing removal of stocks, the supply of rising stocks is constrained by biophysical limits on growing and maintaining standing forests (λ).

Building on standard bio-economic (forestry and fishery) models, it is assumed that such a biological growth constraint follows a logistic pattern [3]. So as to hook the economy to its natural strings, emission flows ($\hat{}ht$) must be capped (Kh) rather than the growth of removal forest stocks

($\hat{G}(Xt)$), which already are by nature. Capping emissions from deforestation, though, implies that, at some former time T, when deforestation was negligible, there was a maximum level of removal stocks, XK, associated with that least emission release (Kh).

Table 1. Algebraic emission-removal matrix

t periods (emission sources)	$j = 1,\dots,2$ removal sinks* ($\mathbf{u} < \mathbf{v}$)		X	Z	λ
	$\mathbf{u}$	$\mathbf{v}$			
1	x_{11}	x_{12}	X_1	Z_1	λ_1
⋮	⋮	⋮	⋮	⋮	⋮
m	x_{m1}	x_{m2}	X_m	Z_m	λ_m
V	V_1	V_2	$V = X$	Z	λ
M	M_1	M_2	M	ε	
k	k_1	k_2	k		

(*) $j = \mathbf{u}$ = the smallest carbon biomass stock sink or the *deforested* sink; $j = \mathbf{v}$ = the largest carbon biomass stock sink or the *conserved* sink. It is required that $\varepsilon > 0$, because, in biophysical terms, there cannot be a negative accountancy [10]. Therefore, so that $Z > 0$, j sinks must be displayed on an increasing biomass stock order. Likewise, so that $M > 0$, $M_j = |x_{mj} - x_{1j}|$.

Table 2. Variables and equations of the BESF model

Dimension	Variable	Description	Equation
Space	X_t	Total removal stock at time t	(1)
	Z_t	Exports, at time t, of removing services across the space and j sinks	(2)
	λ_t	Bio-diversity ratio of deforested to conserved forestland at time t	(3)
	λ	Source-sink system's bio-diversity ratio	(4)a, (4)b
	V	Total spatial bio-economic value of the j removal sinks	(5)
	M	Total spatial imports (across all sinks)	(6)
	ψ_t	†Biophysical overshoot rate at time t	(7)
Time	X	Total removal stock over time	(8)
	Z	Total exports over time	(9)
	V_j	Bio-economic value of sink j arising from its long-run supply of biomass stocks for emission removal	(10)
	M_j	Imports (loans) of removing services per j sink through time	(11)
	k_j	Deforestation rate at sink j	(12)
	k	Source-sink system's deforestation rate	(13)a, (13)b, (13)c
Space-time	ε	Bio-economic exchange rate	(14)
	ψ	Long-run overshoot rate	(15)

(†) k_t = variable k (Eqs. (13)a-c) at time t; k^* = optimal value for k.

Source: [3]

$$X_t = \sum_{j=1}^{n} x_{tj} \tag{1}$$

$$Z_t = v_t - u_t = x_{tn} - x_{t1} \tag{2}$$

$$\lambda_t = v_t / u_t = x_{tn} / x_{t1} \tag{3}$$

$$\begin{aligned} \lambda &= 1 + (Z / V_1) \quad \text{(a)} \\ \ln \lambda &= Z / V_1 \quad \text{(b)} \end{aligned} \tag{4}$$

$$V = \sum_{j=1}^{n} V_j \tag{5}$$

$$M = \sum_{j=1}^{n} M_j = \Delta X = | X_m - X_1 | \tag{6}$$

$$\psi_t = k_t / k^* \tag{7}$$

$$X = \sum_{t=1}^{m} X_t \tag{8}$$

$$Z = \sum_{t=1}^{m} Z_t = \Delta V = | V_n - V_1 | \tag{9}$$

$$V_j = \sum_{t=1}^{m} x_{tj} \tag{10}$$

$$M_j = | x_{mj} - x_{1j} | \tag{11}$$

$$k_j = x_{mj} / x_{1j} \tag{12}$$

$$\begin{aligned} k &= 1 + (M / X_1) && \text{(a)} \\ \ln k &= M / X_1 && \text{(b)} \\ k &= X_{t-1} / X_t && \text{(c)} \end{aligned} \tag{13}$$

$$\varepsilon = \frac{M}{Z} = \frac{\ln k \times X_1}{\ln \lambda \times V_1} = \frac{\Delta X}{\Delta V} \tag{14}$$

$$\psi = \sqrt[m]{\prod_{t=1}^{m} \psi_t} \tag{15}$$

The parameter *Kh* is an algebraically found macro-scale bound to emissions. Actually, it is the value taken on by the emission demand function ˆ*ht* when *Xt* = *XK*. Hence, the first step to set *Kh* is to find *XK*, which is arrived at through vector algebra (Eq. (16)). Eq. (16) fulfils a twofold ideal requirement for sustainability:

a. *Maximum economic efficiency* (*kj* = *k*), so that the rates of economic growth or return even out across the sinks [3][24];

b. *Perfect ecological efficiency* (ε = 1), so that neither further reallocation nor redistribution of emissions removal takes place across the sinks [3][15][25].

TABLE 3. Forest conservation and deforestation in the Brazilian Legal Amazon (1988-2012)

Time periods (t)	Years	Deforest. rates (10^3 km²)	Deforested area (10^3 km²)[a]	Deforest. of the original forest[b]	Forest remnant (10^3 km²)[c]	v_t (GtC)[d]	u_t (GtC)[e]
1	1988	21.05	376.7	8.4%	4107.3	190.58	16.35
2	1989	17.77	401.4	9.0%	4082.6	189.43	17.42
3	1990	13.73	415.2	9.3%	4068.8	188.79	18.02
4	1991	11.03	426.4	9.5%	4057.6	188.27	18.51
5	1992	13.79	440.2	9.8%	4043.8	187.63	19.10
6	1993	14.90	440.2	9.8%	4043.8	187.63	19.10
7	1994	14.90	470.0	10.5%	4014.0	186.25	20.40
8	1995	29.06	497.1	11.1%	3986.9	184.99	21.57
9	1996	18.16	517.1	11.5%	3966.9	184.06	22.44
10	1997	13.23	530.1	11.8%	3953.9	183.46	23.01
11	1998	17.38	547.5	12.2%	3936.5	182.65	23.76
12	1999	17.26	564.7	12.6%	3919.2	181.85	24.51
13	2000	18.23	583.0	13.0%	3901.0	181.01	25.30
14	2001	18.17	601.1	13.4%	3882.8	180.16	26.09
15	2002	21.39	622.5	13.9%	3861.4	179.17	27.02
16	2003	25.25	647.8	14.4%	3836.2	178.00	28.11
17	2004	27.42	675.2	15.1%	3808.8	176.73	29.30
18	2005	18.85	694.0	15.5%	3789.9	175.85	30.12
19	2006	14.11	708.2	15.8%	3775.8	175.20	30.73
20	2007	11.53	719.7	16.1%	3764.3	174.66	31.23
21	2008	12.91	732.6	16.3%	3751.4	174.06	31.79
22	2009	7.46	740.1	16.5%	3743.9	173.72	32.12
23	2010	7.00	747.1	16.7%	3736.9	173.39	32.42
24	2011	6.42	753.5	16.8%	3730.5	173.09	32.70
25	2012	4.57	758.0	16.9%	3725.9	172.88	32.90

(a) According to [21], until 1997, and thereafter calculated from the previous deforested area plus the yearly deforestation rate. (b) Figures in this column are obtained by dividing the deforested area by estimates that consider 4483.972×10^3 km² = 4.483972×10^6 km² as, approximately, the original forested area of Brazil's 5×10^6 km² Legal Amazon [22]. (c) Figures obtained by calculating 4483.972×10^3 km² – deforested area $\times 10^3$ km². (d) "For the region's forests as a whole, the mean biomass loading (t/ha) for pre-logging biomass (weighted by the area of each eco-region present) is estimated at 464 t/ha" [22]. So the figures in this column result from 464 t/ha × forest remnant $\times 10^3$ km² = 464 t/10^{-2} km² × forest remnant $\times 10^3$ km² = (464 × forest remnant $\times 10^5$) t = (464 × forest remnant $\times 10^5$) 10^{-9} Gt. (e) The loading (biomass per hectare) for pre-logging biomass of forests cleared in 1990 (weighted by the deforestation rate in each state) is calculated at 434 t/ha. As before, the figures in this column result from 434 t/ha × forest remnant $\times 10^3$ km² = 434 t/10^{-2} km² × forest remnant $\times 10^3$ km² = (434 × forest remnant $\times 10^5$) t = (434 × forest remnant $\times 10^5$) 10^{-9} Gt [22].

Source: [23]

Theoretically, these conditions not only allow the source-sink system to simultaneously reach its economic and ecological sustainability, but also require it to remain sustainably stable. Therefore, the stock level XK represents the "bio-economic cost" of achieving a stable state of sustainability. Rather than a target to be complied with, it translates into the space-time needed to make k stable ($kj = k$) and $\varepsilon = 1$ [3].

$$2\begin{pmatrix} 1 & \frac{\bar{k}^{-1}+\bar{\lambda}^{-1}}{2} \\ \frac{\bar{k}+\bar{\lambda}}{2} & 1 \end{pmatrix}\begin{pmatrix} -\bar{u}_t \\ \bar{v}_t \end{pmatrix}=\begin{pmatrix} 0 \\ 0 \end{pmatrix} \quad (16)$$

In Eq. (16), the bar over the symbols stands for the corresponding mean values in the last row of Table 4. However, since the distribution of forestland (λ) is known for every period and the release of emissions from deforestation (k) instead is to be limited, $\bar{k}$ must be found by taking $\bar{\lambda} = 7.975$, $\bar{u}t = 24$ and $\bar{v}t = 182.40$ out of Table 4. By so doing, it turns out that $\bar{k} \approx 7.23$ and $\bar{k}^{-1} \approx 0.1383$. Next, by substituting these values into Eq. (17), in which, again, $\bar{u}t = 24$ and $\bar{X}t = 206.40$ come from Table 4, $vK = 1319.20$ and $XK = 1492.78$ are finally found.

$$\begin{pmatrix} 1 & \bar{k}^{-1} \\ \bar{k} & 1 \end{pmatrix}\begin{pmatrix} \bar{u}_t \\ v_K \end{pmatrix}=\begin{pmatrix} \bar{X}_t \\ X_K \end{pmatrix} \quad (17)$$

5. REMOVAL DEMAND FUNCTION ($\hat{H}T$)

In an emissions-saving (low carbon) economy, a seesaw balance between deforestation and conservation of carbon stocks is expected to hold. At lower deforestation rates, the savings (conservation) of biomass stocks outstrip their consumption. But as economic growth increases, conservation savings of biomass stocks decline, while their consumption by deforestation goes up. Thus, in order to account for these offsetting effects at any time, the observed removal demands for deforestation and conservation are respectively determined by:

$$h_t(u_t)=\bar{u}_t - u_t \tag{18}$$

$$h_t(v_t)=\bar{v}_t - v_t, \tag{19}$$

where ˉ*ut* and ˉ*vt* are mean values for either biomass stock, displayed in the last row of Table 4.

By summing Eqs. (18) and (19), the total observed demand for removal per period is obtained:

$$h_t=\bar{x}_t - x_t, \tag{20}$$

where $x_t = u_t + v_t$ and $\bar{x}_t=\bar{u}_t+\bar{v}_t$.

The underlying assumption of Eqs. (18)-(20) is the *Permanent Income Hypothesis* (PIH), put forward by the American economist Milton Friedman in the late 1950. According to this hypothesis, consumption is smoothed in the long run, so as to handle short-run fluctuations in income. Income fluctuates because of its transitory component, which consists of unforeseen additions or subtractions likely to cancel out in the long run [26][27]. In an emissions avoiding world, where carbon storage sounds like the economy's earnings, conservation of carbon stocks adds to income, whereas deforestation lessens it. So, Eq. (20) amounts to the permanent component of income in the long run, which accounts for a smoother path based on consumption expectations (ˉ*xt*) rather than on current consumption (*xt*). The regression[11] - of the results for Eq. (20) on the observed values for *Xt*, in Table 5, yields the estimated long-run mean of removal demand (Eq. (21)).

$$\begin{array}{lll} \hat{h}_t = & 206.121 - & 0.999X_t \\ \text{t-stat.} & 497.034 & -496.808 \\ \text{sig.t.} & 0.0000 & 0.0000 \end{array} \tag{21}$$

TABLE 4. Observed deforested, conserved and demanded biomass in the Brazilian Legal Amazon (1988-2012)

Time periods	Years	Deforestation (GtC)	Conservation (GtC)	Total biomass stock (GtC)	Deforest. demand (GtC)	Conserv. demand (GtC)	Total biomass demand (GtC)	Obs. Exports	Obs. Imports	ϵ	Econ. growth rate	Bio-div. ratio
(t)		(u_t)	(v_t)	(X_t)	Eq. (18)	Eq. (19)	Eq. (20)	(Z_t) Eq. (2)	(M) Eq. (6)	Eq. (14)	(k) Eq. (13c)	(λ_t) Eq. (3)
1	1988	16.35	190.58	206.93	7.65	-8.18	-0.53	174.23	—	—	1.00036	11.657
2	1989	17.42	189.43	206.85	6.57	-7.03	-0.45	172.01	2.2181	12.9	1.00020	10.874
3	1990	18.02	188.79	206.81	5.98	-6.39	-0.41	170.77	1.2392	7.3	1.00016	10.477
4	1991	18.51	188.27	206.78	5.49	-5.87	-0.38	169.77	1.0058	5.9	1.00020	10.174
5	1992	19.10	187.63	206.74	4.89	-5.23	-0.34	168.53	1.2392	7.4	1.00000	9.821
6	1993	19.10	187.63	206.74	4.89	-5.23	-0.34	168.53	0.0000	0.0	1.00043	9.821
7	1994	20.40	186.25	206.65	3.60	-3.85	-0.25	165.85	2.6760	16.1	1.00039	9.131
8	1995	21.57	184.99	206.57	2.42	-2.59	-0.17	163.42	2.4336	14.9	1.00029	8.575
9	1996	22.44	184.06	206.51	1.55	-1.66	-0.11	161.62	1.7960	11.1	1.00019	8.202
10	1997	23.01	183.46	206.47	0.99	-1.06	-0.07	160.45	1.1674	7.3	1.00025	7.974
11	1998	23.76	182.65	206.41	0.23	-0.25	-0.02	158.89	1.5610	9.8	1.00025	7.687
12	1999	24.51	181.85	206.36	-0.51	0.55	0.04	157.34	1.5499	9.9	1.00027	7.420
13	2000	25.30	181.01	206.31	-1.31	1.40	0.09	155.71	1.6367	10.5	1.00026	7.154
14	2001	26.09	180.16	206.25	-2.09	2.24	0.14	154.07	1.6312	10.6	1.00031	6.906
15	2002	27.02	179.17	206.19	-3.02	3.23	0.21	152.15	1.9212	12.6	1.00037	6.632
16	2003	28.11	178.00	206.11	-4.12	4.40	0.28	149.89	2.2672	15.1	1.00040	6.331
17	2004	29.30	176.73	206.03	-5.31	5.68	0.37	147.42	2.4626	16.7	1.00027	6.031
18	2005	30.12	175.85	205.97	-6.13	6.55	0.42	145.73	1.6924	11.6	1.00021	5.838
19	2006	30.73	175.20	205.93	-6.74	7.20	0.47	144.46	1.2670	8.8	1.00017	5.700
20	2007	31.23	174.66	205.90	-7.24	7.74	0.50	143.43	1.0356	7.2	1.00019	5.592
21	2008	31.79	174.06	205.86	-7.80	8.34	0.54	142.27	1.1594	8.1	1.00011	5.475
22	2009	32.12	173.72	205.84	-8.12	8.68	0.56	141.60	0.6703	4.7	1.00010	5.409
23	2010	32.42	173.39	205.82	-8.43	9.01	0.58	140.97	0.6286	4.5	1.00009	5.348
24	2011	32.70	173.09	205.80	-8.71	9.31	0.60	140.39	0.5763	4.1	1.00007	5.293
25	2012	32.90	172.88	205.78	-8.90	9.52	0.62	139.98	0.4105	2.9	—	5.255
Mean		**24.00**	**182.40**	**206.40**							**1.00023**	**7.975**

TABLE 5. Observed forest data and estimated demand (ˆht) and supply (ˆ$G(Xt)$) functions of the BESF model

Time periods	Years	X_t (GtC)	h_t (GtC)	$\hat{h}_t$ (GtC)	$g(v(X_t))$ (GtC)	$\hat{F}(X_t)$ (GtC)	$G(X_t)$ (GtC)	$\hat{G}(X_t)$ (GtC)
(t)			Eq.(20)	Eq.(21)	Eq.(23)	Eq.(25)	Eq.(26)	Eq.(27)
T	?	1492.78	—	-1285.17	0.0000	0.00	-1492.78	1.25×10^7
1	1988	206.93	-0.53	-0.5983	0.5370	1431.25	1224.33	1215.00
2	1989	206.85	-0.45	-0.5242	0.5376	1252.76	1045.91	1039.41
3	1990	206.81	-0.41	-0.4829	0.5379	1153.21	946.40	941.33
4	1991	206.78	-0.38	-0.4493	0.5382	1072.51	865.73	861.74
5	1992	206.74	-0.34	-0.4080	0.5386	973.18	766.45	763.70
6	1993	206.74	-0.34	-0.4080	0.5386	973.18	766.45	763.70
7	1994	206.65	-0.25	-0.3187	0.5393	759.12	552.47	552.05
8	1995	206.57	-0.17	-0.2374	0.5400	564.94	358.38	359.66
9	1996	206.51	-0.11	-0.1775	0.5405	421.94	215.44	217.72
10	1997	206.47	-0.07	-0.1385	0.5409	329.13	122.67	125.48
11	1998	206.4139	-0.02	-0.0864	0.5413	205.20	-1.22	2.17
12	1999	206.36	0.04	-0.0347	0.5418	82.34	-124.02	-120.23
13	2000	206.31	0.09	0.0199	0.5422	-47.19	-253.50	-249.45
14	2001	206.25	0.14	0.0744	0.5427	-176.08	-382.33	-378.21
15	2002	206.19	0.21	0.1385	0.5433	-327.61	-533.80	-529.81
16	2003	206.11	0.28	0.2141	0.5439	-506.05	-712.16	-708.66
17	2004	206.03	0.37	0.2963	0.5446	-699.41	-905.44	-902.84
18	2005	205.97	0.42	0.3528	0.5451	-832.02	-1037.99	-1036.24
19	2006	205.93	0.47	0.3951	0.5455	-931.14	-1137.07	-1136.09
20	2007	205.90	0.50	0.4296	0.5458	-1012.07	-1217.97	-1217.69
21	2008	205.86	0.54	0.4683	0.5461	-1102.58	-1308.43	-1309.03
22	2009	205.84	0.56	0.4907	0.5463	-1154.85	-1360.68	-1361.82
23	2010	205.82	0.58	0.5117	0.5465	-1203.84	-1409.66	-1411.33
24	2011	205.80	0.60	0.5309	0.5466	-1248.73	-1454.53	-1456.72
25	2012	205.78	0.62	0.5446	0.5468	-1280.69	-1486.47	-1489.04

6. REMOVAL SUPPLY FUNCTION (ˆ*G*(*XT*))

By substituting, in Eq. (21), XK = 1492.78, found by Eqs. (16) and (17), for Xt, the least level of emissions demand Kh = -1286.16 comes out. This value is used as the lower limit of a logistic function (Eq. (23)) in the following constrained optimisation problem:

Objective function:

$$\min_{X_t} \sum_t S_t = \min_{X_t} \sum_t [g(v(X_t)) - \hat{h}_t]$$

(22)

Constraints:

$$\hat{h}_t = 206.121 - 0.999 X_t$$

Constraint I:

ˆ*ht*=206.121-0.999*Xt*

Constraint II:

$$g(v(X_t)) = \frac{1}{(1/K_h) + b_0 b_1^{X_t}}$$

$$g(v(X_t)) = \frac{1}{(1/-1286.16) + 0.07075414 \times 1.01593216^{X_t}}$$

23)

Constraint III:

$$g(v(X_t)) \geq \hat{h}_t,$$

(24)

where *St* stands for the instantaneous surplus arising from the gap between removal growth rates (*g*(*v*(*Xt*))) and removal consumption rates (ˆ*ht*). Because of Eq. (24), *St* ≥ 0, that is, at any time, the rate at which removal stocks are supplied must be greater than or equal to the rate at which they are demanded. In Eq. (23), the parameter *Kh* is found by Eqs. (16), (17) and (21), whereas the parameters b_0 and b_1 are provided by GAMS-IDE (General Algebraic Modelling System – Integrated Development Environment), version 24.1.2 (http://www.gams.com/download/), in which the optimisation programme described by Eqs. (21)-(24) was run.

The optimal values for *g*(*v*(*Xt*)) are displayed in Table 5. They are now employed to estimate the future supply of removal stocks (ˆ*F*(*Xt*)) given the existing ones (*Xt*), used up in the present. The estimated variable ˆ*F*(*Xt*) stands for an *outflow-inflow ratio*, defined by logistically constrained rates of demand (numerator) and supply (denominator) of removal stocks (Eq. (25)). The difference between ˆ*F*(*Xt*) and *Xt* is the removal supply per time period (*G*(*Xt*)). Starting from the values for *G*(*Xt*), found by Eq. (26) and displayed in Table 5, it is possible to estimate the removal supply function (Eq. (27)), whose values are also displayed in Table 5.

$$\hat{F}(X_t)=\frac{\partial \hat{h}_t / \partial X_t}{g(v(X_t))}=\frac{\hat{h}_t(K_h-\hat{h}_t)}{g(v(X_t))} \tag{25}$$

$$G(X_t)=\hat{F}(X_t)-X_t \tag{26}$$

$$\begin{array}{lll} \hat{G}(X_t) = & 5.727X_t^2 & -244006.276 \\ \text{t-stat.} & 588.143 & -588.763 \\ \text{sig.t.} & 0.0000 & 0.0000 \end{array} \tag{27}$$

7. AVOIDING DEFORESTATION *VERSUS* STOCK MAINTENANCE APPROACH

All removal market functions are drawn in Figure 3. They look like those in Figure 2.c, and their interplay shows how long emissions from deforestation can be removed by natural forest stocks. In deforestation-conservation settings, the cost of maintaining a high economic growth rate (ln *k*) would be an abrupt fall in the rate at which natural removal stocks grow (ln λ). As demonstrated by Eq. (3), in Table 2, and by the figures in the last column of Table 4, the value of λ indirectly defines the quantities of removal consumption (*ht*) from deforestation (*ut*) or removal savings from conservation (*vt*). Thus, ˆ*ht* (Eq. (21)) hinges on that sink distribution, and so does ˆ*G*(*Xt*) (Eq. (27)), whose estimation ultimately relies on *Kh* — the lower bound of removal demand. As a matter of fact, λ is the critical variable drawing the bottom line in carbon removal markets.

The impacts of λ on the equilibrium of removal markets are shown in Table 6. The equilibrium scenarios checked, but the BESF one, follow the standard fishery analysis [28]. In the context of deforestation and forest conservation, though, there are important remarks regarding equilibrium conditions. To begin with, although the MSY equilibrium might apply to unnatural, even-aged stands, it is not likely to suit the dynamics of conservation of natural forests and deforestation. It is known that in natural forests, the wide frequency and age range of tree species points to a biomass yielding function that does not reach a maximum sustainable yield. Mathematically, this is translated by taking the first derivative of Eq. (27) and making it equal to zero. Thus, it is found that X_{MSY} = 21303.15 GtC, which is a prohibitively high level of removal stocks, provided their logistic upper bound is XK = 1492.78. The same holds for the restricted access (RA) stock level, which is found to be slightly lower (X_{RA} = 21303.06 GtC).

Although in standard bio-economic (fishery) analysis, the latter is expected to be higher than the former, it must be borne in mind that, in a *compensation approach* — such as *avoiding deforestation* —, the demand for removal stocks grows with deforestation (Figure 4.a). Therefore, conservation requires $X_{MSY} > X_{RA}$, since a smaller *compensating* stock means that less deforestation occurred.

On the other hand, Figure 3 shows that the greater the stock, the smaller its demand is, because the stock can only grow when its depletion is low. Thus, as is clear so far, Figure 3 illustrates a rather different standpoint, namely, a *stock maintenance approach* (Figure 4.b). It has been argued that such a *conservation approach* favours the largest forest countries, like Brazil. Unlike in small forest countries, with only tiny remnants of forest left, in countries where large expanses of forest remain standing, stock maintenance represents a much greater carbon service than does avoiding deforestation [21].

In this regard, the equilibrium conditions in Table 6 had to be adjusted to fall within a stock maintenance rationale. Since, in Figure 3, the slopes of the functions $\hat{G}(Xt)$ (Eq. (27)) and $\hat{h}t$ (Eq. (21)) are never expected to be equal — as the columns labelled "rate of return" and "rate of depletion" in Table 6 also show —within the feasible region ($X_{0A} \leq Xt^{*} \leq XK$), RA equilibrium only calls for maximising some positive level of Yt, which could otherwise be warranted if the functions $\hat{G}(Xt)$ and $\hat{h}t$ were to have the same slopes somewhere. However, as the outcomes in Table 6 show, a positive Ytcan be accomplished with a negative value for $\hat{G}(Xt)$, provided that, in absolute terms, this is smaller than that for $\hat{h}t$. Since $\hat{G}(Xt) < 0$ is environmentally threatening, BESF equilibrium becomes a more stringent condition, because it calls both for positive Yt and $\hat{G}(Xt)$. By minimising this positive level of Yt, then, the BESF equilibrium ensures that, for a removal demand function like ht, in Figure 4.b, and $\hat{h}t$, in Figure 3, the stock level satisfying this will lie slightly beyond (greater than) K — the stock level at which $\hat{G}(Xt) = 0$ — and further beyond (greater than) X_{OA} — where $\hat{G}(Xt)$ becomes equal to $\hat{h}t$, but turns out to be negative.

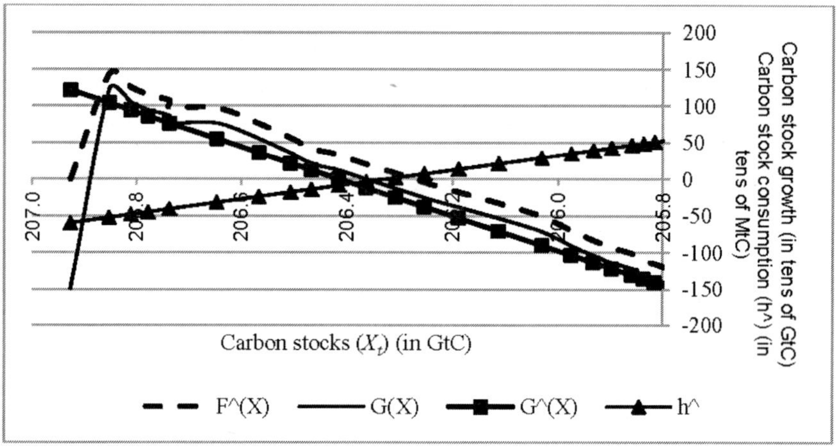

Figure 3. Removal market functions

The open access stock level (X_{OA}) is the smallest, yet just slightly smaller than the others. As expected from standard theory on renewable resources, the economic rent at this level is zero. At any other equilibrium point, it is non-zero and positive. But it is the highest at BESF, which minimises the throughput of maintaining stocks by requiring the supply of their emissions removal services ($\hat{G}(Xt)$) always to be positive.

Finally, by comparing the optimal stock levels (Xt^*), in Table 6, with the observed ones (Xt), in Table 5, it can be inferred when each equilibrium scenario must have occurred. It is worrying to ascertain that all of them are already gone somewhere between 1998 and 1999.

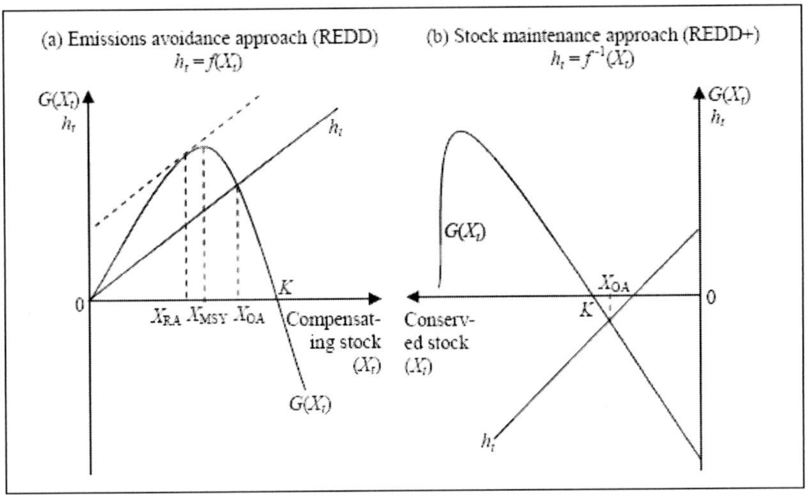

Figure 4. . The BESF model functions and the REDD approaches

Table 6. Removal market equilibrium analysis

Equilibria[a]	Equilibrium conditions[b]	Optimal[c] stock (X_t^*) (in GtC)	Removal supply ($\hat{G}(X_t^*)$) Eq. (27) (in GtC)	Removal demand ($\hat{h}_t^*$) Eq. (21) (in GtC)	Rate of return[d] $\left(\frac{d\hat{G}(X_t^*)}{dX_t^*}\right)$	Rate of depletion[e] $\left(\frac{d\hat{h}_t^*}{dX_t^*}\right)$	Economic rent ($Y_t = \hat{G}(X_t^*) - \hat{h}_t^*$)
BESF	$\min Y_t > 0$ $\hat{G}(X_t) > 0$ $X_{OA} \leq X_t^* \leq X_K$	206.412933	$\rightarrow$ 0.000	-0.086	2.364	0.999	> 0.086
K	$\hat{G}(X_t) = 0$	206.412933	0.000	-0.086	2.364	0.999	0.086
MSY	$\frac{d\hat{G}(X_t)}{dX_t} = 0$ $X_{OA} \leq X_t^* \leq X_K$	—	—	—	—	—	—
RA	$\max Y_t > 0$ $X_{OA} \leq X_t^* \leq X_K$	206.412933	-3.49×10^{-10}	-0.086	2.364	0.999	< 0.086
OA	$\hat{G}(X_t) = \hat{h}_t$	206.412898	-0.08	-0.08	2.364	0.999	0.000

(a) BESF = Bio-Economic carbon Sequestration by Forests; K = steady-state equilibrium; MSY = Maximum Sustainable Yield; RA = Restricted Access equilibrium; OA = Open Access equilibrium. (b) In standard fishery models, however, X_{RA} is found where $d\hat{G}(X_t)/dX_t = d\hat{h}_t/dX_t$. (c) Provided by GAMS-IDE 24.1.2. (d) First derivative of Eq. (27) = $11.454X_t$. However, so that $\hat{G}(X_t)$ and $\hat{h}_t$ can be plotted together, the barter ratio between them is 10 GtC of $\hat{G}(X_t)$ per 10 MtC of $\hat{h}_t$, as the vertical axis of the graph in Figure 3 indicates. More simply, this barter ratio can be expressed as 1 GtC : 1 MtC, which means 10^3 MtC : 1 MtC. Therefore, so that rates of return and depletion can be compared with one another, the former must be multiplied by 10^{-3}. (e) First derivative of Eq. (21).

8. AGGREGATE EMISSIONS DEMAND OR REMOVAL SUPPLY FUNCTION (ˆΛ(ˆE))

Perhaps these worries could have been dismissed before, if the variable regulating land-use changes (λ) had not been overlooked. As seen, ε (Eq. (14)) — the bio-economic exchange rate — is defined as the ratio of ecological debt — excess demand for removal services (supply of emissions) — to ecological credit — excess supply of removal services (demand for emissions). Although it is an underlying variable, it stands for the *shadow price* measuring, along an optimal path through time, the marginal bio-economic value of the forestland asset [19]. When the speed *k* of economic activity drives deforestation, forestland shrinks and thus ε is expected to rise (depreciate).

Since ε critically and ultimately depends on λ and *k*, it must, to begin with, be expressed in terms of them. Methodologically, this can be first accomplished by fixing *k*=¯*k*=1.00023, given in the last row of Table 4. Then this rate is assumed to hold for every year according to the following rule:

$$X_{t-1} / X_t' = \bar{k} = 1.00023, \tag{28}$$

where *Xt'* is the stock level that would be observed in column *Xt* of Table 5, if *k*=¯*k*= 1.0023.

As λ changes, so will ε — whose calculation follows Eqs. (11) and (6), for *M* (imports); (2) and (9), for *Z* (exports); and (14), for ε proper. These outcomes must be ordered pair-wise, according to increasing values of λ. The objective of this disposition is to check how ε is affected by changes in λ. Next, an equation for ε, as a long-run function (thus, bearing no *t* index) of λ, is estimated:

$$\hat{\varepsilon}(\lambda) = \exp\left(-0.131 + \frac{1.096}{\ln \ln \lambda}\right)$$

$$\begin{array}{lcc} \text{t-stat.} & -243.522 & 1048.581 \\ \text{sig.t.} & 0.0000 & 0.0000 \end{array} \tag{29}$$

Finally, the results for $\hat{\varepsilon}(\lambda)$, in Eq. (29), are used as inputs (independent variables) to arrive at an equation for λ as a long-run function of ε:

$$\ln\hat{\lambda}(\hat{\varepsilon}) = \exp\left(-1.127 + \frac{2.766}{\hat{\varepsilon}}\right)$$

$$\begin{array}{lcc} \text{t-stat.} & -133.522 & 214.563 \\ \text{sig.t.} & 0.0000 & 0.0000 \end{array} \tag{30}$$

Eq. (30) stands for the *aggregate emissions demand* or *removal supply* function in the long run.

9. AGGREGATE EMISSIONS SUPPLY OR REMOVAL DEMAND FUNCTION ($\hat{K}(\hat{E})$)

By a similar procedure, the functional relationship between ε and *k*, as well as between *k* and ε, can then be calculated. This time, though, the variable made fixed is $\lambda = \bar{\lambda} = 7.975$, given in the last row of Table 4. This value is kept unchanged for each two consecutive years, to either of which Eqs. (1) and (3) apply:

$$x_{t,j} + x_{t,j+1} = X_t \tag{1)a}$$

$$x_{t,j+1} / x_{t,j} = \bar{\lambda} = 7.975 \tag{3)a}$$

By substituting Eq. (3)a into Eq. (1)a, it turns out that:

$$X_t = (1 + \bar{\lambda})x_{t,j}, \tag{1)b}$$

which can be correspondingly replaced in Eq. (13)c to find the new value of *k*, at each two consecutive periods, when $\lambda = \bar{\lambda} = 7.975$ and thence remains constant. Again, the resulting ε requires Eqs. (11), (6), (2), (9) and (14). Also, like before, so as to inquire into the effects of changing *k* on ε, these variables are taken pair-wise on an increasing order of *k* values.

Lastly, an equation for ε, as a long-run function (thus, bearing no *t* index) of *k*, can be estimated:

$$\begin{aligned} &\hat{\varepsilon}(k) = \ 6434.098 \ln \ln k \\ &\text{t-stat.} \ 3921.720 \\ &\text{sig.t.} \ \ 0.0000 \end{aligned} \tag{31}$$

Now, the outcomes for ˆε(*k*), in Eq. (31), are used as inputs (independent variables) to obtain an equation for *k* as a long-run function of ε:

$$\begin{aligned} &\hat{k}(\hat{\varepsilon}) = 0.99999998 + 1.554575 \times 10^{-4}\hat{\varepsilon} \\ &\text{t-stat.} \quad - \qquad\qquad - \\ &\text{sig. t.} \quad - \qquad\qquad - \end{aligned} \tag{32}$$

Eq. (32) stands for the *aggregate emissions supply* or *removal demand* function in the long run.

10. MACRO-BIO-ECONOMIC SCENARIOS

Based on Eqs. (30) and (32), REDD, CDM and BAU scenarios are tested to understand how well conservation (REDD) and compensation (CDM) strategies can do as compared with business-as-usual (BAU) ones (Table 7 and Table 8). The most useful results shown by Table 8 are those displayed in its last two columns. They make clear how much the natural forest and the economy are expected to grow annually, through 25 years, in each scenario.

It is noteworthy that a stringent conservation scenario, such as REDD1, requires an optimal value for λ (ˆλ*REDD1= 5.787) that is not too far from its observed mean value, displayed in the last row and column of Table 4 (¯λ= 7.975). Thus, the allowed annual deforestation rate through 25 years (Eq. (33)) is 3.18% p.a. (last column of Table 8). This figure might sound startling when compared, for instance, with the deforestation rate in the Brazilian Legal Amazon for a single year: just between August 2012 and

August 2013, this rate reached 20% [31]! However, neither would more "economic growth-driven" strategies (CDM and BAU's) stand such a high annual deforestation rate. Nearly all of them (last four rows and last column of Table 8) would allow for a yearly deforestation rate of about 7.8%. On the other hand, a 100% rate of deforestation reduction, even spread over 25 years (scenario REDD2), would render no more than an economic growth rate as low as 0.00149% (last column of Table 8). These numbers help shed some light on the feasibility of the targets set by deforestation reduction programmes [29] (Figure 7). Figure 5 shows that the supply of emissions ($\hat{k}$) is nearly *perfectly inelastic* to the shadow price ε, whereas the removal of them ($\hat{\lambda}$) dramatically falls with the rise of ε. Although at some high value of ε, the demand for removal also becomes almost inelastic to price changes, this only happens at very low levels of existing removal stocks, when thus $\hat{\lambda} \rightarrow 1$ (Eq. (3)) and the share of forest conservation approaches that of deforestation ($v \rightarrow u$) or becomes even smaller ($v < u$).

$$\delta^* = \left[\sqrt[25]{(\bar{\lambda} - \hat{\lambda}^*)} - 1\right] \times 100, \text{ with } \bar{\lambda} > \hat{\lambda}^* \tag{33}$$

Table 7. Scenario analysis

Scenario	Constraints	Objective-function (W or W')[a]
REDD1	$\widehat{\ln\lambda} > \ln\hat{k};\ \widehat{\ln\lambda} = 1.755682$[b]	$\max W = \widehat{\ln\lambda} - \hat{k}$
REDD2	$\widehat{\ln\lambda} > \ln\hat{k};\ \widehat{\ln\lambda} = 1.028114$[c]	$\max W = \widehat{\ln\lambda} - \hat{k}$
CDM	$\widehat{\ln\lambda} = \ln\hat{k}$	$\max/\min W = \widehat{\ln\lambda} - \hat{k}$ or $W' = \hat{k} - \widehat{\ln\lambda}$
BAU1	$\widehat{\ln\lambda} < \ln\hat{k};\ \hat{k} = 1.8092514$[d]	$\max W' = \hat{k} - \widehat{\ln\lambda}$
BAU2	$\widehat{\ln\lambda} < \ln\hat{k};\ \hat{k} = 2.6658363$[e]	$\max W' = \hat{k} - \widehat{\ln\lambda}$
BAU3	$\widehat{\ln\lambda} < \ln\hat{k};\ \hat{k} = 3.3863549$[f]	$\max W' = \hat{k} - \widehat{\ln\lambda}$

(a) Because, as shown by Table 4, k is much smaller than λ, the greater of them must be scaled down through logarithms to make them comparable. (b) Growth rate for λ that would smooth, over 25 years (1988-2012), the accumulated deforestation reduction rate defined by a 7-year programme, from 2007 to 2015, for reducing deforestation in the Brazilian Amazon [29]. The annual reduction rates for every period $t = 1, \ldots, 7$ are, respectively, 25%, 25%, 30%, 40%, 50%, 75% and 100%. The "capitalisation" (multiplication) of all these rates yields 7.4648438, which amounts to the full figure to be reached in 7 years. This 7-year time is factored into a 25-year period, thereby yielding 3.5714286 sub-periods, over which the deforestation reduction rate accumulated during 7 years is spread according to its geometric mean $\left(\sqrt[\frac{25}{7}]{7.4648438}\right)$. (c) Growth rate for λ that would amount to a 100% deforestation reduction in 25 years. It is calculated by $\sqrt[25]{2}$, where 2 is a rate of growth that is worth 100%. (d) Value needed to yield an annual rate of economic growth of 2.40% during 25 years. This rate is the one projected for Brazil's GDP growth in 2013 by the Brazilian Central Bank, in its latest Focus Report [30]. (e) Value needed to yield an annual rate of economic growth of 4.00% during 25 years. (f) Value needed to yield an annual economic growth rate of 5.00% during 25 years.

Table 8. Optimal (*) results‡ from scenario analysis

Scenario	$\hat{\varepsilon}^*$ (× 10⁻⁴)	$\hat{\lambda}^*$	$\ln\hat{\lambda}^*$	$\hat{k}^*$	$\ln\hat{k}^*$	$\hat{\lambda}^*$% p.a.†	$\hat{k}^*$% p.a.†	Allowed def. rate δ^*% p.a. (Eq.(33))
REDD1	1.637	5.7874	**1.755682**	1.000254	0.000254	7.27520	0.00102	3.1804
REDD2	2.395	2.7958	**1.028114**	1.000372	0.000372	4.19819	0.00149	6.7997
$\ln\hat{\lambda}=\hat{k}$	2.453	2.7193	**1.000381**	**1.000381**	0.000381	4.08267	0.00153	6.8623
CDM	2464.70	1.3832	**0.324368**	1.383156	**0.324368**	**1.30592**	**1.30592**	7.8350
BAU1	5205.61	1.3829	0.324176	**1.809251**	0.592913	1.30515	**2.40000**	7.8352
BAU2	10715.70	1.3828	0.324087	**2.665836**	0.980518	1.30479	**4.00000**	7.8353
BAU3	15350.53	1.3827	0.324062	**3.386355**	1.219754	1.30469	**5.00000**	7.8353

(†) Growth rates per annum (p.a.). Figures under these column captions respectively come from $\left[\left(\sqrt[25]{\hat{\lambda}^*}\right) - 1\right] \times 100$ and $\left[\left(\sqrt[25]{\hat{k}^*}\right) - 1\right] \times 100$. (‡) Provided by GAMS-IDE 24.1.2.

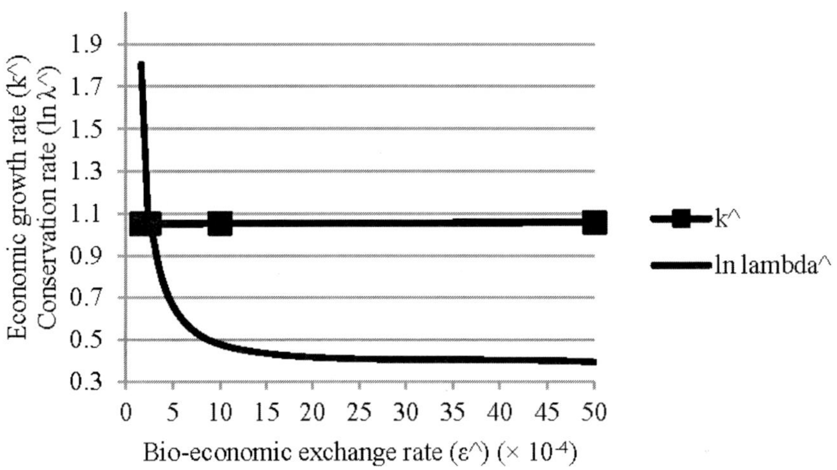

Figure 5. Aggregate equilibrium for conservation and deforestation in the Brazilian Legal Amazon

11. REMOVAL TRADE

Now, the optimal values for ε brought to light in Table 8 can be used in Table 9 to define, as in Figure 2.b, the amount of removal loans across the space (Z) and over time (M). Of course, this previously requires that both Z and M are estimated as long-run functions of ε. However, as shown in Table 4, it is precisely ε that arises from observable Z and M; not the other way around. So, what must be estimated first and foremost is an equation in which ε works as a long-run function of observable Z and M.

$$\begin{array}{ll} \hat{\varepsilon}(Z) = \exp(1.428 \times Z) & \\ \text{t-stat.} & 26.544 \\ \text{sig.t.} & 0.0000 \end{array} \tag{34}$$

$$\begin{array}{lll} \hat{\varepsilon}(M) = 6.643 M^{0.926} & & \\ \text{t-stat.} & 67.364 & 35.514 \\ \text{sig.t.} & 0.0000 & 0.0000 \end{array} \tag{35}$$

Then, the outcomes for ˆε(*Z*) and ˆε(*M*), in Eqs. (34) and (35), are used as inputs (independent variables) to obtain the equations for Z and M as long-run functions of ε:

$$\begin{array}{lll} \hat{Z}(\hat{\varepsilon}) = \exp\left(0.895 - \dfrac{4.154}{\hat{\varepsilon}}\right) & & \\ \text{t-stat.} & 259.616 & -133.481 \\ \text{sig. t.} & 0.0000 & 0.0000 \end{array} \tag{36}$$

$$\begin{array}{lll} \hat{M}(\hat{\varepsilon}) = 0.129 \hat{\varepsilon}^{1.08} & & \\ \text{t-stat.} & - & - \\ \text{sig. t.} & - & - \end{array} \tag{37}$$

Eqs. (36) and (37) define, respectively, the optimal path of *conservation* (REDD) and *compensation* (CDM) to be loaned out in the long run (Figure 6.a). Since conservation is the surplus of carbon removal services, they can be *exported* (lent) to somewhere else. On the other hand, when these missing services have nevertheless been already used, some compensation is due. However, compensating for environmental services that were already lacking before implies *importing* (borrowing) them from somewhere. In this sense, *conservation* (exports) defines an *excess supply* of removal services, while *compensation* (imports) is caused by an *excess demand* for removal services. If these services are to be loaned out, the sink yielding conservation *lends* them, whereas the sink owing compensation *borrows* them.

Table 9. Scenario analysis[b] for conservation and compensation loans

Scenario	$\hat{\varepsilon}$ ($\times 10^{-4}$)	$\ln \hat{Z}$ ($\times 10^{2}$)	$\ln \hat{M}$	$\beta = \ln \hat{Z} - \ln \hat{M}$	$\hat{Z}$ (GtC) Eq.(36)[a]	$\hat{M}$ (GtC) Eq.(37)	$B = \hat{Z} - \hat{M}$
Min B	0.940	-3.523403	-2.114586	-1.408817	0.0295	0.1207	**-0.0912**
REDD1	1.637	-1.642841	-1.515764	-0.127077	0.1934	0.2196	-0.0262
B = 0	**1.799**	-1.413446	-1.413446	0.000000	**0.2433**	**0.2433**	**0.0000**
REDD2	2.395	-0.839176	-1.104520	0.265344	0.4321	0.3314	0.1007
$\ln \hat{\lambda} = \hat{k}$	**2.453**	-0.798110	-1.078638	0.280528	0.4502	0.3401	0.1101
Max B	**5.454**	0.133360	-0.215882	0.349242	1.1427	0.8058	**0.3368**
B = 0	**10.621**	0.503872	0.503872	0.000000	**1.6551**	**1.6551**	**0.0000**
CDM	2464.70	0.893315	6.386668	-5.493353	2.4432	593.87	-591.43
BAU1	5205.61	0.894202	7.194149	-6.299947	2.4454	1331.62	-1329.17
BAU2	10715.70	0.894612	7.973880	-7.079267	2.4464	2904.10	-2901.66
BAU3	15350.53	0.894729	8.362075	-7.467345	2.4467	4281.57	-4279.12

(a) Actually, these figures are in hundreds of GtC, so that, as indicated by Table 4, their order of magnitude can be compared to that of $\hat{M}$. (b) Results provided by GAMS-IDE 24.1.2.

In any event, the demand for these loans, as shown in Figure 2, lies behind disaggregate (Figure 3) — periodical — and aggregate (Figure 5) — long run — removal needs. As long as *B* (last column of Table 9) stands for the balance of carbon trading loans, it is interesting to note that a stringent scenario, such as REDD1 (Table 7), yields more ecological debt ($\hat{M}$) than ecological credit ($\hat{Z}$). This balance only turns out to be positive in the interval where $1.799 \leq \hat{\varepsilon} \leq 10.621$. *B* reaches its maximum in this interval, when $\hat{\varepsilon} = 5.454$. Later, it starts dropping, reaching zero again. From then on, it becomes increasingly negative, and neither the CDM nor BAU scenario can turn it over anymore.

Actually, CDM and BAU1 are but very loose scenarios. When they come into play, the ecological credit region ($1.799 \leq \hat{\varepsilon} \leq 10.621$) has already been left behind. Therefore, in spite of the high values for $\hat{\varepsilon}$ set by CDM and BAU scenarios, these prices appear to be too low to prevent $B < 0$. Although CDM requires that $\ln\lambda = \ln k$ (Table 7 and Table 8), existing ecological conditions demand, rather, that $\ln\lambda > \ln k$.

If these conditions are overlooked, no further rise in $\hat{\varepsilon}$, as in the BAUs scenarios, can prevent *B* from keep falling. That likely occurs because the environmental capacity of providing removal services has been already overshot. As it turns out, $\hat{}\varepsilon$ cannot be set like usual prices can; instead, it is set by ecological conditions, as though ecology were guiding the economy [32].

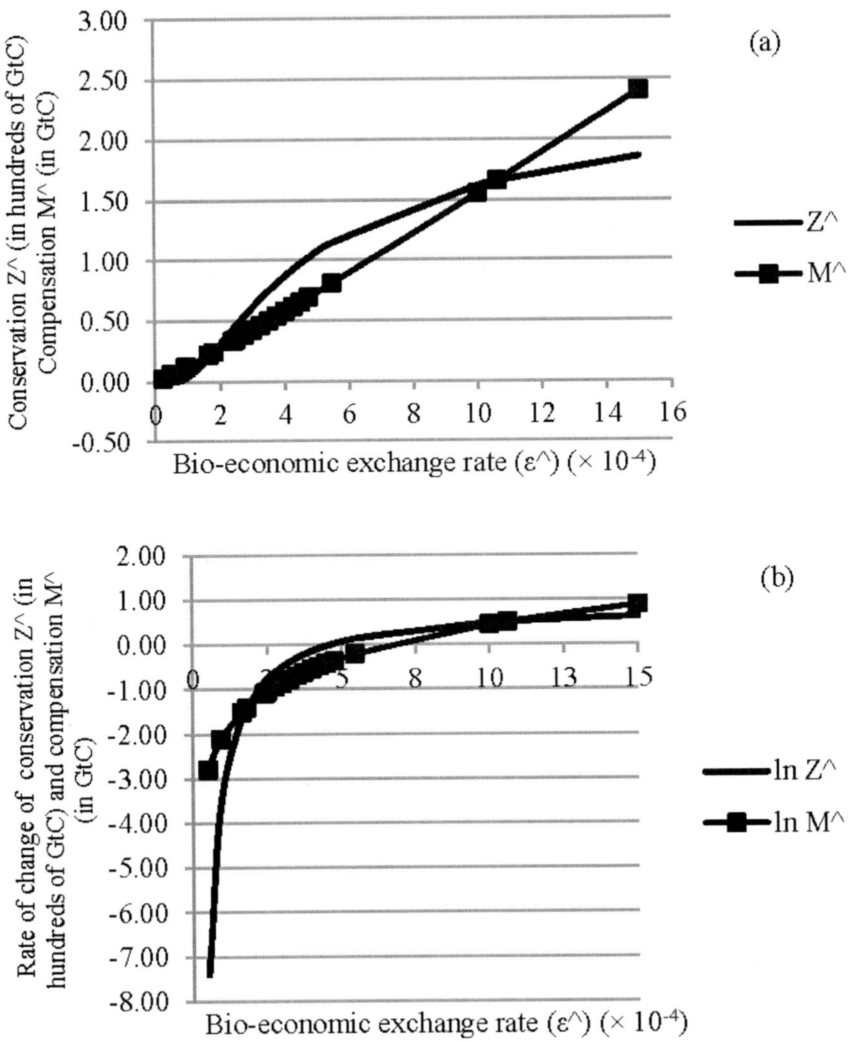

(†) Determined by GAMS-IDE 24.1.2.

FIGURE 6. Optimal path[†] for conservation ($\hat{}Z$) — REDD — and compensation ($\hat{}M$) — CDM — loans (a) and rates (b) in the long run

12. TARGET SETTING

Table 10 focuses on scenario REDD1, from Table 8, in order to demonstrate how an economy can be ecologically guided. Of course, the figures in Table 10 do not account for the real picture. Rather, they relate to a 25-year deforestation period (1988-2012), to show how things would look if it had followed out the optimal path suggested by the conservation scenario REDD1. The percentages in the last column of Table 10 were merged in such a way that the 25-year period is divided into 7 time lags. By so doing, REDD1 reduction path, although applying to a different time period (1988-2012), can be compared to the same time length of the deforestation reduction programme tailored by [29]. The comparison is shown in Figure 7.

Nevertheless, it can still be asked why the percentage deforestation rate along the optimal path in Table 10 (15.59%) does not match that in the last column of Table 8 (3.1804%). Even though the former is referred to in deforested area terms, in km^2, while the latter, in carbon biomass, in GtC, the proper calculation (Eq. (39)) shows that the values would be equal. Accordingly, the percentage deforestation rates displayed in Table 10 can be reckoned either in carbon biomass (GtC) or in area units (km^2). Actually, the underlying reason for the mismatch between deforestation rates in Table 10 and in Table 8 is that the rates in the former are *bounded* (Eq. (33)), whereas those in the latter are *unbounded*. This difference can be grasped from combining Eqs. (1) and (3) to yield:

$$X_t = u_t + \lambda_t u_t = (1 + \lambda_t) u_t \tag{38}$$

From Eq. (38) and from the definition of the deforestation rate (δt) given in note *i* of Table 10, it turns out that:

$$\delta_t = \frac{u_t}{X_t} = \frac{u_t}{(1 + \lambda_t) u_t} = \frac{1}{1 + \lambda_t} \tag{39}$$

If λt, in Eq. (39), takes on whatever optimal value displayed in Table 8, then this equation can be compared with Eq. (33). Yet now, unlike in the former equation, in the latter $\hat{\lambda}^*$ has an upper bound ($\bar{\lambda}$= 7.975). Since this ceiling stands for a mean value for the whole period (1988-2012), it is not surprising that δ^* in Eq. (33) is lower than that arising from Eq. (39) — displayed in the antepenult column of Table 10.

The deforested area in Table 10 is given by Eq. (39). Its size follows closely — yet throughout a 25-year period — the size of forestland that, according to Table 3, has been cut down since 2005 (t = 18). However, while in the observed data of Table 3, the size of deforested land keeps on getting bigger, in the optimal path of Table 10, it increasingly shrinks.

Table 10. Deforestation targets and rates from optimal results in the REDD1 scenario ($\hat{\lambda}^*$=5.7874) (Table 8)

t	Year	u^* (GtC)	v^b (GtC)	X_t (GtC)	Z_t^c (GtC)	M_t^d (GtC)	ε ($\times10^{-4}$)	u^* (10^3 km^2)	v^f (10^3 km^2)	X_t^g (10^3km^2)	Remaining deforest[h] (10^3 km^2)	Def. rate[i] (%)	Def. rate[j] (km^2)	Deforest. targets[k] (%)
1	1988	30.49	176.44	206.93	145.95	0.023890	1.637	702.46	3802.57	4505.03	16806.21	15.59	—	4.01
2	1989	30.48	176.38	206.85	145.90	0.023881	1.637	702.21	3801.21	4503.42	16104.00	15.59	251.55	4.18
3	1990	30.47	176.34	206.81	145.87	0.023877	1.637	702.07	3800.45	4502.52	15401.93	15.59	140.54	4.36
4	1991	30.46	176.31	206.78	145.85	0.023873	1.637	701.96	3799.83	4501.79	14699.98	15.59	114.06	4.56
5	1992	30.46	176.28	206.74	145.82	0.023868	1.637	701.82	3799.07	4500.89	13998.16	15.59	140.54	4.77
6	1993	30.46	176.28	206.74	145.82	0.023868	1.637	701.82	3799.07	4500.89	13296.35	15.59	0.00	5.01
7	1994	30.45	176.20	206.65	145.76	0.023858	1.637	701.51	3797.43	4498.94	12594.83	15.59	303.49	5.28
8	1995	30.43	176.13	206.57	145.70	0.023848	1.637	701.24	3795.93	4497.17	11893.60	15.59	275.99	5.57
9	1996	30.42	176.08	206.51	145.66	0.023841	1.637	701.03	3794.83	4495.86	11192.57	15.59	203.68	5.89
10	1997	30.42	176.05	206.47	145.63	0.023837	1.637	700.90	3794.12	4495.02	10491.67	15.59	132.40	6.26
11	1998	30.41	176.00	206.41	145.59	0.023831	1.637	700.72	3793.16	4493.88	9790.94	15.59	177.03	6.68
12	1999	30.40	175.96	206.36	145.55	0.023825	1.637	700.55	3792.21	4492.75	9090.40	15.59	175.77	7.16
13	2000	30.40	175.91	206.31	145.52	0.023818	1.637	700.36	3791.20	4491.56	8390.04	15.59	185.62	7.70
14	2001	30.39	175.87	206.25	145.48	0.023812	1.637	700.18	3790.20	4490.38	7689.86	15.59	185.00	8.35
15	2002	30.38	175.81	206.19	145.43	0.023805	1.637	699.96	3789.02	4488.98	6989.90	15.59	217.88	9.10
16	2003	30.37	175.75	206.11	145.38	0.023796	1.637	699.70	3787.63	4487.33	6290.20	15.59	257.12	10.01
17	2004	30.35	175.68	206.03	145.32	0.023786	1.637	699.42	3786.12	4485.54	5590.78	15.59	279.28	11.12
18	2005	30.35	175.63	205.97	145.28	0.023780	1.637	699.23	3785.08	4484.31	4891.55	15.59	191.93	12.51
19	2006	30.34	175.59	205.93	145.25	0.023775	1.637	699.09	3784.30	4483.39	4192.46	15.59	143.69	14.29
20	2007	30.34	175.56	205.90	145.23	0.023771	1.637	698.97	3783.66	4482.63	3493.49	15.59	117.44	16.67
21	2008	30.33	175.53	205.86	145.20	0.023767	1.637	698.84	3782.95	4481.79	2794.65	15.59	131.49	20.00
22	2009	30.33	175.51	205.84	145.18	0.023764	1.637	698.76	3782.54	4481.30	2095.89	15.59	76.02	25.00
23	2010	30.32	175.49	205.82	145.17	0.023762	1.637	698.69	3782.15	4480.84	1397.20	15.59	71.29	33.34

t	Year	u^a (GtC)	v^b (GtC)	X_t (GtC)	Z_t^c (GtC)	M_t^d (GtC)	ε ($\times 10^{-4}$)	u^e (10^3 km²)	v^f (10^3 km²)	X_t^g (10^3km²)	Remaining deforest.[h] (10^3 km²)	Def. rate[i](%)	Def. rate[j] (km²)	Deforest. targets[k] (%)
24	2011	30.32	175.48	205.80	145.16	0.023759	1.637	698.62	3781.80	4480.43	698.58	15.59	65.36	50.00
25	2012	30.32	175.46	205.78	145.15	0.023758	1.637	698.58	3781.55	4480.13	0.00	15.59	46.55	100.00
Total								**17508.67**	**94778.08**					

(a) $u_t = X_t \div (1 + \hat{\lambda}^*)$. (b) $v_t = X_t - u_t$. (c) Eq. (2). (d) Eq. (14): $M_t = \varepsilon \times Z_t$. (e) $(u_t \div 434) \times 10^{-4}$, according to note e in Table 3. (f) $(v_t \div 464) \times 10^{-4}$, according to note *d* in Table 3. (g) X_t (in km²) = u_t (in km²) + v_t (in km²). However unrealistic it may sound, the total area "shrinks" to account for the loss of forestland implied by substituting the carbon stored in deforested sites for that stored in natural forests. (h) For $t = 1$ (1988), $\Sigma_t u_t$ (in km²) = 17508.67 − u_1 (in km²); for $t > 1$, (remaining deforestation)$_{t-1}$ − u_t (in km²). (i) u_t (in km²) ÷ X_t (in km²). (j) $(u_{t-1} - u_t)$ (in km²) × 10³. (k) For $t = 1$ (1988), u_t (in km²) ÷ $\Sigma_t u_t$ (in km²); for $t > 1$, u_t (in km²) ÷ (remaining deforestation)$_{t-1}$.

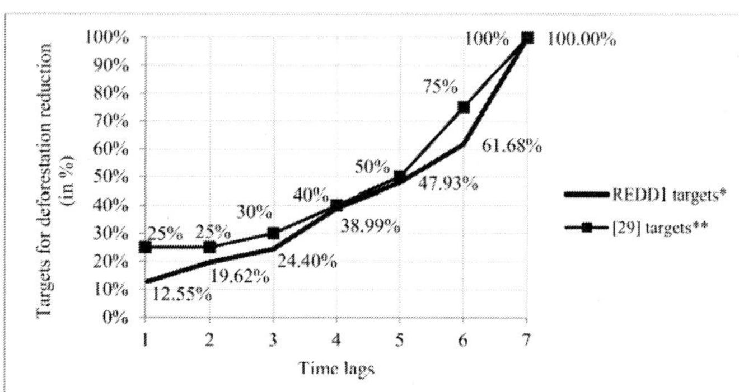

(*) During 1988-2012. (**) From 2007 to 2015 [29].

FIGURE 7. Targets for reducing deforestation in the Brazilian Legal Amazon

13. CONCLUSION

The analysis carried out so far has demonstrated that, where policy climate and deforestation are concerned, carbon conservation (REDD) and compensation (CDM) entail a trade-off that cannot be overcome by monetary mechanisms. Instead of money, the underlying variable which forest value rests upon is the bio-diversity ratio (λ). Although it is typically a space-based measure, this ratio also accounts for forestland distribution over time. When λ is affected by the demand of removal stocks (*h*) set off by the emissions growth from the economy, the amount of compensation (M) for these emissions over time is assumed to feed on the conservation (Z) of carbon savings carried out at each period.

Conservation surplus can be loaned out to afford compensation demands from an economy producing growing emissions. However, such demands must be halted somewhere, otherwise the supply of removal forest stocks ($G(Xt)$) will hit a ceiling and fall short of delivering enough conservation. As shown by Figure 6 and Table 9, it is likely there also is a biophysical limit to conservation, which should prevent this environmental service from being further encouraged by incentives such as monetary payments (PES). When conservation is driven too far, as in the REDD1 scenario (Table 8 and Table 9), it might end up turning the balance of carbon trading loans negative. On the other hand, when it is traded off against compensation, as in the CDM scenario, it deteriorates the balance of carbon trading loans even further. Quite often, economy-wise price setting and policy-making grow apart from ecological conditions.

Figure 2, Figure 3, Figure 5 and Figure 6 show how ecology could economically guide the economy [32]. If this linkage should be fully accomplished, setting targets for reducing deforestation would be grounded in biophysical, rather than in economic or political bearings.

Shifting from a deforestation avoidance approach to a forest stock maintenance one would certainly be a step forward. While the former carries a misleadingly uneconomic meaning, the latter brings forest conservation to the economic foreground. As avoiding deforestation usually implies forgoing profit-making activities, it mistakenly underlies monetary rewards. Thus, a great deal of the REDD mechanism draws heavily upon them. However, from a bio-economic standpoint, they are most likely to become romantic red roses, whose purchase will, rather than mend a broken heart, make money melt into thin air.

NOTES

[1] - The first time REDD came into the UNFCCC (United Nations Framework Convention on Climate Change) agenda was in 2005, when a mechanism for reducing deforestation and forest degradation was proposed by Papua New Guinea, Costa Rica and eight other countries. Two years earlier, Brazilian researchers had already come up with a similar tool backed by international payments for reduced deforestation. Between the Conference of the Parties (COP) in Bali, Indonesia, in 2007 (COP-13), and in Copenhagen, Denmark, in 2009 (COP-15), different REDD proposals emerged. As the scope of the mechanism was getting wider, its abbreviation

was getting longer with the addition of "D's" and plus (+) signs. Chronologically, it started with RED — short for Reducing Emissions from Deforestation; next, it became REDD — when forest Degradation was added; then, it turned out to be REDD+ — so as to encompass conservation and enhancement of forest stocks through Sustainable Forest Management (SFM) techniques; at present, it has often been labelled REDD++, including reforestation and afforestation — that is, reforestation of non-forested areas [1]. Throughout this chapter, simply REDD will be employed, regardless of its several chronological meanings. However, afforestation and reforestation will be considered to fall into forestry-CDM — the branch of the Clean Development Mechanism suited to unnatural forests.

[2] - In monetary economics, this relationship is known as the "quantity theory of money". "The quantity theory is a mechanistic proposition strangely alien to the assumptions of rational maximising behaviour on which classical and neoclassical theories generally rely ... It ignores the effects of the returns to holding money on the amounts economic agents choose to hold ... Money holdings depend ... on the opportunity costs, the expected changes in the value of money and the real yields of other assets into which the same funds could be placed" [11]. Of course, these remarks follow a Keynesian theoretical tradition, within which money is thought to be an asset rather than a token of bank liabilities to current account holders. Taking money as liabilities dates back to the days when bank-notes were "promises to pay", handed over at once as a receipt to depositors who had voluntarily given up gold to the bank, which, in turn, promised to repay them on demand [10].

[3] - Dating back to classical times, the labour theory of value paradoxically stated that only when land (natural resources) is (are) running out is the maximum value reached. Therefore, any natural resource that gets into the market and thus acquires exchange value is dying out or close to extinction. So, exchange value lays bare that there is less of that natural resource than there was when it held no value at all! Later on, the neoclassical utility theory displaced the economic value to an even odder container: consumption. In the "utility world", the economic value was placed in the individual preferences for commodities. As preferences were a subjective matter, the economic value then turned out to rest upon the abstract basis of utility. From then on, the economic analysis has been cut off from its biophysical roots [14]. But as early as 1883, S. Podolinsky pioneered the idea of associating energy with value, as claimed by energy theories of value. These theories aimed at substituting energy for money as the only source of value. However, money is not particularly correlated with energy content, because there is no direct equivalence between low entropy and economic value [13][15]. For instance, the monetary value per unit energy content of a diamond is extremely large compared to the monetary value

per unit energy content of a lump of coal. Nonetheless, if all indirect energy flows were to be tracked down and accounted for, the discrepancy between diamonds and coal might not be so great [15]. By and large, economists have rebutted these energy evaluation methods because of the fear that economics might end up turning into a branch of thermodynamics [14]. Furthermore, as Georgescu-Roegen [13] once pointed out, the economic process has only two flows: an input flow of low entropy and an output flow of high entropy, namely, waste. If the balance sheet of value should be set on the basis of these inputs and outputs, one "would arrive at the absurd conclusion that the value of the low entropy flow on which the maintenance of life itself depends is equal to the value of the flow of waste, that is, zero" (p. 284). This paradox only vanishes by acknowledging that the true "product" of the economic process is not a material flow, but a psychic flux — the enjoyment of life. Thus, the economic value has both psychic (neoclassical) and physical (classical) roots. An entropy-oriented, energy theory of value would only account for the supply side of the process and neglect the corresponding demand side [9]. That is why the proposition of a shadow price for natural resource inputs, on one hand, and for the waste sinking capacity of the environment, on the other, results from an economic rather than from an energy theory of value [14].

[4] - Although nature might well have other values — existence and bequeath — than use ones, non-use values are harder to estimate. Obviously, the use value is the one arising from the real — direct, indirect or optional — use of a given resource, whether in the present or in the future. On the other hand, the existence value is simply related to the existence of specific riches. The bequeath value measures the benefit that any individual obtains from knowing that, in the future, other people will also be able to benefit from the resource they have been left [17]. First and foremost, both non-use values essentially depend on estimating the preferences of future generations, which is not that easy to foreshadow. Moreover, non-use values resemble very much the intrinsic value of nature, which was claimed by Deep Ecology followers. According to them, nature had to be preserved for itself, rather than for satisfying the well-being of present and future generations. The intrinsic value is fully separated from any use value, even in the future [14]. Once these non-use concepts are difficult to apply, environmental goods and services are taken up thereafter in their usefulness sense.

[11] - All statistical estimations were performed in SPSS Statistics 17.0.

REFERENCES

1. Linhares-Juvenal T. REDD and the Challenge of Protecting the Global Forest Cover. In: Serôa da Motta R., Hargrave J., Luedemann G., Gutierrez MBS. (eds.) Climate Change in Brazil — Economic, Social and Regulatory Aspects. Brasília: IPEA; 2011. p319-328.
2. Kanninen M., Murdiyarso D., Seymour F, Angelsen A., Wunder S., German L. Do Trees Grow on Money? The Implications of Deforestation Research for Policies to Promote REDD. Bogor, Indonesia: CIFOR; 2007.
3. Giacomelli Sobrinho V. Carbon Bio-Economics and Forests — Getting the BESF out of Climate Policy. In: Kheradmand H. (ed.) Climate Change — Socioeconomic Effects. Rijeka: InTech; 2011. p401-424.
4. Martin RM. Deforestation, Land-Use Change and REDD. Unasylva 2008; 59(230): 3-11.
5. Angelsen A., editor. Analyzing REDD+ — Challenges and Choices. Bogor, Indonesia: CIFOR; 2012.
6. Polanyi K. A Grande Transformação — As Origens de Nossa Época. Rio de Janeiro: Campus; 2000.
7. Whiteman A. Money Doesn't Grow on Trees: A Perspective on Prospects for Making Forestry Pay. Unasylva 2003; 54(212): 3-10.
8. Prince's Charities' International Sustainability Unit. Interim REDD+ Finance — Current Status and Ways Forward for 2013-2000. London: The Prince of Wales' Charitable Foundation; 2012. http://www.pcfisu.org/the-princes-rainforests-project/interim-redd-finance (accessed 8 September 2013).
9. Daly HE., Farley, J. Economia Ecológica — Princípios e Aplicações. Lisboa: Instituto Piaget; 2004.
10. Soddy F. The Rôle of Money — What It Should Be, Contrasted with What It Has Become. London: Routledge; 1934.
11. Tobin J. Money. In: Durlauf SN., Blume LE. (eds.) Monetary Economics. 2nd. ed. Hampshire, UK: Palgrave Macmillan; 2010. p224-241.
12. Instituto do Homem e Meio Ambiente da Amazônia. IMAZON: Mapas: Desmatamento Acumulado na Amazônia Legal 2007-2012: Desmatamento na Amazônia Legal até 2012. http:// www.imazon.org.br/mapas/desmatamento-acumulado-2007-2010/desmatamento-amazonia-legal-acumulado-ate-2012/view (accessed 25 September 2013).
13. Georgescu-Roegen N. Entropy, Value, and Development. In: Georgescu-Roegen N. The Entropy Law and the Economic Process. Cambridge, US, London, UK: Harvard University Press; 1999. p276-315.
14. Faucheux S., Noël JF. Economia dos Recursos Naturais e do Meio Ambiente. Lisboa: Instituto Piaget; 1995.
15. Ayres RU. On the Life Cycle Metaphor: Where Ecology and Economics Diverge. Ecological Economics 2004; 48: 425-438.
16. Costanza R. Value Theory and Energy. Encyclopedia of Energy 2004; 6: 337-346.
17. Pearce D., Moran D. O Valor Económico da Biodiversidade. Lisboa: Instituto Piaget; 1994.
18. Daly HE. The Economics of the Steady State. The American Economic Review 1974; 64: 15-21.
19. Clark CW. Mathematical Bio-Economics — the Mathematics of Conservation. 3rd. ed. Hoboken, US: John Wiley & Sons; 2010.
20. Wackernagel M., Rees W. Our Ecological Footprint — Reducing Human Impact on the Earth. Gabriola Island, Canada: New Society Publishers; 1996.
21. Fearnside PM. Forests and Global Warming Mitigation in Brazil: Opportunities in the Brazilian Forest Sector for Responses to Global Warming under the "Clean Development Mechanism". Biomass and Bioenergy 1999; 16: 171-189.

22. Fearnside PM. Greenhouse Gases from Deforestation in Brazilian Amazonia: Net Committed Emissions. Climatic Change 1997; 35: 321-360.
23. Instituto Nacional de Pesquisas Espaciais. INPE: Amazonia: PRODES: Taxas Anuais 1988 a 2012. http://www.obt.inpe.br/prodes/prodes_1988_2012.htm (accessed 20 September 2013).
24. Common M. Environmental and Resource Economics — an Introduction. 2nd. ed. New York: Addison Wesley Longman; 1996.
25. Ayres RU. The Second Law, the Fourth Law, Recycling and Limits to Growth. Ecological Economics 1999; 29: 473-483.
26. Houthakker HS. The Permanent Income Hypothesis. The American Economic Review 1958; 48: 396-404.
27. Meghir C. A Retrospective on Friedman's Theory of Permanent Income. The Economic Journal 2004; 114: F293-F306.
28. Perman R., Ma Y., McGilvray J., Common M. Natural Resource and Environmental Economics. 3rd. ed. Harlow, England: Pearson Education; 2003. http://rapidlibrary.com/files/ perman-natural-resource-and-environ mental-economics-3rd-edition-pdf_ulczebmccni89on.html (accessed 19 November 2012).
29. Instituto Socioambiental. ISA, Greenpeace, Instituto Centro de Vida. ICV, The Nature Conservancy. TNT, Conservação Internacional, Amigos da Terra - Amazônia Brasileira, Instituto do Homem e Meio Ambiente na Amazônia. IMAZON, WWF-Brasil. Pacto pela Valorização da Floresta e pelo Fim do Desmatamento na Amazônia Brasileira. http://www.socioambiental. org/banco_imagens/pdfs/doc-pacto%20desmatamento%20zero%20SUM%20ONGs%20FINAL. pdf (accessed 25 September 2013)
30. Banco Central do Brasil. BCB. Sistema de Metas para a Inflação: Sala do Investidor: Focus: Focus — Relatório de Mercado. http://www.bcb.gov.br /pec/GCI/PORT/readout/R20130920.pdf (accessed 25 September 2013).
31. Martins H., Fonseca A., Souza Jr. C., Sales M., Veríssimo A. Boletim do Desmatamento (SAD) Agosto de 2013. Boletim Transparência Florestal da Amazônia Legal 2013: 12. http://www. imazon.org.br/publicacoes/transparencia-florestal/transparencia-florestal-amazonia-legal/boletim-do-desmatamento-sad-agosto-de-2013-1 (accessed 25 September 2013).
32. Giacomelli Sobrinho V. Bio-economics and ecosystem services: how ecology can economically guide the economy. In: Bürgi M. (ed.) Frontiers in Historical Ecology: Abstracts: proceedings of the International Conference of the International Union of Forest Research Organisations (IUFRO), Swiss Federal Institute for Forest, Snow and Landscape Research (WSL), 27 August - 2 September 2011, Birmensdorf, Switzerland. Birmensdorf: WSL; 2011. http://www.wsl.ch/ publikationen/ pdf/11114.pdf

CHAPTER 16

Risk Factors among Greenhouse Farmers in Gaza Strip

Yasser El-Nahhal

Faculty of Science, The Islamic University of Gaza, Gaza, Palestine

ABSTRACT

Greenhouse farmers are exposed to many risk factors during work. This paper characterized the risk factors and discussed their health impacts among greenhouse farmers. Risk factors under greenhouse conditions emerged from pesticide application and greenhouse gas emission from soil during day. Risk factors among farmers were determined by recording toxicity symptoms, acute and chronic obstructive pulmonary disease (COPD). This study targeted 67 greenhouses in Gaza Strip and found 169 farmers doing many agricultural activities without protective clothes. We documented nine highly toxic insecticides and five moderately toxic fungicides being applied for insect and fungal control. A large number of farmers (55%) revealed that they visited respiratory health clinic quite two to three times a month immediately after insecticide application. About 8.88% (n = 15) of farmers stated that they had unrecoverable lung disease due to working in greenhouses during winter season. Personal health records of farmers showed variety of cholinergic symptoms, severe reduction on acetyl choline esterase activity (ACHE), many cases of COPD and some pneumonia cases. Classification of farmers according to risk factors indicates that 18% of farmers are at high risk due to extreme to insecticides. It is recommended to seasonally investigate the activity of ACHE of greenhouse farmers and to routinely check their chests and lungs to avoid any potential health risk factors due to works in greenhouse environment.

Keywords: Risk Factors, Insecticides, Acetylcholine Esterase, Chronic Obstructive Pulmonary Disease

1. INTRODUCTION

A large number of farmers are doing many agricultural activities in greenhouses in Gaza Strip to meet the population needs from vegetables and ornamentals. Greenhouse construction in Gaza Strip is progressively increased due to ornamental and medical plant production. Farmers in greenhouses are doing the following agricultural activities: manure application, manual pesticide spray, fruit collection sorting and separation, storage and post-harvest activities. It has been shown that application of pesticides in greenhouse caused many environmental problems [1] [2] [3] [4] . Moreover, health risks such as cancer cases [5] , health disabilities [6] and biochemical changes [7] have been documented with pesticides application. In addition, drinking of contaminated water [8] [9] , ingestion of contaminated food [3] , working with soil components [10] and inhalation of contaminated air [11] may become extreme health risk due to indirect exposure to pesticides. Application of pesticides was found to create risks to non-target organism such as cyanobacteria [12] [13] [14] [15] , plant [16] , fish [17] [18] . Many authors tried hard to reduce health risks associated with pesticide application. This included development of ecologically acceptable organoclay formulations of pesticides [19] - [29] . Health risk among farmers may be emerged due to occupational and/or accidental exposure to direct insecticide solution or vapor. In addition, misuse or violation with safety measures of insecticides may lead to appearance of health risk among farmers. So far, health risk among farmers may be indicated by measuring the toxicity symptoms, activity of acetylcholine esterase in blood serum, chest and pulmonary disease. Several authors [30] [31] [32] indicated the reduction of erythrocyte acetylcholinesterase as a potential health risk in greenhouse workers due to occupational pesticide exposure. Similarly, they found methamidophos-treated groups had significantly lower butyrylcholinesterase and paraoxonase activities. In a different study, Liu et al. [33] evaluated the prevalence of farmer's lung disease (FLD) and the potential risk factors for FLD among Chinese greenhouse farmers. They revealed that prevalence of FLD among these farmers was 5.7% (308/5,420) and the risk factors for

FLD were years of age, shorter time interval for re-entry greenhouse, ventilation frequency of greenhouse. Earlier, Jurewicz et al. [34] reviewed the evidence on the association between working in greenhouses and the occurrence of and respira- tory disorders. They indicated that greenhouse exposure was associated with an increased risk of respiratory disorders, sensitization to allergens and skin reactions. Previously, Terho et al. [35] studied the prevalence and incidence of chronic bronchitis and farmer's lung with respect to age and sex. They found that the prevalence of chronic bronchitis was 8.0%, and the mean annual incidence was 2017 per 100,000 farmers. So far, it is necessary to characterize the risk factors among greenhouse farmers in Gaza Strip. Accordingly, this study was initiated to answer the following questions: 1) what are risk factors among greenhouse farmers in Gaza? 2 what are health impacts of these factors? and 3) what can be done to reduce the impacts of risk factors?

2. MATERIALS AND METHODS

2.1. Data Collection

About 67 greenhouses were visited for data collection and 169 farm workers were interviewed in the field via direct contact, face to face interview. The data collection included knowledge of risk factors, working conditions under greenhouse, temperature of greenhouses, humidity, manure application and pesticide knowledge, toxicity symptoms and protective measures related to pesticide application in greenhouse.

2.2. Determination of Risk Factor Associated with Pesticide Application

The greenhouse farmers were convinced to donate blood samples (10 ml each) in a private clinic. Blood samples were collected in heparin containing tubes and analyzed in the same day for Acetylcholine esterase activity (ACHE) using a spectrophotometer according to previous method [36] .

2.3. Analysis of Personal Health Records of Farmers

Personal health records of farmers having acute and chronic lung disease were collected and carefully summarized.

2.4. Statistical Analysis

Insecticides and fungicides applied in greenhouse were collected along with their properties and presented in a separate table. Toxicity symptoms were arranged according to locations, total and percentage values of symptoms were calculated. Risk factors according to toxicity among farmers were categorized into four cases according to the activity of Acetyl cholinesterase. Average and standard deviation of each group was calculated. These groups are extreme risk, high risk, and moderate risk.

3. RESULTS

Risk Factors Emerged from Pesticide Application

The study targeted 67 greenhouses distributed in different governorate in Gaza Strip and found 169 farmers doing all agricultural activities. Moreover, about 20 non-farmer individuals were taken as control sample. It was found that majority of farmers (70%, n = 118) are male in the age group of 20 - 35 years old, having 12 - 14 years of education. Farmers indicated that they have about 5 years working experience with greenhouses. About 65% (n = 110) of farmers indicated that insecticides are not risky for human beings. In addition, 60% (n = 100) of farmers stated that they did not comply with safety measures such as wearing protective clothes, gloves, glasses or long shoes during pesticides application. Majority of greenhouse farmers revealed that insecticides and fungicides were applied while the greenhouse is totally closed and the application rate was 1.5 folds higher than the recommended rate. They also indicated that pesticide spray processes were applied mechanically using high volume techniques at the several cases of pest control whereas manual applications were occasionally applied by farmers using low volume technique. Farmers revealed that both techniques resulted in face and skin contamination. Some farmers (31%, n = 52) revealed that their internal clothes have the same odor of pesticide solution at the end of working day. We found about 11 insecticides are commonly used in greenhouses for insects, nematodes and mites control, with application rates ranged from 1 - 2.5 kg/ha (Table 1). The applied insecticides included Nemacur, Methomyl and Vertimic. These insecticides are extreme risky materials due to their high toxicity and low LD_{50} values (6 - 34 mg/kg) [37] . Moreover, Pirate, Malathion, and Pyrethrine are moderately risky insecticides due to higher LD_{50} values than Nemacur, Methomyl and Vertimic [38] . These applied insecticides belonged to organphosphorus compounds (OPs) carbamate compounds, and organo-chlorine compounds (OC). Application of OPs and carbamate compounds are the most risky compounds due to irreversible binding to ACHE. In addition fungicides with different chemical groups (diazole, and triazoles) were applied in greenhouses to control Botrytis cinerca, Oidiumspp, Oidiopsissicula, Alternariacucamerina, Pseuduperonasporacubenis of cucumber, melon and tomato. These fungicides y have LD_{50} values ranged from 700 - 5000 mg/kg and considered as unsafe materials [37] . More risk of pesticides application may be emerged from their water solubility which ranged from 0.1 - 2000 mg/l (Table 1). This factor enables applied pesticides to move quickly in soil profile and reach the ground water creating threats to human life. Leaching of pesticides in soil has been studied [39] [40] [41] [42] . These reports stated high risk to ground water. In addition, it was found that greenhouse temperature usually in the range of 23°C - 35°C and high relative humidity (50% - 65%) during winter season due to close

Table 1. List of insecticides and fungicides applied in greenhouses.

Serial #	Pest	Pesticides	Rate g/ha	Freq.	LD_{50} mg/kg	Sol. mg/l	$K_{ow}logP$
1	*Tetranychus urtica*	[1]Smash (quinalphos)	150 - 200	57	71	17.8	4.44
		[2]Pirate (chlorfenapyr)	40	18	441	Ins	4.83
2	*Aphidoidae*	[1]Malathion	200	66	1375	145	2.75
		[3]Methomyl (lannate)	50	56	34	*57.9	0.093
3	*Bemisia tabaci*	[4]Pyrethrins	60 - 100 ml		1030	0.2	5.9
4	*Trips tabaci*	[1]Nemacur (fenamiphos)	200	10	6	400	3.3
		[4]Cymbush (cypermethrin)	150	37	250	0.004	6.6
5	*Spodoptra littoralis*	[1]Dursban (chlorpyrifos)	200	56	135 - 165	1.4	4.7
		[5]Vertimic (abamectin)	150	29	10	7-10	4.4
		[1]Dimethoate	200	37	387	*23.3	0.704
6	*Botrytis cinerca* Oidium spp. *Oidiopsis sicula*	Rovral (iprodione)	100	12	2000	13	3
		Antracol (2-naphthyloxyacitic acid)	250	25	1000	-	-
7	*Pseudoperonaspora cubensis* *Pseudomonas syringae*	Daconil (chlorothalonil)	150	25	5000	0.81	2.92
		Bayfidan (triadimenol)	50	100	700	33	3.28
8	*Alternaria cucumerina*	Copper sulfate (tribasic)	10	12	100	Ins	-

1= Organo-Phosphorus compounds; 2= Organo-Chlorine compounds; 3= Oxim carbamate compounds; 4= Pyrothriods; 5= Bio-pesticides; *= g/l.

1= Organo-Phosphorus compounds; 2= Organo-Chlorine compounds; 3= Oxim carbamate compounds; 4= Pyrothriods; 5= Bio-pesticides; *= g/l.

conditions. Under this condition the applied pesticides tend to evaporate and attacking the respiratory enzymes in farmer's lung creating lung disease. This suggestion can be supported by our results which indicated about 26% (n = 36) farmers revealed that they had acute respiratory disease immediately after long working day under greenhouse condition.

Farmers also stated that they recovered after medical treatment. In addition 27.3% (n = 38) of farmers revealed that they have unrecoverable disease during winter season whereas the symptoms disappeared in summer time. Investigation of clinical health records indicated that 18% (n = 25) farmers have pneumonia and 11.5% (n = 16) farmers have chronic obstructive pulmonary disease (COPD). More support to our results can be obtained from the results of Jurewicz et al. [34] who reviewed the evidence on the association between working in greenhouses and the occurrence of and respiratory disorders and indicated that greenhouse exposure is associated with an increased risk of respiratory disorders, sensitization to allergens and skin reactions. Further supports to our hypothesis come from the work of Liu et al. [33] who evaluated the prevalence of farmer's lung disease (FLD) and revealed that prevalence of FLD among these farmers was 5.7% (308/5420) and the risk factors for FLD were years of age, shorter time interval for re-entry greenhouse, ventilation frequency of greenhouse. In addition, the risk factors emerged from insecticide application can be classified to four cases according to ACHE activity: case 1, extreme risk group included farmers got slight ataxia and having the lowest ACHE activity (538 ± 302), case 2, high risk, included farmers got convulsions tremors and having ACHE activity (1829 ± 1048), case 3, moderate risk, included farmers got dizziness, and headache and having ACHE activity (6000 ± 1414) case 4, Low risk, included farmers got vomiting, and diarrhea, Nozai and having (8389 ± 1622) whereas the normal range of ACHE activity was (14100 ± 2798) included farmers not exposed to pesticides in greenhouse (control sample) more details are shown in Table 2.

Further risks in greenhouse may be emerged from the biochemical degradations of applied manures or from the organic matter that originally exists in soil. This process may result in excess emission of greenhouse gases such as NO_x, SO_2, CO_2 and CH_4. It can be suggested that at winter season where the greenhouse windows are usually closed and air exchange between outside and inside are not usually done at a suitable way. This may result in accumulation of humidity, under this condition NO_x, SO_2 and CO_2 tended to react with water molecules (accumulated humidity) and form corresponding acids that may reach farmer's body and make some damages to them. Furthermore, at low relative humidity NO_x, SO_2, and CO_2 may reach the lung and react with the blood producing corresponding acids that increase the acidity in lung. This situation may result considerable inhibition of respiratory enzymes and creating lung disease. These results agree with previous report [43] who found increased acidity in lung due to accumulation of acid gas in human building. This disease may not be recovered by medical treatment as farmers revealed. In summer time the windows of greenhouses are usually opened and air exchange between inside and outside of greenhouse would be at most. Under this condition there are no accumulations of greenhouse gas and farmers do not expose to high risk.

Table 2. Cholinergic and non-cholinergic symptoms among framers exposed to insecticides in greenhouse.

Toxicity symptoms	Locations				Total	%	Ave ± Std (u/L)	State of risk
	Rafah	Kh Y	Gaza	North				
Tremors, Convulsions, Slight ataxia	9	7	13	10	39	23.1	538 ± 302	Extreme
Dizziness, Headache,	14	18	25	23	80	47.35	1829 ± 1048	high
Diarrhea, Vomiting	7	8	11	12	38	22.48	6000 ± 1414	Moderate
Nozai	3	3	3	3	12	7.07	8389 ± 1622	Low
Control group	5	5	5	5	20	00	14100 ± 2798	Normal range

4. DISCUSSION

The presented results in Table 1 clearly shows the applied insecticide or fungicides have different solubility limits in water, different K_{ow}, different vapor pressure (Hunry constant), different application rate and frequency. These data indicate that applied pesticides have different behavior under greenhouse conditions. Moreover, risk factors emerged from toxicity symptoms (Table 2) showed cholinergic and non- cholinergic effects. The explanation of these results is that applied insecticides contained OP and oximcarbamate compounds that regard as strong acetylcholine esterase inhibitors as indicated by WHO [37] and fungicides have high LD_{50} values, indicating non- cholinergic effects. Moreover, toxicity symptoms associated with ACHE levels (Table 2) indicated that farmers exposed to extreme risk due to toxic substances. The explanation of these results is that under greenhouse conditions, high temperature and humidity, farmers may be exposed to pesticides in different ways such as inhalation, ingestion and/or skin absorption. Moreover, farmers revealed that they used high and low volume techniques during pesticides spray process. Under high volume technique, farmers may ingest few drops of spray solution whereas at low volume technique farmers may become in contact with micro-droplets of pesticides active ingredients that penetrate the skin and adipose tissues according to K_{ow} and reach the active site in the body faster than usual and make toxic symptoms as in Table 2. These multi-exposure techniques may enhance the toxicity of pesticides. Moreover, application of insecticides followed by fungicides may expose farmers to binary mixture toxicity. This agrees with recent reports [7] [16] [30] [31] [32] [44] [45] [46] who found decreased levels of acetyl cholinesterase in greenhouse workers due to pesticide exposure. Nevertheless, it can be suggested that greenhouse conditions enhance the evaporation of pesticides due to high temperature and exposing farmer lungs to toxic vapors. This situation might enhance appearance of oxidation stress accordingly elevated levels of liver enzymes were observed. Some authors found elevated liver enzyme in farmers worked in greenhouse condition. Moreover, recent published work [7] found elevated levels of liver enzymes in farmers having long term exposure to pesticides. Nevertheless, exposure to insecticides under greenhouse condition elevated the oxidation stress in liver [47] [48] [49] and result in elevation of ALT and AST levels. It can be suggested that irregular metabolic reaction of amino acid may occur resulting in accumulation of uric acid in the blood.

5. CONCLUSION

Working at greenhouse conditions exposes farmers to different types of health risk. This includes risks from pesticides application and their vapors. More risk factors emerged from greenhouse gas emission and exposing lungs to acidic gas such as NO_x, SO_2 and CO_2 which generate corresponding acids when reacting with blood, lowering its pH and creating lung diseases. Application of pesticides generates 4 types of healthy risks due to level of ACHE activity. Lung disease was the 2nd type of risk factors in greenhouse conditions.

ACKNOWLEDGEMENTS

Special thanks to the Alexander von Humboldt foundation for a Research Fellowship at Leipzig University and BAM institute Germany. I would also like to thank my students at faculty of science for helping me in collecting field data.

ETHICAL STATEMENT

This study was not funded by any organization.

Compliance with Ethical Standards Conflict of Interest: Author declares that he has no conflict of interest.

The study complies with the international ethics issues. Consent form was filled by each farmer participating in the study. Helsinki human right ethics were received before conducting the study.

CITE THIS PAPER

El-Nahhal, Y. (2017) Risk Factors among Greenhouse Farmers in Gaza Strip. Occupational Diseases and Environmental Medicine, 5, 1-10.

REFERENCES

1. Schecter, A., Papke, O., Isaac, J., Hrimat, N., Neiroukh, F., Safi, J. and El-Nahhal, Y. (1997) 2,3,7,8 Chlorine Substituted Dioxins and Dibenzofuran Congeners in 2,4-D, 2,4,5-T and Pentachlorophenol. Organohalogen Compounds, 32, 51-55.
2. Schecter, A., Papke, O., Ryan, J., Furst, P., Isaac, J., Hrimat, N., Neiroukh, F., Safi, J., El-Nahhal, Y., Abu El-Haj, S., Avni, A., Richter, E., Chuwers, P. and Fischbein, A. (1997) Dioxins, Dibenzofurans and PCBs in Human Blood, Human Milk and Food from Israel, The West Bank and Gaza. Organohalogen Compounds, 33, 457-461.
3. Safi, J., Abu Foul, N., El-Nahhal, Y. and El-Sebae, A. (2002) Monitoring of Pesticide Residues on Cucumber, Tomatoes and Strawberries in Gaza Governorates, Palestine. Nahrung/Food, 46, 34-49.
4. El-Nahhal, Y. (2004) Contamination and Safety Status of Plant Food in Arab Countries. Journal of Applied Science, 4, 411-417.
5. Safi, J.M., El-Nahal, Y.Z., Soliman, S.A. and EL-Sebae, A.H. (1993) Mutagenic and Carcenogenic Pesticides Used in Agricultural Environment of Gaza Strip. The Science of the Total Environment, 132, 371-380.
6. El-Nahhal, Y. and Radwan, A. (2013) Human Health Risks: Impact of Pesticide Application. Journal of Environment and Earth Science, 3, 199-209.
7. El-Nahhal, Y. (2016) Biochemical Changes Associated with Long Term Exposure to Pesticide among Farmers in the Gaza Strip. Occupational Diseases and Environmental Medicine, 4, 72-82.
8. El-Nahhal, Y. (2006) Contamination of Groundwater with Heavy Metals in Gaza. Proceeding of the 10th International Water Technology Conference, Alexandria, 20-25 August 2006, 1139-1150.
9. El-Nahhal, Y. and Harrarah, S. (2013) Contamination of Groundwater and Associated Disease: Case Study from Khan Younis Governorate, Gaza, PNA. Journal of Environment and Earth Science, 3, 147-153.
10. Heinze, S., Chen, Y., El-Nahhal, Y., Hadar, Y., Jung, R., Safi, J., Safi, M., Tarchitzky, J. and Marschner, B. (2014) Small Scale Stratification of Microbial Activity Parameters in Mediterranean Soils under Freshwater and Treated Wastewater Irrigation. Soil Biology & Biochemistry, 70, 193-204.
11. Bornstein, R., Safi, J., El-Nahhal, Y., Isaac, J., Rishmawi, K., Luria, M., Mahrer, Y., Ranmar, D. and Weinroth, E. (2001) Transboundary Air-Quality Effects from Urbanization. UJSU Report to USAID-MREC.
12. El-Nahhal, Y., Awad, Y. and Safi, J. (2013) Bioremediation of Acetochlor in Soil and Water Systems by Cyanobacterial Mat. International Journal of Geosciences, 4, 880-890.

13. Safi, J., Awad, Y. and El-Nahhal, Y. (2014) Bioremediation of Diuron in Soil and Cyanobacterial Mat. American Journal of Plant Sciences, 5, 1081-1089.
14. EL-Nahhal, Y., Kerkez, M.F.S. and Abu Heen, Z. (2015) Toxicity of Diuron, Diquat and Terbutryn Cyanobacterial Mats. Ecotoxicology and Environmental Contamination, 10, 71-82.
15. Ma, J., Qin, W., Lu, N., Wang, P., Huang, C. and Xu, R. (2005) Differential Sensitivity of Three Cyanobacteria (Anabaena flos-aquae, Microcystisflos-aquae and Mirocystis aeruginosa to 10 Pesticide Adjuvants. Bulletin of Environmental Contamination and Toxicology, 75, 873-881.
16. El-Nahhal, Y. and Hamdona, N. (2015) Phytotoxicity of Alachlor, Bromacil and Diuron as Single or Mixed Herbicides Applied to Wheat, Melon, and Molokhia. SpringerPlus, 4, 1-19.
17. L-Nahhal, Y., EL-Najjar, S. and Afifi, S. (2015) Toxicity of Carbaryl, Chlorpyrifos and Diuron to Different Aquatic Organisms. Toxicology International, 22, 45-53.
18. El-Nahhal, Y. and El-Dahdouh, N. (2015) Toxicity of Amoxicillin and Erythromycin to Fish and Mosquito. Ecotoxicology and Environmental Contamination, 10, 13-21.
19. El-Nahhal, Y., Nir, S., Polubesova, T., Margulies, L. and Rubin, B. (1997) Organo-Clay Formulations of Alachlor: Reduced Leaching and Improved Efficacy. Proceedings of Brighton Crop Protection Conference, 1, 21-26.
20. Rubin, B., El-Nahhal, Y., Nir, S. and Margulies, L. (2001) Slow Release Formulations of Pesticides. Patent No. US6261997 B1.
21. El-Nahhal, Y. (2003) Adsorption Mechanism of Chloroacetanilide Herbicides to Modified Montmorillonite. Journal of Environmental Science and Health Part B, 38, 591-604.
22. El-Nahhal, Y. (2003) Persistence, Mobility, Efficacy and Safety of Chloroacetanilide Herbicide Formulation under Field Conditions. Environmental Pollution, 124, 33-38.
23. El-Nahhal, Y. (2003) Adsorptive Behavior of Acetochlor on Organoclay Complexes. Bulletin of Environmental Contamination and Toxicology, 70, 1104-1111.
24. El-Nahhal, Y. and Safi, J. (2004) Adsorption Behavior of Phenanthrene on Organoclays under Different Salinity Levels. Journal of Colloid and Interface Science, 269, 265-273.
25. El-Nahhal, Y. and Safi, J. (2004) Stability of an Organo Clay Complex: Effects of High Concentrations of Sodium Chloride. Applied Clay Science, 24, 129-136.
26. El-Nahhal, Y. and Safi, J. (2005) Adsorption of Benzene and Naphthalene to Modified Montmorillonite. Journal of Food, Agriculture & Environment, 3, 295-298.
27. El-Nahhal, Y., Lagaly, G. and Rabinovitz, O. (2005) Organo-Clay Formulations of Acetochlor: Effect of High salt. Journal of Agricultural and Food Chemistry, 53, 1620-1624.
28. Nir, S., Undabeytia, T., Yaron, D., El-Nahhal, Y., Polubesova, T., Serban, S., Rytwo, G., Lagaly, G. and Rubin, B. (2000) Optimization of Adsorption of Hydrophobic Herbicides on Montmorillonite Preadsorbed by Monovalent Organic Cations: Interaction between Phenyl Rings. Environmental Science & Technology, 34, 1269-1274.
29. Nir, S., El-Nahhal, Y., Undabeytia, T., Rytwo, G., Polubesova, T., Mishael, Y., Rabinovitz, O. and Rubin, B. (2006) Clays and Pesticides. Developments in Clay Science, 1, 677-691.
30. García-García, C.R., Parrón, T., Requena, M., Alarcón, R., Tsatsakis, A.M. and Hernández, A.F. (2016) Occupational Pesticide Exposure and Adverse Health Effects at the Clinical, Hematological and Biochemical Level. Life Sciences, 145, 274-283.
31. El-Nahhal, Y., EL-Dahdouh, N., Hamdona, N. and Alshanti, A. (2016) Toxicological Data of Some Antibiotics and Pesticides to Fish, Mosquitoes, Cyanobacterial Mats and to Plants. Data in Brief, 6, 871-880.
32. Araoud, M., Neffeti, F., Douki, W., Khaled, L., Najjar, M.F., Kenani, A. and Houas, Z. (2016) Toxic Effects of Methamidophos on Paraoxonase 1 Activity and on Rat

Kidney and Liver and Ameliorating Effects of Alpha-Tocopherol. Environmental Toxicology, 31, 842-854.

33. Liu, S., Chen, D., Fu, S., Ren, Y., Wang, L., Zhang, Y., Zhao, M., He, X. and Wang, X. (2015) Prevalence and Risk Factors for Farmer's Lung in Greenhouse Farmers: An Epidemiological Study of 5,880 Farmers from Northeast China. Cell Biochemistry and Biophysics, 71, 1051-1057.
34. Jurewicz, J., Kouimintzis, D., Burdorf, A., Hanke, W., Chatzis, C. and Linos, A. (2007) Occupational Risk Factors for Work-Related Disorders in Greenhouse Workers. Journal of Public Health, 15, 265-277.
35. Terho, E.O., Husman, K. and Vohlonen, I. (1987) Prevalence and Incidence of Chronic Bronchitis and Farmer's Lung with Respect to Age, Sex, Atopy, and Smoking. European Journal of Respiratory Diseases—Supplement, 152, 19-28.
36. Ellman, G.L., Courtney, K.D., Andreess, V. and Featherstone, R.M. (1961) A New and Rapid Colorimetric Determination of Acetylcholinesterase Activity. Biochemical Pharmacology, 7, 88-95.
37. WHO Library Cataloguing-in-Publication Data (2009) World Health Organization Recommended Classification of Pesticides by Hazard and Guidelines to Classification.
38. Tomlin, C. (2000) The Pesticide Manual. British Crop Protection Council, UK.
39. El-Nahhal, Y., Nir, S., Polubesova, T., Margulies, L. and Rubin, B. (1998) Leaching, Phytotoxicity and Weed Control of New Formulations of Alachlor. Journal of Agricultural and Food Chemistry, 46, 3305-3313.
40. El-Nahhal, Y., Undabeytia, T., Polubesova, T., Golda Mishael, Y., Nir, S. and Rubin, B. (2001) Organo-Clay Formulations of Pesticides: Reduced Leaching and Photodegradation. Applied Clay Science, 18, 309-326.
41. El-Nahhal, Y., Abadsa, M. and Affifi, S. (2013) Adsorption of Diuron and Linuron in Gaza Soils. American Journal of Analytical Chemistry, 4, 94-99.
42. El-Nahhal, Y. (2014) Development of Controlled Release Formulations of Thiabendazole. Journal of Agricultural Chemistry and Environment, 3, 1-8.
43. El-Nahhal, Y. (2013) Alcohol like Syndrome: Influence of Increased CO2 Concentration in the Respiration Air. Journal of Environment and Earth Science, 3, 222-227.
44. El-Nahhal, Y., Wheidi, B. and El-Kurdi, S. (2016) Development of Ecologically Acceptable Chlorpyrifos Formulation for Effective and Safe Application. Journal of Encapsulation and Adsorption Sciences, 6, 91-108.
45. Al-Arifi, S.N., Al-Agha, R.M. and El-Nahhal, Z.Y. (2013) Environmental Impact of Landfill on Groundwater, South East of Riyadh, Saudi Arabia. Journal of Natural Sciences Research, 3, 222-242.
46. Al-Arifi, S.N., Al-Agha, R.M. and El-Nahhal, Z.Y. (2013) Hydrogeology and Water Quality of Umm Alradhma Aquifer, Eastern Saudi Arabia. Journal of Environment and Earth Science, 3, 118-127.
47. Riu, E., Monsó, E., Marin, A., Magarolas, R., Radon, K., Morera, J., Andreo, F. and Nowak, D. (2008) Occupational Risk Factors for Rhinitis in Greenhouse Flower and Ornamental Plant Growers. American Journal of Rhinology, 22, 361-364. https://doi.org/10.2500/ajr.2008.22.3186
48. Yang, J., Cao, J., Sun, X., Feng, Z., Hao, D., Zhao, X. and Sun, C. (2012) Effects of Long-Term Exposure to Low Levels of Organophosphorous Pesticides and Their Mixture on Altered Antioxidative Defense Mechanisms and Lipid Peroxidation in Rat Liver. Cell Biochemistry and Function, 30, 122-128.
49. El-Nahhal, Y. (2016) Acute Toxicity among Greenhouse Farmers in Gaza Strip. IOSR Journal of Environmental Science, Toxicology and Food Technology, 10, 56-64.

CHAPTER 17

Expert System for Greenhouse Production Management

Yongguang HU[1], Jizhang WANG[1] and Pingping LI[1]

[1] *Jiangsu University, China*

1. INTRODUCTION

Greenhouse is a kind of agricultural building structure with glass or plastic roof, where the vegetables or flowers can grow well year round by controlling the inside climatic environment. Fig. 1 shows commercial multispan greenhouses in China. The closed environment of a greenhouse has its own unique requirements, compared with outdoor production. Pests and diseases, and extremes of heat and humidity, have to be controlled, and fertigation is necessary to provide water and nutrients. Significant inputs of heat and light may be required, particularly with winter production of warm-weather vegetables. So greenhouse cost-effectiveness can not be ignored.

Figure 1. Modern commercial greenhouses.

Greenhouses are increasingly important in the food supply of high latitude countries and regions, especially Northern China, because they can allow certain crops to be grown year round such as lettuce and tomato.

Greenhouses protect crops from too much heat or cold, shield plants from dust storms and blizzards, and help to keep out pests or diseases. Light and temperature control allows greenhouses to turn inarable land into arable land. Hydroponics can be used in greenhouses as well to make the most use of the interior space.

Therefore, greenhouse industry is one of the most important parts and symbol of modern agriculture, since it can provide abundant fresh products stably and sustainably. However, greenhouse production process is much more complicated, compared with outdoor production. Precision management is badly needed under such controlled space. Generally, greenhouse system consists of the following elements – surrounding structures, covering, cooling and heating system, shading and light-supplementary system, CO_2 enrichment system, culture facilities, fertigation and automatic control system. Crop cultivation techniques and related equipment are also necessary.

In a word, faced with so huge amount of knowledge and techniques greenhouse engineering and its production management are challengeable, which leads to the emerging of expert system for greenhouse production management.

The application of expert system on greenhouse production management has been emphasized by many researchers, especially on greenhouse cultivation and environment control. A decision analysis and expert system model was applied to the problem of individual nutrient selection for cucumber growth in a controlled environment (Fynn et al., 1994). A computer-aided decision-making expert system for greenhouse cucumber was developed, which included cultivation management, pest and disease diagnosis and prevention (Chen & Li, 2001). VEGES was developed for diagnosis of nutrient deficiency, pest and disease on greenhouse vegetable (Yialouris et al., 1997). The humidity and temperature of a laboratory scale greenhouse are controlled using a rule-based expert system (Türkay, 1994). Artificial neural network was adopted to determine the temperature setpoint for greenhouse tomato by data collection rather than growers' experiences. Compared with conventional expert system it has the advantage of short control time and higher accuracy (Seginer et al., 1996). An expert system based on optimal fuzzy control strategy fulfilled real-time temperature control by on-line adjustment of setpoint (Jacobson et al., 1989). GHMES was developed to carry out the real-time environment control based on crop model and cultivation management (Hu et al., 2006).

The chapter deals with expert system development and application for greenhouse vegetable production management, including general description of greenhouse production system, design on expert system for greenhouse production management, integration of environment control subsystem with the expert system, software fulfilment of the expert system and two application cases in greenhouse vegetable production practice.

2. DESIGN ON EXPERT SYSTEM FOR GREENHOUSE PRODUCTION MANAGEMENT

2.1. System Functional Structure

According to the analysis of greenhouse and its production process, the expert system is composed of four subsystems (in Fig. 2), which are cultivation techniques, consultation of pest/disease and nutrient deficiency, diagnosis of pest/disease and nutrient deficiency, and environment control decision. Under each subsystem, there are still several subsystems.

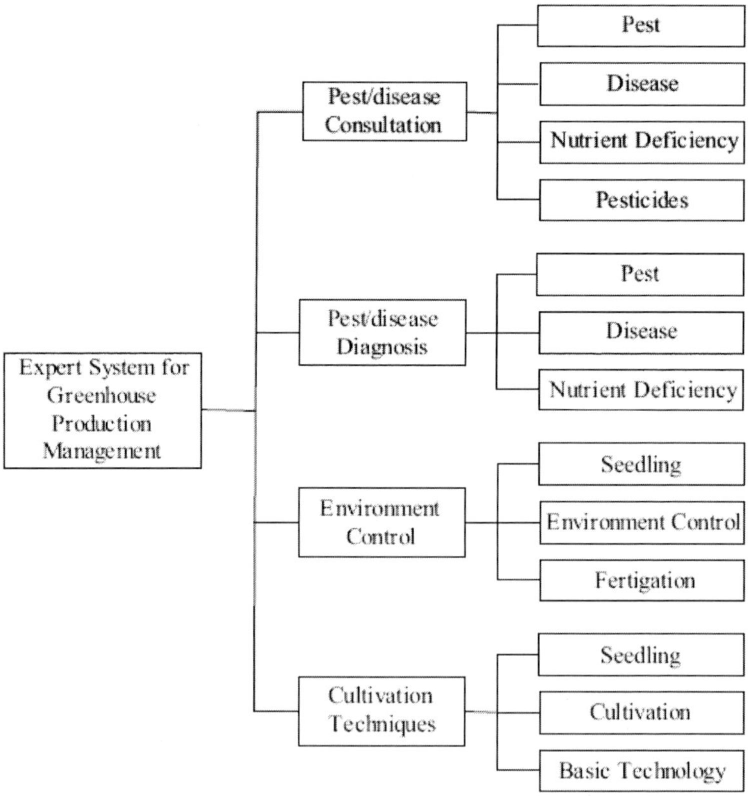

Figure 2. System structure.

2.2. System Configuration

Modular design method is adopted to configurate the expert system and every independent module is combined by master module. Main modules of the expert system include inference diagnosis module, database management module, cosultation module, decision on environment control, text browsing module and help module. Correspondingly, system software structure is shown as Fig. 3.

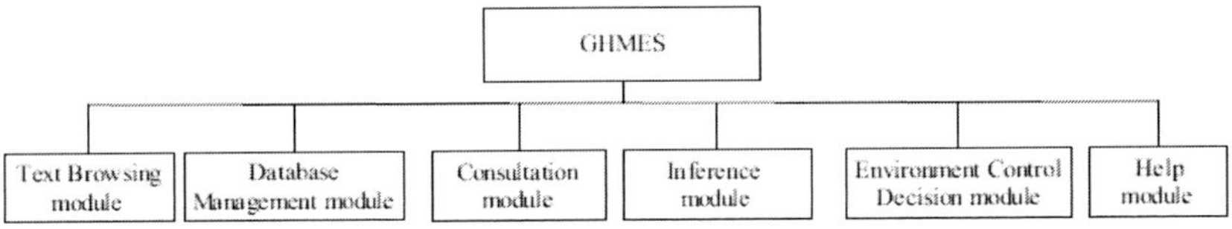

Figure 3. System software structure.

1. Inference and Diagnosis Module

This is the core part of the expert sytem, which deals with the diagnosis of common pest/disease and nutrient deficiency for greenhouse crops. It can draw a diagnosis conclusion step by step and provide prevention and treantment mothed for user's reference.

2. Database Management Module

The module is used for creating, editing and modifying knowledge and database. The known data, knowledge and rules of the expert system are also created, edited and modified through the module.

3. Text Browsing Module

The module describes the botanical and physiological features of several typical greenhouse crops in text. Cultivation techniques and requirements of environment are also involved.

4. Consultation Module

It includes both pest/disease and nutrient deficiency consultation modules. The module provides users with the type of pest/disease or nutrient deficiency and the expert system goes into the details such as symptom, occurrence regularity, prevention and treatment, and related pictures.

5. Decision of Environment Control Module

There are two decision modes for greenhouse environment control, suiting for different occasions. One is on-line decision mode and the other is off-line decision mode. Under the on-line mode the expert system is integrated with the environment information collection system, based on which real-time environment control strategy can be worked out to complete the adjustment of greenhouse environment for good growth of crops. Under the off-line mode users input the present environment and crop information by hand and the expert system gives proper decision alternatives for user's reference.

3. FULFILLMENT OF EXPERT SYSTEM FOR GREENHOUSE PRODUCTION MANAGEMENT

3.1. Database Management

Among consultation module, cultivation techniques is called through text mode, and pest/disease, nutrient deficiency and chemicals are based on data dictionary, which result from large amount of searched basic data. For example, intuitive description of disease symptom and its pictures can be given by the system according to user's selection of different diseases.

Moreover, diagnosis module is also based on symptom, but the symptom is represented by particular knowledge. Here production rule knowledge representation based on relational database is adopted, and the knowledge is represented through database tables.

First, E-R model is established, which is an abstraction of real world. E-R model extracts common characteristics and neglects nonessential details, and describes the characteristics accurately using all kinds of concepts. There are three abstractions.

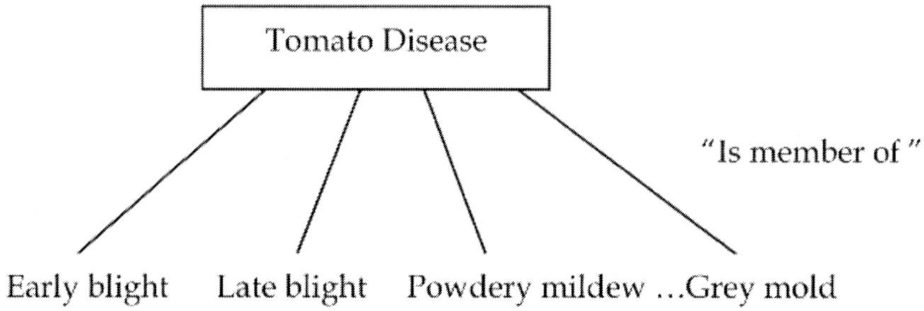

Figure 4. Classification model.

a. Classification defines a certain concept as a type of a group of objects, which have the common features and behaviours. Take tomato as an example, classification model is described in Fig. 4.

b. Aggregation is defined as a component part of a type. An entity is the aggregation of several attributes. In Fig. 5, the relationship between symptom and different parts is an aggregation.

c. Generalization is defined as a certain subset relationship between different types, which includes subtype and supertype. Generalization model is described in Fig. 6.

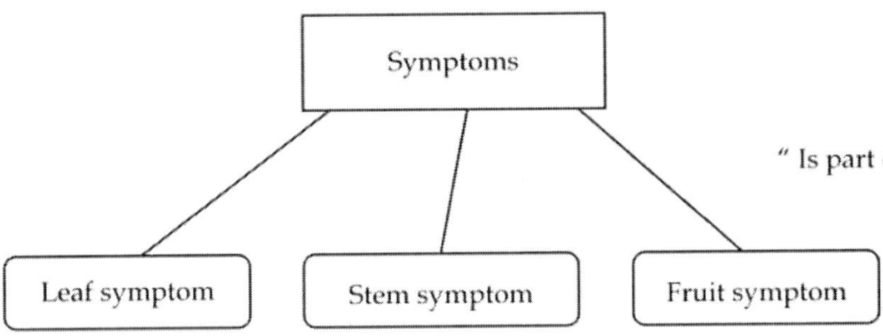

Figure 5. Aggregation model.

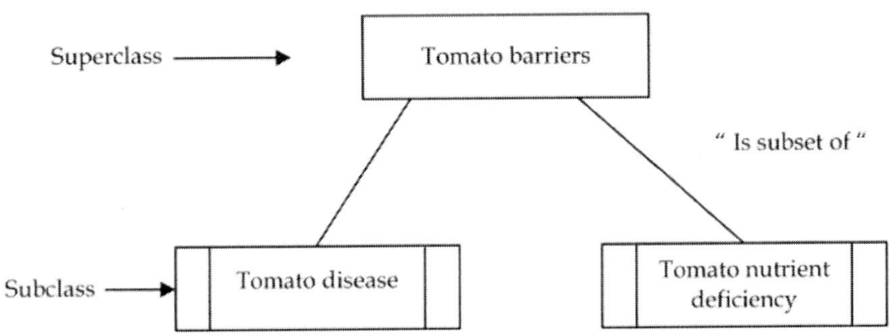

Figure 6. Generalization model.

Therefore, E-R model of the expert system is as follows in Fig. 7.

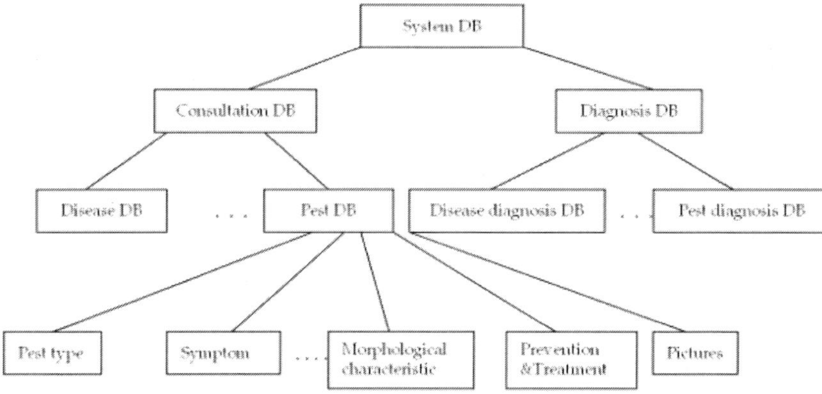

Figure 7. E-R model.

Then, E-R model is transformed into relational model, following the principle: one entity type is transformed into one relational schema, so entity attribute is relation attribute and entity code is relation code. The transformed data model is listed below.

Diseases:

Title	Symptom recognition	Occurrence regularity	Prevention/treatment	Pictures

Pests:

Title	Damage symptom	Morphological feature	Prevention/treatment	Pictures

Finally, the memory size of each data table field is determined, data and knowledge is recorded, and database is built. When using the database, maintenance of adding, deletion and updating is needed.

3.2. Design of Knowledge Base

3.2.1. Acquisition of knowledge

The source of knowledge is kinds of books, scientific reports and pictures. The reliability of the expert system depends on the quality of knowledge, so accuracy of fact description need to be guaranteed. The knowledge of greenhouse tomato growth barriers is briefly illustrated as follows.

1. Symptom description with natural language

Tomato barrier symptom is described with simple natural language, and a piece of knowledge table is made. For example, as for the case of phosphorus deficiency, symptom and diagnosis is described as follows.

Crop title: Tomato

Symptom: older leaves etiolating with purple brown spots

Older leaves dropping off easily

Short small plant

Thin stem

Olive green leaves

Violet Leaf Venation on the back

Diagnosis: Phosphorus deficiency

Simple language helps to purify and order the knowledge, and forms a description and diagnosis table. Even though there is large number of tomato growth barriers, the most common symptoms and easily-recognized symptoms are listed in Table 1, with which a series of targets (conclusion) can be decided for every barrier. All described symptoms and important facts are linked to the targets, and then proper diagnosis rules are obtained.

2. Rules of knowledge base

Representation of system symptom is based on the form of Object-Attribute-Value (OAV), which is suitable for developing any expert system based on rules. Most of the symptoms have one or several attributes and each attribute has at least one value. All the possible attributes and values of each symptom are collected and prepared.

Knowledge in description/diagnosis table is used to relate disease with its symptom. In the table the columns represent diagnosis and the rows represent symptom. Rules conformed to given column or row units are found out from rules of knowledge base. In addition, if a disease needs to be recognized by several rules, the attribute of the disease will appear in several columns or rows of the table. When coding or decoding the table, rules are sorted out according to the etiopathology or inducement of the disease in Table 2. The rules of Table 2 are described as,

Table 1. Type of tomato diseases, pests and nutrient deficiency.

Disease	Pest	Nutrient deficiency
Grey mold	Whitefly	Potassium
Late blight	Liriomyza sativae blanchard	Nitrogen
Leaf mold	Polyphagotarsonemus latus	Phosphorus
Spot blight	Spodoptera litura	Ferrite
Virus disease	Helicoverpa armigera	Calcium
Cercospora leaf spot		Magnesium
Blossom-end rot		Boron
Bacterial wilt		
Wilt		
Powdery mildew		
Sclerotinia sclerotiorum		

If growth limited with light green/yellow older leaves then Nitrogen deficiency;

If yellow or light yellow young leaves with green texture then Ferrite deficiency.

Table 2. Relation between tomato symptom and nutrient deficiency.

Symptom	Nutrient deficiency				
	N	N	Fe	Fe	……
Plant:					
Growth limited	X				
Wilting		X			
Color of older leaf:					
Light green	X				
Light yellow		X			
Green-purple					
Color of young leaf:					
Light yellow			X		
Green texture among nervations				X	
…………					

3.2.2. Knowledge Representation Method

Knowledge base is essential to expert system and its structure and performance influence the accuracy and efficiency of problem solving. At the same time proper knowledge representation method may enhance the performance of knowledge base.

The knowledge within the expert system is represented and organized by the following types: descriptive knowledge, data knowledge and rule knowledge. Production knowledge representation based on database is used to strengthen the relation among knowledge and to reduce the number of production rules.

Rule base is the knowledge memory of specific field, and its representation is If < premise > then < conclusion > λ which means the reliability of conclusion is λ when premise comes into existence. $\lambda=1$ for accurate inference, and λ varies within [0,1] for inaccurate inference. < premise > is the conjunction of facts and assertion.

Semantic model of the rules is expressed as Fig. 8.

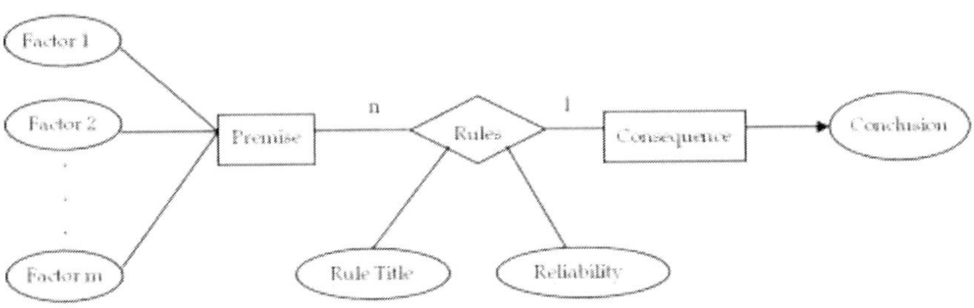

Figure 8. Semantic model of the rules.

Based on above semantic model, relational database and knowledge inference, rules are represented by relational database. Rule premise and conclusion are separated, and put into different database. Rules and related information is stored in rule premise base, and its structure is as follows.

Number of rules	Rule premise	Meet or not?

Rule conclusion and related information is stored in rule conclusion base, and its structure is as follows.

Number of rules	Rule conclusion	Number of premise	Number of auxiliary	contentment

Take a complete production rule as an example,

if < premise 1 >, < premise 2 >, < premise 3 > then < conclusion >

which means number of premise is three and conclusion can be made only when all the three premises meet the requirements.

3.2.3. Establishment Of Knowledge Base

1. Knowledge base of pest/disease diagnosis

Besides rule premise base and rule conclusion base, dictionary base is designed to make knowledge base management easier by storing premise, conclusion and code of the rule base. Its structure is dictionary (fact number, facts, known or not) Knowledge base is essential to expert system and its structure and performance influence the accuracy and efficiency of problem solving. At the same time proper knowledge representation method may enhance the performance of knowledge base. The above three bases are structured through a table in Fig. 9.

2. Knowledge base of environment control decision

① Knowledge base of crop growth

It includes the requirement of growth on temperature, humidity and light intensity during different stages, which is acquired from expert's experiences, experiments and literature.

② Knowledge base of environment control

It covers the basic requirement on inside and outside environment control, priority of control parameter selection, actuator requirement under different climate, diagnosis and control strategy under abnormal situations. For example, roof window must be closed when it rains, or rolling screens must be closed if wind speed is too high.

③ Other auxiliary knowledge base

A. knowledge base of open-close of greenhouse

When the greenhouse actuators are in operation the open-close status of greenhouse is very important. For example, greenhouse must be closed when heating or CO_2 enrichment, or south and north rolling screens must be open and east and west rolling screens must be closed when cooling with sprinkling.

B. knowledge of environmental influence on crop growth

For example, higher temperature is needed to promote the growth of cucumber during the early growth stage, while too low humidity could cause its physiological barrier during the development stage.

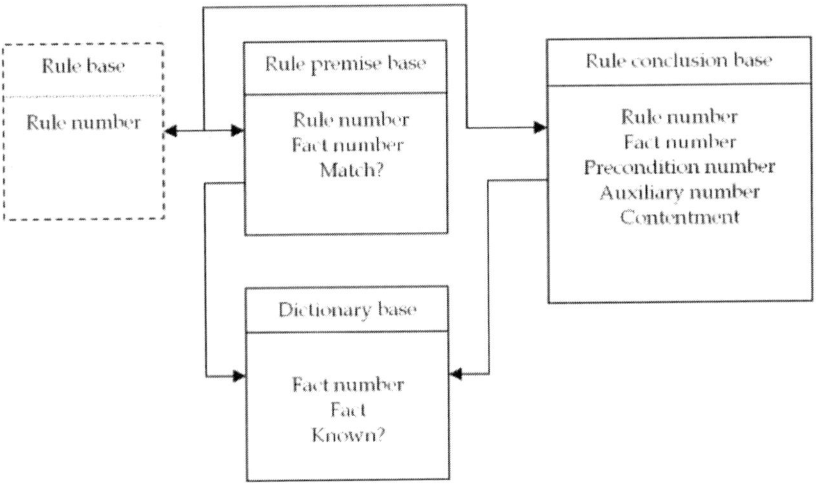

Figure 9. Table-structured knowledge base.

3.2.4. Organization of Knowledge Base

Knowledge base is organized by database management mechanism of SQL Server 2000, since knowledge is represented through relational database table. This kind of organization has the following advantages.

1. Knowledge base and inference engine are mutually independent, and the internal variation of organizational structure of knowledge base does not affect inference engine.
2. Expansion, maintenance and modification of knowledge are easily done with the help of database.
3. Rule files can be transformed into database files so that the normative and flexibility of knowledge representation are improved.
4. Multimedia knowledge representation is achieved through setting fields of picture, sound and animation in the conclusion base.

3.3. Design of Inference Engine

Inference engine may control and coordinate the whole expert system for greenhouse production management. The knowledge used for inference is accurate, and inference conclusion is definite.

3.3.1. Diagnosis of pest/disease and nutrient deficiency

Occurrence and development of pest/disease and nutrient deficiency are restricted by many factors, and here cropping system, growth season, temperature and humidity, and time are considered. Diagnosis of pest/disease and nutrient deficiency is based on symptom. The diagnostic flow chart is shown in Fig. 10.

The inference engine is built up with structured query language and forward inference is adopted. According to the given facts, rule premise database is searched to find out the matched record. The matching process is shown in Fig. 11.

For example, when users input information (pupae; oval shape), auxiliary number of rule is not reduced to 0 after inference. The inference conclusion is: greenhouse whitefly and liriomyza sativae blanchard, and detailed information of the two pests is listed. Final decision will be made by watching and judging the further features.

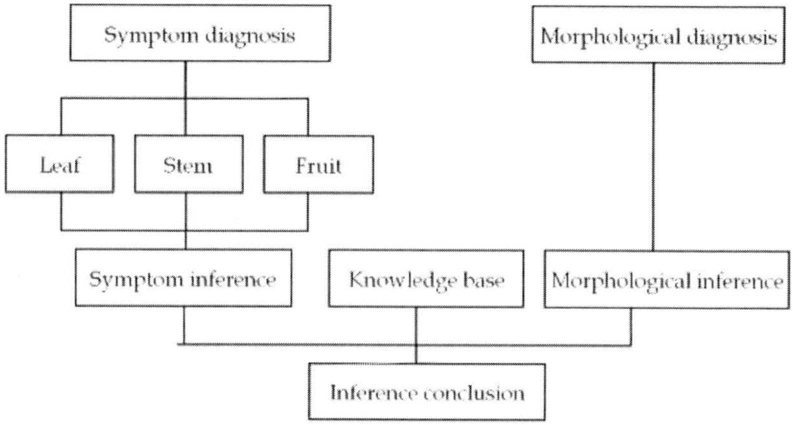

Figure 10. Flow chart of diagnosis.

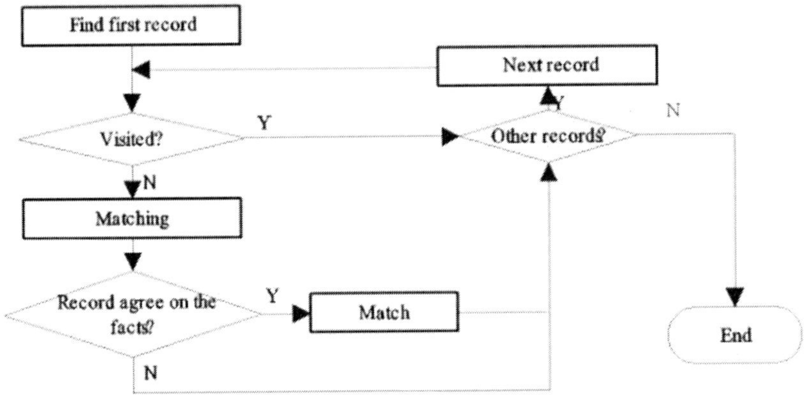

Figure 11. Flow chart of matching record.

3.3.2. Environment control decision based on models

Different combinations of environment control actuators are listed in Table 3 (for summer) and Table 4 (for spring and autumn).

Table 3. Combinations of environment control actuators for summer.

Combination number	Roof window	Rolling screen	Shading	Fan	Wet pad
1	0	0	0	0	0
2	1	0	0	0	0
3	1	1	0	0	0
4	1	1	1	0	0
5	1	1	1	1	0
6	1	1	1	1	1

Table 4. Combinations of environment control actuators for spring and autumn.

Combination number	Half roof window	Roof window	Rolling screen	Shading	Fan
1	0	0	0	1	0
2	0	0	0	0	0
3	1	0	0	0	0
4	0	1	0	0	0
5	0	1	1	0	0
6	0	1	1	1	0
7	0	1	1	1	1

In Table 3 and Table 4, 1 stands for the status "Open" of the actuators, while 0 stands for the status "Closed" of the actuators.

The inference process is described as Fig. 12. First, users input environment information and the status of actuators manually or automatically through greenhouse environment information collection system. With photosynthetic model and knowledge of crop growth appropriate environment parameters can be figured out. At the same time current environment inside the greenhouse is judged to make sure it conforms to the proper range. Then keep the current status of the actuators. Otherwise, inference of environment control decision must be done in order to output the optimal environment control decision for users' reference.

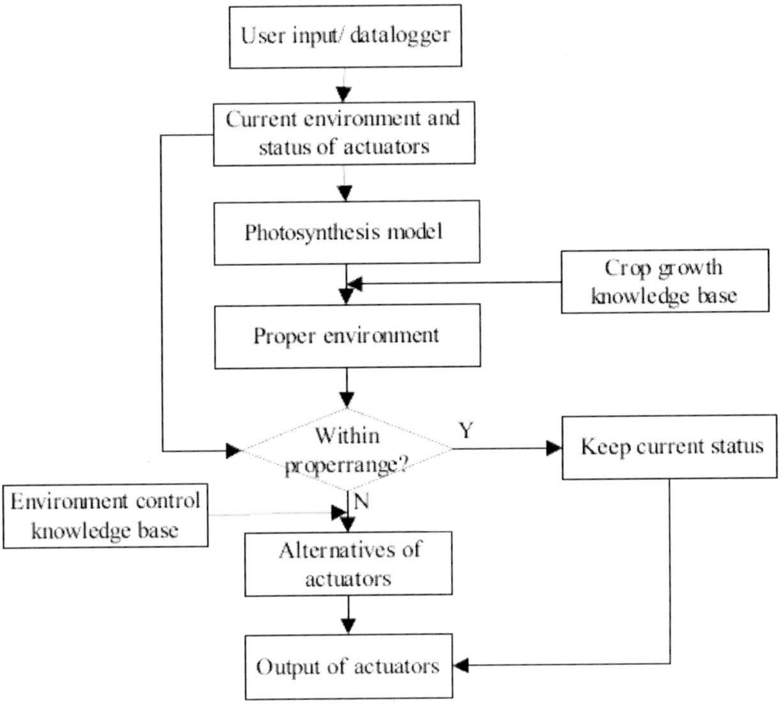

Figure 12. Inference process of environment control decision based on models.

4. INTEGRATION OF ENVIRONMENT INFORMATION COLLECTION WITH EXPERT SYSTEM

To integrate obtained models and knowledge, especially current environment and crop information with greenhouse management expert system for vegetables and make better decisions, real-time environment information collection system is indispensable.

4.1. Overall Design Of Environment Information Collection System

As a sub-system of greenhouse management system, environment information collection system provides expert system with real-time environment information and help to make proper environment control decisions.

4.1.1. Design Requirements

The sub-system can collect environment information inside and outside greenhouses independently. All the information is stored in database, and can be used for greenhouse management expert system. The climatic parameters include temperature, relative humidity, light intensity and data of CO_2concentration wind speed, wind direction and rainfall can also be collected by reserved extensible channels and interfaces for future purpose.

The system needs separate monitoring and unified management

4.1.2. System Functions

For greenhouse environment control, single chip microcomputer receives climatic data from kinds of sensors, carries out logical operation and judgement with expert system, and then makes control decisions to adjust greenhouse environment. Meanwhile, it can help growers to analyze history dada and mine good management strategies. The specific functions are listed as follows:

- Real-time monitoring greenhouse environment;
- Updating the environment information database for expert system;
- Information output and display with text, graph table or printout;
- Alarming in case of extreme occasions or crop stress.

4.1.3. System Configuration

Environment information collection system consists of sensors, data collection module and monitoring software. System configuration is shown in Fig. 13.

Each of the sensors is connected with data collection modules to obtain environment parameters inside and outside greenhouses. All data collection modules communicate with monitoring system (PC) through RS232/485 converter. Monitoring software is developed with configuration software, which is easily customized and interactive.

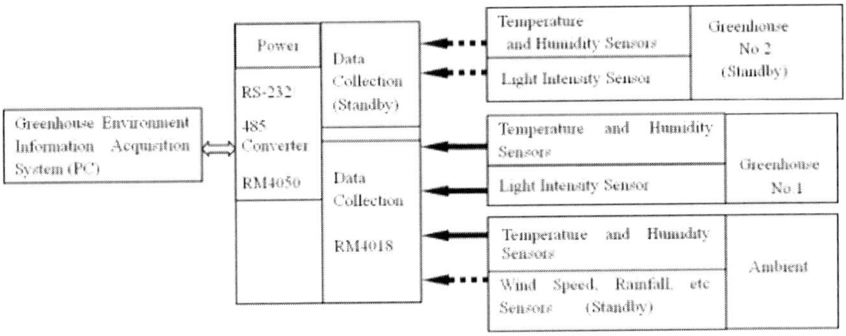

FIGURE 13. Configuration of greenhouse environment information collection system.

4.2. System Hardware Design

Selected sensors are used to measure temperature, relative humidity and light intensity. Integrated temperature and humidity sensor LT/W/S and light intensity sensor LT/G adopt advanced circuit module to transmit signals, and output standard voltage and current, which can be switched by jumping lines. Specification of sensors is shown in Table 5.

8 channels can be linked to data collection module RM4018 and communicate with monitoring computer through the interface of RS485. RM4018 adopts MCU of AT89C2051, 12-bit A/D chip ADS7822 and multiplexer chip MPC508. Photoelectric isolation is used to enhance anti-jamming.

Module RM4050 adopts the techniques of photoelectric isolation and auto reversion to gain high reliability. System supply voltage and current is DC24V/2A and plastic case is self-made.

Table 5. Specifications of sensors.

Sensor type	Range	Output	Accuracy
LT/W/S Temperature Humidity	-20~80°C 0~100%	0~20 mA 4~20 mA 0~5 V	±0.5°C ±3%
LT/G	0~100KLux	0~10 V	±3%

4.3. System Software Design

The software of greenhouse environment information acquisition system is developed on the platform of configuration software PCAuto3.1, which is a kind of special software for data collection and process control. PCAuto3.1 has flexible configuration not programming mode and can provide customers with friendly interface and easy operation.

Configuration software consists of project manager, development system, interface running system, real-time database, I/O driver, network communication program, serial communication program, dialing communication program and web server program.

The software of greenhouse environment information acquisition system includes real-time display, text and graph output, database management and system parameters setting. The main window is shown in Fig. 14.

The database of the software is called by the model base and environment control system, and then control decision can be made based on it.

First, database files of ACCESS is created, entitled data.mdb. Then the database form is correlated with configuration software form by the command of dynamic link library---SQLCreateTable. And one record of environment information is inserted by the command of dynamic link library---SQLInsert.

Finally, expert system directly calls real-time ACCESS database files to be processed for making control decisions.

4.4. System Implementation

According to measuring requirement, sensors were put on proper locations and connected with the data collection modules. And the data collection modules were connected with a computer in which monitoring and control software had been installed

The system automatically recorded environment information at a certain interval and was put into expert system by the ACCESS database files. The system ran for 20 days with no software or hardware failures and successfully supported greenhouse management system with proper decision-making.

Figure 14. Main window of the system software.

5. APPLICATION EXAMPLES OF EXPERT SYSTEM

The operation of the expert system is simple and easy through menu-driven mode. Given functions include,

1. Consultation of basic cultivation information, dealing with botanical characteristics, requirements of growth environment and cultivation techniques during each growth stage.
2. Consultation of frequent pest/disease and nutrient deficiency symptom in greenhouses and chemicals for prevention and treatment.
3. Diagnosis and recognition of pest/disease and nutrient deficiency.

4. Decision on greenhouse environment control for on/off-line operation mode.

Fig. 15 shows the main interface of the expert system, consisting of title bar, menu bar and shortcut bar. Users can select corresponding functions by clicking them.

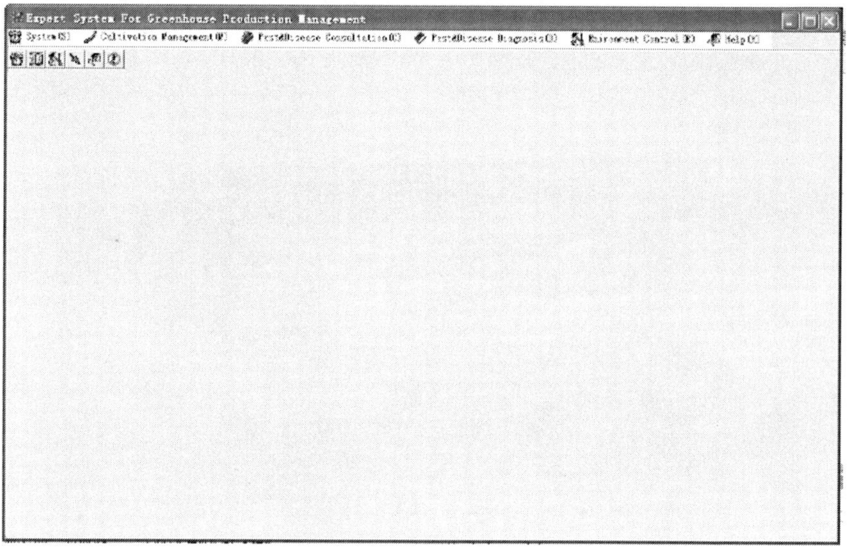

Figure 15. Main interface of the expert system.

5.1. Example of Disease Diagnosis

The general process of diagnosis is briefly described through tomato diseases. Fig. 16 is one of the subforms for diagnosis of tomato diseases. Tomato diseases are diagnosed by their damage symptoms, which have much difference according different growing parts such as leaves, stems and fruits. Users observe the tomato plant on site carefully, and then choose corresponding parts suffering potential diseases in the expert systen. Finally the diseases and their prevetion/treatment motheds are inferred by the expert system.

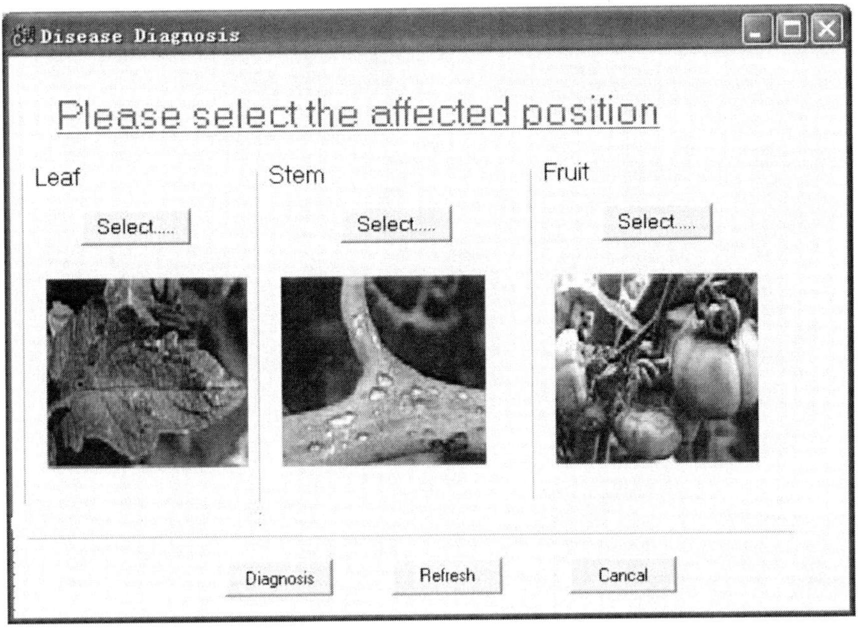

Figure 16. Interface diagnosis for tomato diseases.

Fig. 17, symptom selection dialogue box, will come out when clicking on "Leaf", affected position, in Fig. 16. Then all the disease symptoms related to tomato leaf will appeare in the list box of "selected knowledge or data". After proper selection the knowledge or data will be put into the list box of "selected inference knowledge". Here we choose three pieces of knowledge: "water-stain diseased spots, V-shape extended inwards from leaf edge", "drab or tawny spots" and "with grey mold layer", and save it as a database file.

For the fruit part, the operation is the same as the leaf part (see Fig. 18). At the same time another database file is created.

After the above two-step selection, click the botton "Inferring" and the conclusion will be made in Fig. 19 by calling the knowledge base of disease diagnosis. The inference result shows that the tomato plant suffers grey mold, whose detailed information is listed in Fig. 19.

If only insufficient information is input, several inference conclusions may be obtained for user's reference and selection.

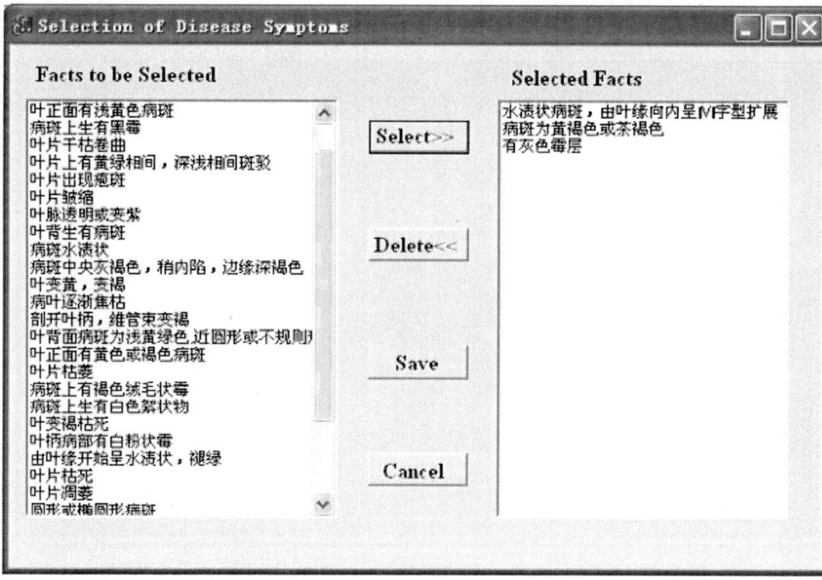

Figure 17. Dialogue box of leaf symptom selection.

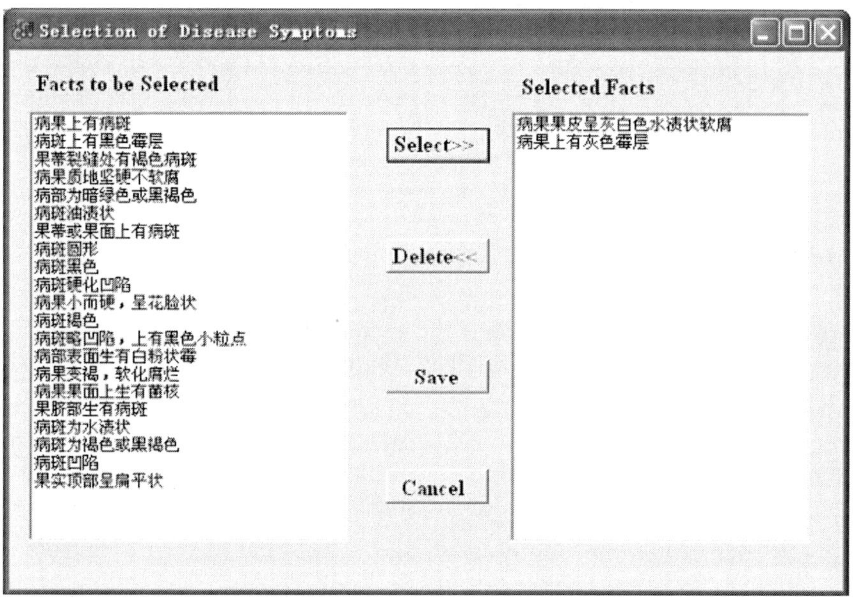

Figure 18. Dialogue box of fruit symptom selection.

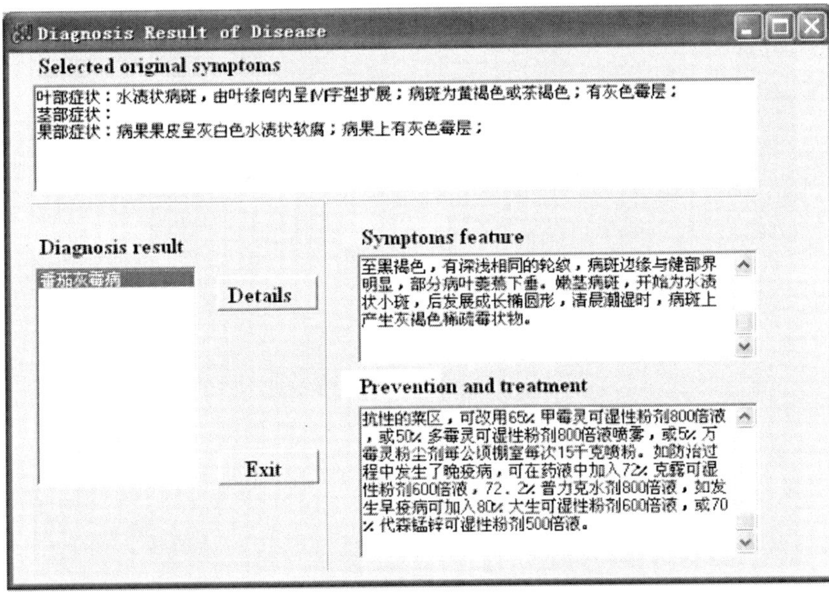

Figure 19. Diagnosis result of tomato diseases.

5.2. Example of Greenhouse Environment Control Decision

5.2.1. On-Line Decision

When users choose on-line decision mode the system collects environmental information inside and outside the greenhouse automatically. Fig. 20 shows both inside and outside environment parameters, including temperature, humidity and light intensity. The information is updated every a certain period. The current status of each environment controlling actuator is also shown. According to the specific crops cultivated in greenhouse, optimal environment conditions, irrigation span and recommended actuating alternatives will be decided through the expert system, wihich is shown in Fig. 21.

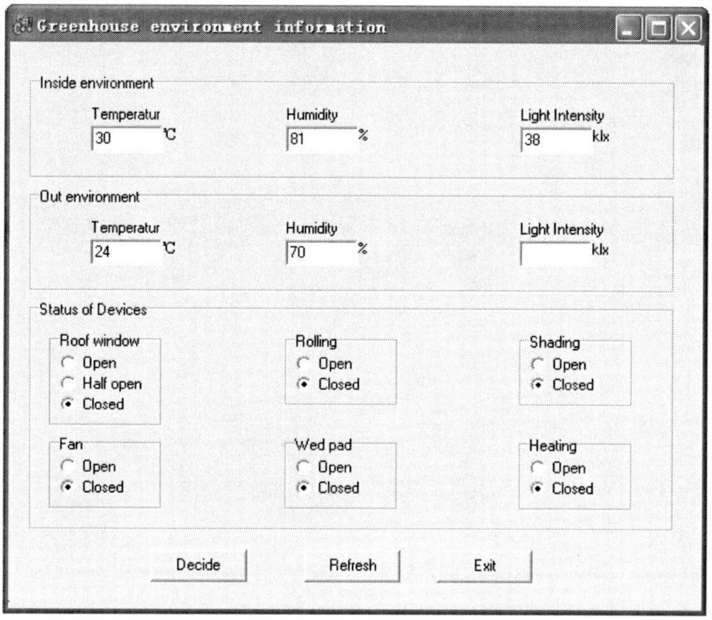

Figure 20. Information output of greenhouse environment and actuator status.

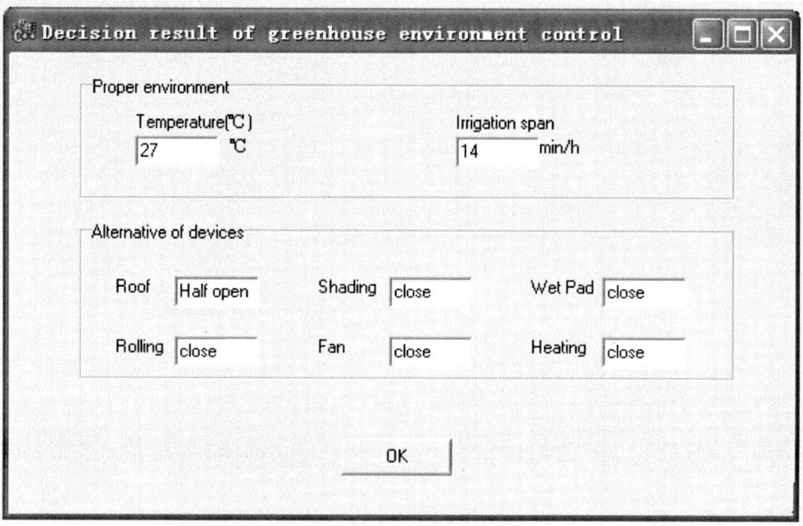

Figure 21. On-line decision result.

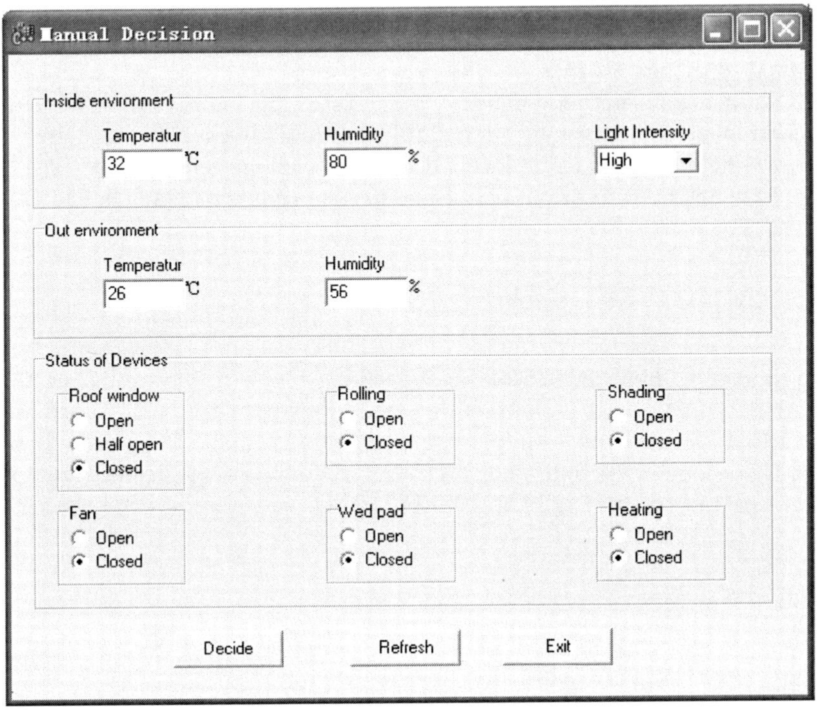

Figure 22. Manual decision.

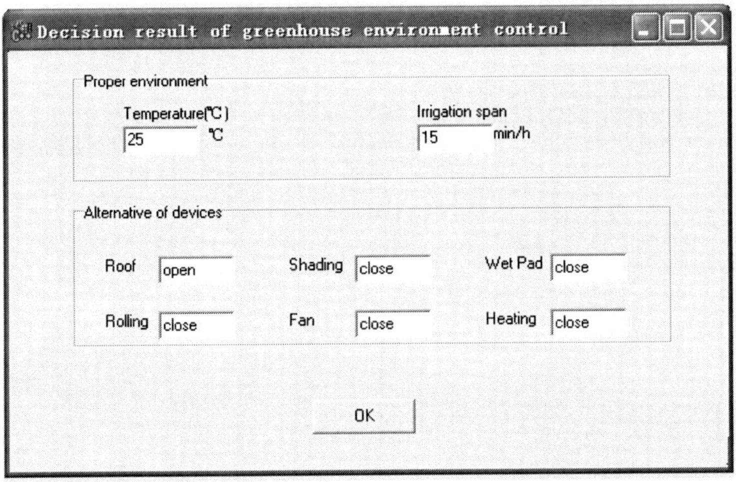

Figure 23. Off-line decision result.

5.2.2. Off-Line Decision

Manual decision is carried out when users input the information of greenhouse environment and actuator status by hand, not through the collection system of environment information. Users may estimate and input light intensity at five levels due to no radiometer available (see Fig. 22). Base on above information off-line decision could be made in Fig. 23.

6. ACKNOWLEDGEMENTS

We sincerely thank Professor Xiliang Zhang and Mr. Zuohua Li for their contributions to the development of greenhouse environment information collection system and pest/disease diagnosis module. The research was also funded by China and Jiangsu Provincial government (Project grant No. 2006BAD11A10, 30771259 and 08KJA21001).

REFERNCES

1. Q. Y. Chen, H. Li, 2001 Study on the expert system of cucumber culture management in greenhouse, Transactions of the CSAE, 17 6 142-146, 1002-6819
2. R. P. Fynn, W. L. Bauerle, W. L. Roller, 1994 Implementing a decision and expert system model for individual nutrient selection, Agricultural Systems, 44 2 125-142, 0308-521X
3. Y. G. Hu, P. P. Li, X. L. Zhang, J. Z. Wang, L. F. Chen, W. H. Liu, 2007 Integration of an environment information acquisition system with a greenhouse management expert system, New Zealand Journal of Agricultural Research, 50 5 855-860, 0028-8233
4. B. K. Jacobson, P. H. Jones, J. W. Jones, J. A. Paramore, 1989 Real-time greenhouse monitoring and control with an expert system, Computers and Electronics in Agriculture, 3 4 273-285, 0168-1699
5. I. Seginer, Y. Hwang, T. Boulard, J. W. Jones, 1996 Mimicking an expert greenhouse grower with a neural-net policy, Transactions of the ASAE, 39 1 299-306, 0001-2351
6. O. S. Türkay, 1994 An expert system to control humidity and temperature of a laboratory scale greenhouse, Mechatronics, 4 6 607-615, 0957-4158
7. C. P. Yialouris, H. C. Passam, A. B. Sideridis, C. Métin, 1997 VEGES-A multilingual expert system for the diagnosis of pests, diseases and nutritional disorders of six greenhouse vegetables, Computers and Electronics in Agriculture, 19 1 55-67, 0168-1699

CHAPTER 18

Heavy Metals Contamination in Greenhouse Soils and Vegetables in Guanzhong, China

Ling Liu[1,2*], Jinyin Lu[2], Zhenwen Zhang[1], Hai Zheng[3], Xiaoqing Gao[1], Wei Zhang[2]

[1]*Shaanxi Provincial Academy of Environmental Sciences, Xi'an, China*

[2]*College of Life Science, Northwest A & F University, Yangling, China*

[3]*Environmental Protection Department of the Northwest Environmental Protection Supervision Center, Xi'an, China*

ABSTRACT

This study used a flame atomic absorption spectrometer (FAAS) and atomic fluorescence spectrophotometer (AFS) to detect the concentrations of chromium (Cr), cadmium (Cd), lead (Pb), hydrargyrum (Hg) and arsenic (As) in soils and three genotypes of vegetables in greenhouse, as well as analyzed the physical and chemical properties of soils, including soil pH, soil organic matter (OM), basic nutrients, electrical conductivity (EC) and cation exchange capacity (CEC) in Guan- zhong areas, Shaanxi province, China. The results showed that comparing to subsoil, the sampled topsoil is enriched in Cr, Cd, Pb, As and Hg. Cd (0.83 - 3.17 mg·kg^{-1}) and Hg (0.40 - 1.44 mg·kg^{-1}) are exceeding the limited value stated in "the 2006 Greenhouse Vegetable Producing Environmental Quality Evaluation Standards" of 0.40 mg·kg^{-1} and 0.35 mg·kg^{-1} respectively. However, Nanzhuang greenhouse soil is within the limits. The heavy metal pollution index (HPI) of soil in Sanyuan (8.10) is the highest and in Dongzhang (4.23) is the lowest. The contents of Pb (0.201 - 0.376 mg·kg^{-1}) were exceeding the limited value (0.20 mg·kg^{-1}) in vegetables species, and Cd (0.0363 - 0.0572 mg·kg^{-1}) in some place were also exceeding the limited value (0.05 mg·kg^{-1}). Greenhouse soils were becoming acidified year after year; the ratios of N, P and K in soil were seriously imbalanced. According to the impacting factors, OM, pH, available P, EC and CEC have obviously effected the accumulation of Cr and Hg. However, there was not enough evidence for the effects of available nitrogen and available potassium.

Keywords: Heavy Metals, Soils, Vegetables, Greenhouse

1. INTRODUCTION

Heavy metals are toxic to animals and human through the food chain. Metal pollution of agricultural soils is a major environmental problem that can affect plant productivity, food quality and human health [1] -[3] . Elevated levels of metals in agricultural soils were the results of atmospheric deposition, wastewater irrigation, sludge amendment, and fertilizer application, as well as industrial activities, in particularly the metallurgical industries [4] [5] . The uptake and bioaccumulation of heavy metals in vegetables are influenced by a number of factors such as climate, atmospheric depositions, the concentrations of heavy metals in soil, the nature of soil on which the vegetables are grown and the degree of maturity of the plants at the time of harvest [6] [7] . Furthermore, it is a direct basis for rational fertilization. Many phosphate rocks contain heavy metals such as Pb and Cd and high application rates of phosphorus (P) fertilizer may not only increase soil P but also lead to accumulation of the metals above the maximum limited values [8] -[10] . Jinadasa et al. [11] found that Cd concentrations in vegetables could exceed the limited value if the soil Cd concentration reached 0.3 mg·kg^{-1} Cd.

Guanzhong plain lies in the middle of Shaanxi province of China, where the land is fertile with plenty agriculture products, and is one of the principal birthplaces of Chinese Yellow River Valley civilization. The agricultural history is long with well developed irrigation farming, concentrating 52% of cultivated land and 75% of effective irrigated areas of Shaanxi [12] [13] , and serving also as the main provincial base for suburban vegetable plot in Qin River. However, most farmers apply a large number of organic fertilizers and mineral manures in greenhouse, leading to heavy metals accumulation in soil and vegetables, increasing potential health risks. At present, as the worldwide food pollution problem becomes more and more serious and the environmental consciousness improves continuously, the limiting measures to the agricultural products, fruits and vegetables markets are becoming gradually complete, and there are many studies about greenhouse soil acidification, nutrient accumulation, salinization, biotic population and enzymatic activity at home and abroad [14] . To our knowledge, there are few systematic published studies on metal contamination in Chinese vegetable-growing regions, particularly in greenhouse. In recent years, with the development of establishment agriculture, areas with greenhouses rapidly increased in Shannxi province of China. However up to now, heavy metal contents of agricultural soil in Guanzhong and its pollution problem have been studied very little [15] , especially there is no systematical investigation for heavy metals in greenhouse soil and vegetables, as well as soil nutrition status in Shaanxi Guanzhong district. For heavy metals in soil which have a long residence time and are potentially dangerous, it is very important to study the condition of heavy metals accumulation in soil and the heavy metal pollution index (HPI) in greenhouse soil. It is also vital to pay attention to food safety, which will bring up potential adverse effects to us through vegetables polluted by heavy metals. In recent years, with the development of establishment agriculture, the areas of greenhouses rapidly increased in Shannxi province of China. At present, our study focused on the contents of heavy metals in the greenhouse soils and the main plant species of vegetables in Guanzhong district, Shannxi province. However, the content of heavy metals in greenhouse soils and vegetables in other districts in Shannxi province, as well as other species of vegetables should be tested in order to ensure safe production of establishment agriculture.

In this study, field survey method includes sampling and analysis, the system investigated on greenhouse heavy metals, including Cd, Pb, Cr, Hg and As in soil and vegetables, as well as soil nutrition status in Guan- zhong district of Shaanxi province. The aims of this study: 1) to detect the concentrations of five metals in greenhouse soil and three main species of greenhouse vegetables, and to examine the relationship between them; 2) to assess environmental risks of heavy metals pollution in the greenhouse soils and vegetables by using pollution index; 3) and to research the contents of basic nutrients, as well as the relationships between basic nutrient and metal concentrations in greenhouse soils so that it can provide a scientific basis for harmless cultivation.

2. MATERIALS AND METHODS

2.1. Site Description

The study area is located in Shaanxi province, including Xi'an city, Xianyang, Yangling and Baoji, these districts is called Guanzhong areas. We choose seven greenhouses as our main study places which were built 10 - 12 years ago in Guanzhong areas. Xi'an is the capital city of Shaanxi province in central China, we selected Xiwang and Wenchang village in Xi'an as the sampling places. The last three cities are located at the suburb of Xi'an city. Baoji is the second largest city of Shaanxi province, is situated at the western end of the Guanzhong (Wei River) valley about 150 km west of the provincial capital city Xi'an, we selected village Dongbai as the sampling place. Sanyuan, Fanyao and Dongzhang were also chosen as the sampling place in Xianyang city, as well as Nanzhuang village in Yangling district. Detailed locations of the sampling sites are shown in Figure 1. A total of 7 topsoil (0 - 20 cm) samples, 7 subsoil (20 - 40 cm) samples, and 21 vegetable samples, including tomato (Lycopersicum esculentum Mill), cucumber (Cucumis sativus L.) and celery (Apium graveolens) were collected from the study areas. Topsoil, subsoil and vegetable samples were taken from the same sites.

2.2. Soil and Vegetable Sampling

From each plot topsoil and subsoil samples were collected with a stainless steel shovel. Five random samples from each soil plot were taken and bulked together as one composite sample, approximately 200 g in total weight. At the same time, vegetable samples were collected on the plot where soil sample was taken. Samples of soil and vegetables were stored in polyethylene bags in the field, and were transferred to the laboratory within three hours for sample processing. The soil samples were air-dried at room temperature, and mechanically ground using a wood roller to pass through a 100-mesh sieve. This fine material was used to determine the total metal contents. Vegetables were washed and the inedible parts were removed immediately. The edible parts of the vegetables were washed with deionized water, blotted dry with tissue paper and weighed. All the vegetable samples were then dried in an oven at 70°C - 80°C for 24 h to constant weight, weighed again, and ground using a smash instrument and a homogenizer [16] .

2.3. Sample Analysis

A 2.0 g sample of soil was digested with perchloric acid and nitric acid (1:4) solution, and taken to 25 ml with distilled water for metals determination. A 0.5 g sample of plant powder was also digested with perchloric acid and nitric acid (1:4) medium, and taken to 10 ml with distilled water for metals determination. Three replicates of Soil and plant for each sample. Concentrations of Cd, Cr and Pb in all samples were determined by a atomic absorption spectrometer (Perkin-Elmer AA800) with an autosampler (combined with flame atomic absorption spectrometer (FAAS). And the before concentrations of Hg and As were determined by atomic fluorescence spectrophotometry (AF-640A) (AFS) according to standard analytical procedures. Samples of soil and plant with certified concentration of metals (GSS-15, GSD-8 and GSV-2, respectively, China National Center for Standard Materials) were included for quality assurance.

2.4. The Basic Nutrient in Soil

Prior to total elemental analysis, soils were air dried (25°C, 14 d) and mechanically ground using a wood roller to pass through a 100-mesh sieve. The basic nutrients of soil were determined by the Bao's method [17] .

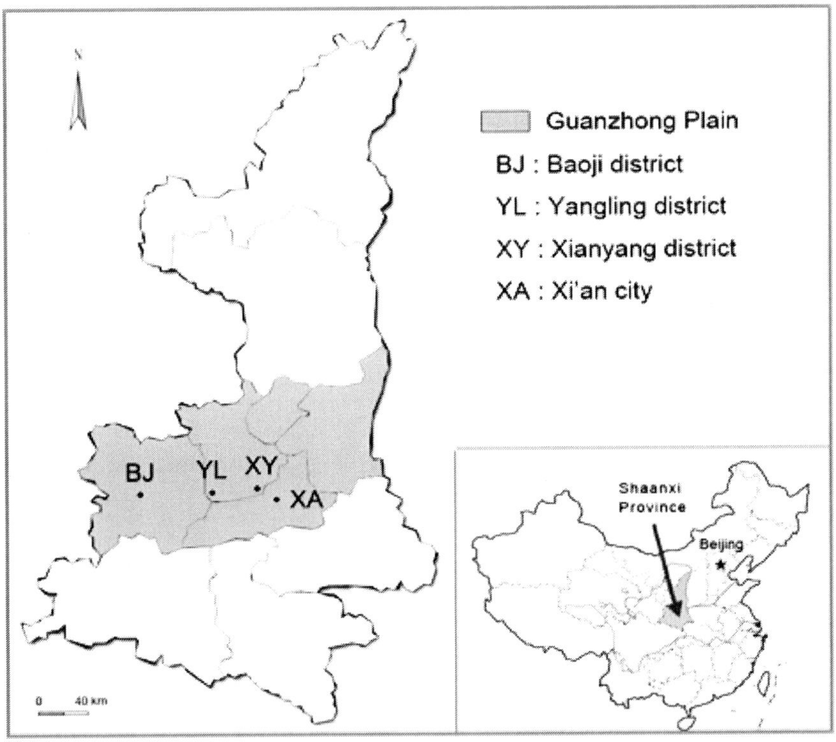

Figure 1. The locations of the greenhouse and sampling sites.

The soil pH (209 pH meter, Hanna Instruments) and electrical conductivity (EC) (4010 EC meter, Jenway) were measured by a glass electrode in a 1:2.5 soil/water suspension. The organic matter in soil was determined by the

$K_2Cr_2O_7$-H_2SO_4 oxidation method, the 0.3 g soil sample was acceded to 10 ml of 1.0 mol·L^{-1} $K_2Cr_2O_7$- H_2SO_4 solution, entering a 170°C - 180°C paraffin oil bath pot for 5 min. After cooling, affiliating with 2 - 3 drops of ferrous indicator, using standard $FeSO_4$ titration. Soil NH_4^+-N was extracted from 1:5 (w/v) dry soil samples with 20% NaCl solution, shaken for 1 h on a reciprocating shaker at 250 rev·min^{-1}, acceded to 1.0 ml sodium tartrate solution, and determined at 390 nm by a spectrometer. Soil NO^{-3}-N was extracted from fresh soil samples with 0.5 mol·L^{-1} $CaSO_4 \cdot 2H_2O$ solution, added 0.05 g $CaCO_3$, 2.0 ml phenol disulfonic acid solution and 1:1 NH_4OH, and determined at 420 nm by a spectrometer. Olsen P was extracted by 0.5 mol·L^{-1}$NaHCO_3$ and a ratio of 1:20 (w/v) fresh sample extractant and shaken for 1 h on a reciprocating shaker at 250 rev·min^{-1} and then determined colourimetrically. Soil available potassium was extracted by 1:5 (w/v) sample extractant ratio of 1.0 mol·L^{-1} NH_4OAc, buffered to pH 7.0, shaken for 30 min on a reciprocating shaker at 250 rev·min^{-1} and was then analysed by flame spectrophotometer. Cation exchange capacity (CEC) was extracted by CH_3COONa, 4.0 g soil sample was added to 33 ml of 1.0 mol·L^{-1} (pH 8.2) CH_3COONa; after shaking 5 min, the solution was discarded and repeated distilling for 4 times, alcohol was used in washing the sample 3 times in the same way, and using 1.0 mol·L^{-1} CH_3COONH_4 washed twice, collected the solution for 100 ml, determined Na^+ concentration by a flame spectrophotometer.

2.5. Statistical Analysis

To compare the total heavy metal load through soils and vegetables at different sampling locations, heavy metal pollution index (HPI) was calculated using the equation given by Usero et al. [18] .

$$HPI = (Cf_1 \times Cf_2 \times Cf_3 \times \cdots \times Cf_n)^{1/n}$$

where Cf is the concentration of n heavy metals in soil samples or vegetable samples. Original data was processed and correlation analysis was done in Excel 2003 and SPSS 16.0. The results are the means ± SD of the three replicates.

3. RESULT AND DISCUSSION

3.1. The Concentration of Heavy Metals in Topsoil and Subsoil

The total concentrations of metals (Cr, Cd, Pb, As, Hg) in greenhouse topsoil and subsoil and the limited values for metals are presented in Table 1. The results indicated that the total concentrations of metals were elevated in surface layers for all samples relative to the underlying subsoil. It demonstrated that the contents of metals in the order: topsoil (0 - 20 cm) > subsoil (20 - 40 cm), but it was not significant. The concentrations of Cd were exceeding the limits in soils of 7 districts. Furthermore, Cd contents (2.9 - 3.2 $mg \cdot kg^{-1}$) in Dongbai soil samples were 7 to 8 times higher than the limited level (0.4 $mg \cdot kg^{-1}$), other districts' Cd contents (0.83 - 2.13 $mg \cdot kg^{-1}$) were 2 to 5 times higher than the limited level. Except in Nanzhuang, Yangling district, the Hg concentration was exceeding the limit in soil of 7 districts. Hg content (1.44 $mg \cdot kg^{-1}$) in Xiwang was significantly higher than others, and 3 to 5 times higher than the limited level. Meanwhile, Cr, Pb and As contents in all soil samples were not exceeding the limits. According to seven villages in Guanzhong district, HPI in the soil samples were all up by 1.0, the total heavy metal load in Sanyuan was the highest. Cd and Hg caused the most heavy metal pollutions in greenhouse soils in Guanzhong districts.

Greenhouse arable soil layer is generally 0 - 40 cm; artificial disturbance frequency, may be the main reasons for the small difference in heavy metal contents between the 0 - 20 cm and 20 - 40 cm layers. Heavy metal accumulation in soil is generally caused by elevated fertilization, pesticides, sewage irrigation and agricultural residues, such as thin-film. At present, the study areas are using groundwater irrigation, except for Xiwang, and have a history of sewage irrigation in Xi'an city. And there are certain restrictions on pesticides, therefore, irrigation and pesticides in greenhouse were not the main reasons for heavy metal accumulation in soil. Study found that in recent years, some feeds contained large amounts of heavy metals; therefore, the defecation of livestock that ingest these feeds is another source of heavy metal pollution in vegetable field [19] . Fertilization led to the occurrence of a large number of heavy metal accumulation in soil throughout the world [20] . Application of large number of poultry droppings, such as chicken defecates, and calcium magnesium phosphate fertilizer,

Table 1. The total concentrations and heavy metal pollution index (HPI) in the soil samples.

Sample sites	Soil layers	Heavy metals (mg·kg^{-1})					HPI
		Cr	Cd	Pb	As	Hg	
Xiwang	T	27.92 ± 0.56	1.08 ± 0.29	29.38 ± 2.61	9.78 ± 0.18	1.44 ± 0.05	6.60
	S	21.08 ± 0.95	0.83 ± 0.36	26.67 ± 0.19	8.90 ± 0.51	1.26 ± 0.03	5.55
Wenchang	T	34.21 ± 0.52	1.17 ± 0.14	17.08 ± 1.72	10.24 ± 0.91	0.48 ± 0.02	5.07
	S	31.38 ± 2.10	0.91 ± 0.40	13.58 ± 3.22	9.37 ± 0.50	0.45 ± 0.01	4.39
Fanyao	T	32.13 ± 1.63	1.13 ± 0.33	23.63 ± 1.44	16.52 ± 0.72	0.44 ± 0.02	5.74
	S	28.17 ± 1.86	0.63 ± 0.45	20.29 ± 4.84	15.07 ± 0.43	0.40 ± 0.02	4.65
Sanyuan	T	42.17 ± 0.52	2.13 ± 0.63	20.42 ± 3.21	13.48 ± 0.41	1.41 ± 0.79	8.10
	S	31.50 ± 1.02	1.42 ± 0.44	18.33 ± 1.87	12.34 ± 0.36	1.10 ± 0.06	6.45
Dongzhang	T	26.38 ± 0.78	0.83 ± 0.40	21.58 ± 2.80	11.98 ± 0.62	0.53 ± 0.01	4.96
	S	20.21 ± 1.92	0.71 ± 0.36	18.71 ± 2.42	9.91 ± 1.07	0.51 ± 0.01	4.23
Nanzhuang	T	30.42 ± 1.99	2.08 ± 0.51	22.04 ± 6.47	15.92 ± 0.90	0.34 ± 0.02	5.96
	S	22.21 ± 1.23	0.83 ± 0.07	20.37 ± 2.28	13.88 ± 1.21	0.28 ± 0.02	4.29
Dongbai	T	36.29 ± 2.37	3.17 ± 0.26	19.38 ± 2.55	12.79 ± 0.84	0.67 ± 0.04	7.18
	S	29.13 ± 0.45	2.88 ± 0.13	19.04 ± 2.60	11.12 ± 0.43	0.61 ± 0.02	6.41
Limit (pH > 7.5)		250	0.40	50	20	0.35	

Note: T-topsoil (0 - 20 cm), S-subsoil (20 - 40 cm). Limit: limit values in soil (HJ333-2006); HPI: heavy metal pollution index. The content of heavy metals in the table is based on dry weight of soil, values are means ± SD (n = 3); Limit: the limit values for metals set by the State Environmental Protection Administration of China (GVPEQES, 2006).

which contains a certain amount of heavy metals, may be the main sources of heavy metal accumulated in greenhouse soils, in Guanzhong districts. Compared with the vegetable field in Guanzhong district, heavy metals contents in greenhouse soils had been increasing with farming year after year. So people should pay more attention to it.

3.2. The Concentrations of Heavy Metals in Vegetables

The total heavy metal concentrations and the limited values in the vegetables are presented in Table 2. The concentrations of heavy metals in three species of vegetables are in the following order: celery > tomato > cucumber. The contents of Pb (0.201 - 0.376 mg·kg^{-1}) in three species of vegetables were exceeding the limited values (0.20 mg·kg^{-1}) according to the 7 districts in Guanzhong. And the contents of Cd were exceeding the limited values in 4 districts, including Xiwang, Sanyuan, Nanzhuang and Dongbai. In addition, Cd found in celery was only exceeding in Xiwang and Nanzhuang. The other three metals were not exceeding the limited values in three species of greenhouse vegetables. However, they have a high metal content when compared to field vegetables. HPI is the highest in cucumber and tomato in Sanyuan, as well as in Xiwang's celery. Meanwhile, HPI is the lowest in tomato and celery in Dongzhang, as well as in Wenchang's cucumber.

In recent years, celery, tomato and cucumber are the main cultural greenhouse vegetables in Guanzhong district, in Shannxi province, having high numbers of growing areas and the higher yields. Previous studies showed that leafy vegetables have a stronger absorptive capacity for heavy metals in soil, atmosphere and water, which are easily polluted, followed by root vegetables, and fruits vegetables has less absorptive capacity of heavy metals [21] . In our study, we only detected the heavy metals in root vegetables and fruit vegetables, the result showed that celery has the highest concentrations of metals, because celery has the enrichment ability to heavy metals [22] . It is indicated that in the vegetables, Xiwang and Sanyuan have higher metal concentrations, because Xiwang has the waste-water irrigation history, and Sanyuan has a large number of fertilization. Although the source of heavy metals contamination of vegetables were more complex, according to the previous studies, heavy metal pollution in greenhouse vegetables in Guangzhong district was caused by heavy metals polluting agricultural soil [21] . Our studies showed that greenhouse soils were mainly polluted by Cd and Hg in Guan- zhong district, but greenhouse vegetables were polluted by Pb and Cd, this is because Pb and Cd element have the ability of enrichment in vegetables [23] . Farming measures, using many fertilizers and pesticides will cause heavy metal residues in vegetables.

3.3. The Basic Nutrition, pH and Organic Matter (OM) in Greenhouse Soil

The basic nutrients, pH and OM in the greenhouse soils are given in Table 3, respectively. Basic nutrients, pH and OM in the greenhouse soils is in the following order: topsoil (0 - 20 cm) > subsoil (20 - 40 cm). The pH of the upper soil layer was always higher than the bottom layer. The pH (7.0 - 7.7) of greenhouse soils was lower than the normal value of the general field soil (7.9 - 8.3), and with the years passing by the soil became acidified, especially in the topsoil (0 - 20 cm). Soil pH can directly affect metals activity, thereby impacting on metals transfer and accumulate in soil. Usually, the soil pH is much lower, and the metal is easier to transfer. However, when the soil pH is litmus or alkalescent, metal is easy to accumulate in soil. OM in topsoil is the highest in Xiwang, and is the lowest in Fanyao. Commonly, with the growing stage increased, the OM contents in soil declined. However, the content of OM in greenhouse soil was higher than that of growing crops field, because of the usage of chicken defecates. In addition, soil organic matters provided nutrients such as carbon, N, P and K to support plant growth and reduced the metal toxicity to plants. The

Table 2. The total concentrations and heavy metal pollution index (HPI) in vegetables.

Sample sites	Vegetable samples	Heavy metals ($mg \cdot kg^{-1}$)					
		Cr	Cd	Pb	As	Hg	HPI
Xiwang	Cucumber	0.359 ± 0.56	0.0477 ± 0.21	0.219 ± 1.11	0.178 ± 0.18	0.0066 ± 0.02	0.085
	Tomato	0.41 ± 0.52	0.0489 ± 0.14	0.234 ± 1.32	0.185 ± 0.45	0.0078 ± 0.03	0.092
	Celery	0.517 ± 0.95	0.0517 ± 0.27	0.376 ± 1.29	0.196 ± 0.51	0.0153 ± 0.01	0.125
Wenchang	Cucumber	0.224 ± 0.34	0.0421 ± 0.33	0.204 ± 1.55	0.146 ± 0.32	0.0058 ± 0.01	0.070
	Tomato	0.278 ± 0.41	0.0437 ± 0.19	0.212 ± 1.60	0.159 ± 0.34	0.0067 ± 0.02	0.077
	Celery	0.359 ± 0.33	0.0481 ± 0.27	0.227 ± 1.34	0.164 ± 0.41	0.0073 ± 0.02	0.086
Fanyao	Cucumber	0.298 ± 0.71	0.0363 ± 0.12	0.203 ± 1.28	0.165 ± 0.47	0.0057 ± 0.01	0.073
	Tomato	0.301 ± 0.58	0.0387 ± 0.34	0.217 ± 1.62	0.177 ± 0.45	0.0062 ± 0.04	0.077
	Celery	0.379 ± 0.28	0.0419 ± 0.33	0.241 ± 0.98	0.182 ± 0.23	0.0071 ± 0.03	0.087
Sanyuan	Cucumber	0.455 ± 0.72	0.0503 ± 0.17	0.227 ± 0.69	0.196 ± 0.31	0.0084 ± 0.03	0.097
	Tomato	0.482 ± 0.43	0.0512 ± 0.28	0.249 ± 1.32	0.159 ± 0.45	0.0089 ± 0.04	0.097
	Celery	0.522 ± 0.55	0.0545 ± 0.55	0.325 ± 1.01	0.168 ± 0.65	0.0091 ± 0.01	0.107
Dongzhang	Cucumber	0.221 ± 0.51	0.0363 ± 0.78	0.201 ± 1.43	0.199 ± 0.55	0.0063 ± 0.01	0.073
	Tomato	0.247 ± 0.39	0.0379 ± 0.69	0.217 ± 1.71	0.117 ± 0.71	0.0072 ± 0.03	0.070
	Celery	0.298 ± 0.44	0.0434 ± 0.17	0.233 ± 1.82	0.154 ± 0.91	0.0081 ± 0.02	0.082
Nanzhuang	Cucumber	0.376 ± 0.83	0.0467 ± 0.49	0.209 ± 2.02	0.154 ± 0.74	0.0043 ± 0.03	0.075
	Tomato	0.399 ± 0.73	0.0479 ± 0.81	0.211 ± 1.62	0.161 ± 0.32	0.0055 ± 0.02	0.081
	Celery	0.432 ± 1.21	0.0509 ± 0.23	0.224 ± 1.23	0.173 ± 0.52	0.0067 ± 0.01	0.089
Dongbai	Cucumber	0.401 ± 0.99	0.0536 ± 0.71	0.218 ± 1.37	0.171 ± 0.43	0.0058 ± 0.02	0.085
	Tomato	0.437 ± 0.87	0.0554 ± 0.32	0.244 ± 1.24	0.187 ± 0.59	0.0066 ± 0.02	0.094
	Celery	0.475 ± 0.79	0.0572 ± 0.92	0.293 ± 1.44	0.199 ± 0.71	0.0078 ± 0.01	0.104
Safe limit		0.50	0.05	0.20	0.50	0.01	

Note: HPI: heavy metal pollution index. The content of heavy metals in the table is based on dry weight of vegetables, values are means ± SD (n = 3).

Table 3. Basic nutritions ,pH and soil organic matter (OM) in the greenhouse soils.

Sample sites	Soil layers	pH	OM (%)	NO_3^--N ($mg \cdot kg^{-1}$)	NH_4^+-N ($mg \cdot kg^{-1}$)	NH_4Ac-K ($mg \cdot kg^{-1}$)	Olsen-P ($mg \cdot kg^{-1}$)	EC ($mS \cdot cm^{-1}$)	CEC Me/100g
Xiwang	T	7.04	3.14	59.80	65.8	292.8	91.8	65.5	30.7
	S	7.13	1.28	52.10	58.8	264.1	69.2	61.0	24.2
Wenchang	T	7.42	2.34	690.7	102.0	260.0	92.9	63.9	26.6
	S	7.55	1.50	541.9	90.2	197.6	87.2	60.5	15.9
Fanyao	T	7.32	1.51	783.2	47.5	191.2	78.1	58.4	25.6
	S	7.36	1.24	731.1	44.8	166.2	50.8	56.6	15.7
Sanyuan	T	7.02	2.90	787.3	280.2	465.8	134.2	99.5	34.7
	S	7.19	1.89	738.6	163.6	399.3	109.7	86.5	24.9
Dongzhang	T	7.35	1.58	199.5	191.1	394.7	62.8	57.9	26.5
	S	7.56	1.34	154.9	153.3	232.2	29.8	52.6	21.5
Nanzhuang	T	7.39	2.08	138.2	111.8	185.8	82.5	52.8	27.9
	S	7.59	1.35	97.9	103.1	112.5	53.6	47.2	22.9
Dongbai	T	7.11	2.80	331.9	118.3	332.4	120.5	93.9	30.9
	S	7.38	2.18	249.1	79.91	281.4	115.3	89.4	26.79

Note: T-topsoil (0 - 20 cm), S-subsoil (20 - 40 cm). The content of NO_3^--N in the table was based on fresh weight of soil, others were based on dry weight of soils. values are means ± SD (n = 3).

topsoil contents of basic nutrient, including available soil potassium, available phosphorus, available nitrogen, soil EC and CEC were the highest in Sanyuan.

This high level of soluble nutrients in soil can be attributed to high levels of EC in the soil. Most of greenhouse soil Olsen-P are more than 60 $mg \cdot kg^{-1}$, especially in Sanyuan and Dongbai. In general, vegetables need about 60 - 90 $mg \cdot kg^{-1}$ [19] of P. However, the available P were exceeding the limit by 1 - 2 times in Guanzhong district, it was far beyond the necessity for vegetables. Although P has little mobility in soil profile, when plowed, soil P content is higher than 60 $mg \cdot kg^{-1}$, it may pollute the environment. Applying large amounts of organic fertilizer in the soil may lead to P downward migration to soil profile, thus creating a potential pollution of groundwater [24] . Greenhouse soil nitrate and ammonium nitrogen contents are higher than average vegetable content, especially the soil nitrate content found in Sanyuan. With the greenhouse shedage increasing, available K content increased, but the increasing amount of the content was limited This indicated that the fertilization of greenhouse vegetable production was imbalanced in Guanzhong district where input of fertilizer N and P is too much while fertilizer K is relatively small. The imbalanced fertilization not only lead to wasting many fertilizers, but also result in secondary environment pollution, and nitrate content increasing, finally the quality of vegetables declining. Therefore, people should pay attention to the rational application of fertilizer K in greenhouse soil, in order to keep fertilizer balanced. The level of soil salinity can be measured by soil EC and CEC, which were rising in study areas, this mainly due to a large number of fertilizers in soil.

3.4. Relationships between Basic Nutrients and Metal Concentrations in Soil

According to these impacting factors, OM, pH, available soil phosphorus, EC and CEC have a great impact on heavy metal contents (Table 4). Soil pH and the contents of heavy metals (Cr, Cd, Pb, Hg) were negative correlation coefficients, except for As, and it had a great impact on the Hg, achieving a very significant level ($p < 0.01$). Soil Olsen-P had a great impact on the Cr, by a very significant level ($p < 0.01$). It maybe because the elements can combine with soil P and form phosphate, accumulating with the increasing of available soil phosphorus contents. EC and Cr had a positive correlation of a significant level ($p < 0.05$). CEC and Hg had a positive correlation of a significant level ($p < 0.05$). Available soil nitrogen and available soil potassium were not significant to correlation coefficients of heavy metals. The significant differences between greenhouse soil and growing crop field are the large accumulation of OM and high available nutrient contents.

From different view of the impacting factors, correlation analysis showed that the OM, pH, available P, EC and CEC have more influence on the heavy metals Cr and Hg in soil.

4. CONCLUSIONS

1) Major survey of the seven regions in Shaanxi Province greenhouse soil for heavy metals showed that the contents of Cd and Hg were exceeding the limited value stated in "the Greenhouse Vegetable Producing Environmental Quality Evaluation Standards". The contents of As, Pb and Cr were not exceeding the limited value. The

Table 4. Correlation coefficients of heavy metal contents with properties of greenhouse soils.

Soil property	Cr	Cd	Pb	As	Hg
pH	−0.432	−0.395	−0.394	0.244	−0.895**
OM	0.327	0.118	0.320	−0.546	0.860*
Olsen-P	0.898**	0.724	−0.271	−0.076	0.573
NH_4 Ac-K	0.415	0.157	−0.176	−0.375	0.596
NO_3^- -N	0.718	0.015	−0.314	0.367	0.144
NH_4^+ -N	0.522	0.221	−0.376	−0.040	0.481
EC	0.834*	0.689	−0.292	−0.117	0.565
CEC	0.659	0.556	0.079	−0.178	0.823*

Note: *indicated that the correlation between the two factor to achieve $p < 0.05$ significant level; **indicated that the correlation between the two factor to achieve $p < 0.01$ significant level.

HPI in Sanyuan is the highest, and in Dongzhang is the lowest. The heavy metal contents were increasing in greenhouse soils in Guanzhong districts; maybe it is related to the use of large amounts of the chemical and organic fertilizers and pesticides.

2) The heavy metals for the seven regions of Shaanxi greenhouse vegetables are in the following order: celery > tomato > cucumber. The content of metals in rooted vegetables is higher than that in eggplant vegetables. The contents of Pb were exceeding the limited value in three species of vegetables. Cd in tomato and cucumber were exceeding the limited value in Sanyuan and Dongbai. Cr, Pb and Hg were not exceeding the limited value in three species of vegetables.

3) The contents of soil pH, OM, soil available nitrogen, Olsen-P, available potassium, CEC and EC were higher than common vegetable contents in

field. Moreover, the contents of available nitrogen and Olsen-P were the highest and the ratio of N, P and K in soil was seriously imbalanced.

4) Using the correlation analysis, according to these impacting factors, OM, pH, Olsen-P, EC and CEC have obvious effect on Cr and Hg. However, the effects of available nitrogen and available potassium were not evidence.

At present, our study focused on the contents of heavy metals in the greenhouse soils and the main plant species of vegetables in Guanzhong district, Shannxi province. However, the contents of heavy metals in greenhouse soils and vegetables in other districts in Shannxi province, as well as other species of vegetables should be tested in upcoming years, in order to ensure the safety production of establishment agriculture.

ACKNOWLEDGEMENTS

Ling Liu, Jinyin Lu, Zhenwen Zhang, Hai Zheng, Xiaoqing Gao, Wei Zhang (No. 200803034).

REFERENCES

1. Alloway, B.J. (1990) Cadmium. In: Alloway, B.J., Ed., Heavy Metals in Soils, Blackie and Son Ltd., Glasgow, 100-124.
2. Jorgensen, S.E. (1991) The Environmental Crisis. In: Hansen, P.E. and Jorgensen, S.E., Eds., Introduction to Environmental Management, Elsevier, Amsterdam, 6-9.
3. Kabata-Pendias, A. and Pendias, H. (2001) Trace Elements in Soils and Plants. 3rd Edition, CRC Press, LLC, Boka Raton.
4. Zhong, W.N., Yan, H.M., Xiu, J.H. and Xian, X.L. (2004) Heavy Metal Concentrations in Vegetable Garden Soils from the Suburb of Hangzhou, Peoples Republic of China. Bulletin of Environmental Contamination and Toxicology, 72, 165-169.
5. Kachenko, A.G. and Singh, B. (2006) Heavy Metals Contamination in Vegetables Grown in Urban and Metal Smelter Contamination Sites in Australia. Water, Air, Soil Pollution, 169, 101-123.
6. Scott, D., Keoghan, J.M. and Allen, B.E. (1996) Native and Low Input Grasses—A New Zealand High Country Perspective. New Zealand Journal of Agricultural Research, 39, 499-512.
7. Voutsa, D., Grimanis, A. and Samara, C. (1996) Trace Elements in Vegetables Grown in Industrial Areas in Relation to Soil and Air Particulate Matter. Environmental Pollution, 94, 325-335.
8. Sharpley, A.N. and Menzel, R.G. (1987) The Impact of Soil and Fertilizer Phosphorus on the Environment. Advances in Agronomy, 41, 297-324.
9. Song, F., Guo, Y.W. and Liu, X.Y. (1996) Composite Contamination of Cd, Zn and Pb in Soils. Chinese Journal of Environmental Science, 16, 431-436. (in Chinese with English Abstract)
10. Powlson, D.S. (1997) Integrating Agricultural Nutrient Management with Environmental Objective: Current State and Future Prospects. Proceedings of the Fertiliser Society, International Fertiliser Society, York.

11. Jinadasa, K.B.P.N., Milham, P.J., Hawkins, C.A., Cornish, P.S., Williams, P.A., Kaldor, C.J. and Conroy, J.P. (1997) Survey of Cadmium Levels in Vegetables and Soils of Greater Sydney, Australia. Journal of Environmental Quality, 26, 924-933.
12. Si, Q.Y., Ran, X.Q., Zhou, X.D., et al. (2000) Study of Regional Water Resources Control and Protection of Ecological Environment. China Environment Science Press, Beijing, 1-6, 151-161.
13. Shi, J., Chen, Z.F., Xing, D.W., et al. (2002) The Rational Exploitation of Water Resources and Protection of Ecological Environment in Guanzhong Area. Yellow River Water Conservancy Press, Zhengzhou, 198-206.
14. Ju, X.T., Kou, C.L., Christie, P., et al. (2007) Changes in the Soil Environment from Excessive Application of Fertilizers and Manures to Two Contrasting Intensive Cropping Systems on the North China Plain. Environmental Pollution, 145,497-506.
15. Zheng, G.Z., Yue, L.P., Li, Z.P., et al. (2006) Assessment on Heavy Metals Pollution of Agricultural Soil in Guanzhong District. Geographical Sciences, 16, 105-113.
16. Yan, S., Ling, Q.C. and Bao, Z.Y. (2007) Metals Contamination in Soils and Vegetables in Metal Smelter Contaminated Sites in Huangshi, China. Bulletin of Environmental Contamination and Toxicology, 79, 361-366.
17. Bao, S.D. (2001) Soil Agriculture and Chemistry Analyzes. China Agriculture Press, Beijing.
18. Usero, J., Gonza-Regalado, E. and Gracia, I. (1997) Trace Metals in the Bivalve Molluscs Ruditapes decussatus and Ruditapes philippinarum from the Atlantic Coast of Southern Spain. Environment International, 23, 291-298.
19. Lu, R.K. (1998) Soil-Principles of Plant Nutrition and Fertilization. Chemical Industry Press, Beijing, 412-422.
20. Tayler, M.D. (1997) Accumulation of Cadmium Derived from Fertilizers in New Zealand Soil. Science of the Total Environment, 208, 123-126.
21. Li, G.L., Su, H.X., Duan, M., et al. (2008) Analyses and Assessment of Heavy Metal Pollution of Vegetables in Xi'an. Northwest Plant Journal, 28, 1904-1909.
22. Duan, M., Ma, W.X. and Li, L. (1999) Analysis and Research on Pb, Cr, Cd Amount of 17 Vegetables in Shanxi. Journal of Arid Land Resources and Environment, 13, 74-80.
23. Guo, Z.H., Xiao, X.Y., Chen, T.B., et al. (2008) Heavy Metal Pollution of Soils and Vegetables from Midstream and Downstream of Xiangjiang River. Geographical Journal, 63, 3-11.
24. Chardon, W.J., Oenema, O. and Castilho, P. (1997) Organic Phosphorus in Solutions and Leachates from Soils Treated with Animal Slurries. Journal of Environmental Quality, 26, 372-378.

CHAPTER 19

How Can Multifunctional Agriculture Support a Transition to a Green Economy in Africa? Lessons from the COMACO Model in Zambia

Orleans Mfune [1,2,*], Moses Ngongo Chisola [1] and Ignitius Ziba [1]

[1]*Department of Geography and Environmental Studies, The University of Zambia, P. O. Box 32379, Lusaka 10101, Zambia*

[2]*United Nations University-Institute for Natural Resources in Africa (UNU-INRA) Green Economy Project, International House, University of Ghana, Legon Campus, Accra, Ghana*

ABSTRACT

This paper examines the link between the green economy and multifunctional agriculture. In particular, the paper uses the case of the Community Markets for Conservation (COMACO) initiative, an agro-based enterprise promoting a multifunctional agriculture model in Eastern Zambia, to examine how the potential of smallholder farmers can be harnessed to support a transition towards the green economy. The empirical data on which the paper is based were collected through questionnaire surveys and in-depth interviews conducted with farmers and other actors in the agricultural sector. The results of the study show that a number of elements underpinning the COMACO model including sustainable land management practices, conservation outreach, community markets, value addition, and conservation dividends have great potential to deliver benefits related to the green economy. However, to truly foster a transition towards a green economy, a number of constraints need to be overcome. These include lack of a supportive policy and institutional framework, technological backwardness, and lack of consumer awareness of environmental information instruments such as eco-labelling.

Keywords: multifunctional agriculture; green economy; smallholder farmers; agro-based enterprise; community markets

1. INTRODUCTION

Growing dissatisfaction with the conventional development model driven by massive exploitation of natural capital, a huge appetite for fossil fuel and misallocation of capital is driving the search for alternative development and growth models. More recently, the notion of a green economy has gained ascendancy as the preferred alternative growth model. With the promise of delivering a low carbon, resource efficient, and socially inclusive development, the green economy has been endorsed by governments, inter-governmental organizations, businesses, and scholars alike. One of the major proponents of the green economy, the United Nations Environmental Programme (UNEP), argues that the 'green economy' should be viewed as one whose growth (in income and employment) is driven by investments (public or private) that reduce emissions and pollution, enhances energy and resource efficiency, and prevents loss of biodiversity and ecosystem services [1]. In other words, in a green economy, the generation of wealth should not result in the production of environmental risks or increase in social disparities. Indeed, the multiple crises facing the world today, such as the threats posed by climate change, loss of biodiversity, and increases in inequalities within and between regions, require that the world makes a radical departure from the status quo towards a green economy.

Drawing from the various definitions in literature (e.g., [2,3,4,5]), a green economy is seen in this paper as being characterised by at least six key elements: (a) low carbon economic activity; (b) resource efficiency; (c) social inclusion; (d) green jobs and businesses; (e) pollution and waste reduction; and (f) protection of biodiversity and ecosystem services. Green jobs, here, are viewed as jobs in business that produce goods or provide services beneficial to the environment and conserve natural resources (and includes workers with green skills) [3,6]. As Yi [4] notes, green jobs are multipurpose, providing tax and revenue in addition to employment.

Inevitably, with its focus on change, the green economy has important implications for various sectors that underpin the global economic system. Among these sectors is agriculture. As a source of livelihood for millions of people across the developing world, agriculture is key to sustainability and green economy aspirations. For decades, a model of agriculture based on intensive production systems to support high yielding crops has been promoted in a bid to feed the world's growing population. Over the years, a lot of studies have already drawn our attention to the ecological implications of extensive use of agro-chemicals and tillage practices in such agricultural systems [7,8,9]. Use of agro-chemicals and tillage practices have resulted in negative impacts on natural ecosystems and biodiversity and increased greenhouse gas emissions [10,11,12]. As Pearce et al. [13] note, unchecked abuse of resources (such as fertilizers) in an agricultural system do not only affect the sustainability of agroecosystems but also increase its vulnerability to external shocks and stresses such as market conditions, prolonged dry seasons, and changes in land tenure and others.

While, as already noted, on one hand, agriculture has been associated with an array of environmental externalities, on the other; there is a growing argument that agriculture provides one of the greatest opportunities for change towards a green economy. According to Musvoto et al. [2] agricultural activities potentially offer solutions to a myriad of social, economic, and environmental challenges. However, for agriculture to drive the green economy, it is acknowledged that changes are required within the sector [2]. As Leakey [14] equally argues, agriculture needs to be rethought, to move it from being an environmental culprit to an environmental, social, and economic savior. Here, multifunctional agriculture emerges as one such strategy that is reconfiguring agriculture towards some of the elements associated with a green economy. Traditionally, agricultural spaces have been viewed as primarily production spaces rather than multifunctional spaces [15]. In this regard, agro-based businesses have often ignored other functions that can be provided by agricultural landscapes. The concept of multifunctional agriculture, in this regard, highlights the fact that agriculture can play a significant role in delivering a range of socio-economic and environmental dividends beyond its productive functions [15]. In other words, this is a type of agriculture that embraces various other functions besides food production [8,11,16,17]. As a concept, the term multifunctional agriculture was first used at the Earth Summit in Rio de Janeiro (1992) [18] and further developed by the IAASTD [19] who have been highly instrumental in promoting its principles and practices.

Today, it is evident that there is a growing trend across the world towards multifunctional agricultural systems such as eco-agriculture and conservation agriculture that not only focus on food production but also ecosystem services and biodiversity conservation [7,20,21]. To its proponents, multifunctional agriculture is key to halting deforestation, biodiversity loss, land degradation, loss of ecosystem services in rural environments, and addressing inequalities in society [7,8,17,20]. However, it is important to note here that analyses of multifunctional agriculture have mostly focused on examining the role of agriculture in delivering ecosystem services and addressing environmental externalities in agricultural systems.

While provision of ecosystem services and addressing environmental externalities is a critical element in multifunctional agriculture, the notion of a green economy suggests that much more is required of agriculture. Beyond, addressing environmental externalities, the economic and social dimensions of the green economy requires that agriculture also addresses the inequalities that characterize the sector such as unequal access to farming inputs and markets. A few initiatives in Africa are already showing the importance of going beyond the focus on environmental externalities in delivering multifunctional agriculture. One such notable multifunctional agricultural initiative has been the agroforestry based three-step model implemented in Cameroon [22,23,24,25]. According to the proponents [22,23,24,25], the three step model involves rehabilitation of degraded lands through planting leguminous trees and shrubs as the first step. The second step involves the domestication of indigenous trees to produce salable products and in the third step; products are processed to add value and then marketed. Studies on these initiatives are already generating important lessons about the value of multifunctional agriculture and demonstrating the need to study such initiatives more keenly, especially in the context of the green economy. Moreover, with the concept of the green economy being new, the reality on the ground is that currently, there is insufficient information to aid green economy strategies or policy development [2] especially in African countries such as Zambia. In this regard, there are still some gaps in our understanding of the type of strategies and conditions that offer the most effective approach for supporting agriculture systems that are resilient, provide sustainable livelihoods, and at the same time foster inclusive and sustained economic growth and environmental sustainability.

Given these circumstances, in this paper, we draw on both empirical and theoretical data to analyze a model on the ground that similarly holds promise for fostering a green economy through multifunctional agriculture. In particular, we examine a multifunctional agricultural model being promoted in the Eastern Zambia by an organization called Community Markets for Conservation (COMACO). COMACO runs an agro-based enterprise that seeks to green the agricultural sector through promotion of sustainable land management practices, community markets, and premium pricing. The goal of the COMACO model is to simultaneously improve farmers' production and protect the environment. While being similar than the three-step model, the COMACO is conservation farming based and includes elements of eco-labeling and conservation dividends. By examining this model, we seek to answer one important question: how can multifunctional agriculture in Zambia be used to harness the potential of smallholder farmers to support a transition towards a green economy?

In the COMACO model, the multifunctional agriculture approach adopted is that of conservation farming (often used synonymously with the term conservation agriculture). Conservation farming (CF) is a farming approach that emphasizes zero or minimum tillage, use of permanent planting basins, retention of crop residues on the farm, reduction in inorganic fertilizer use and promotion of animal manure use. Further, conservation farming also encourages integration of nitro-fixing shrubs and other woody plants in crop lands, crop rotation, and mixed cropping. For an extended discussion of the major characteristics of conservation farming, see [11,26].

2. MATERIALS AND METHODS

Our study is premised on the idea that to support a green economy, multifunctional agriculture must not only be underpinned by the motivation to produce food and fibers, but must also be guided by social, economic, and environmental goals. In this regard, our analysis follows a four step approach. In the first step we examine the conventional agricultural model in the study area that provides, to some extent, a basis for comparison with the COMACO model. In the next step, we then examine the COMACO model in terms of how its multifunctional agricultural strategy is translated into practice. This then, allows us, in the third stage, to tease out the most innovative features of the COMACO model that hold promise for a green economy. We thus, with the fourth step, generate lessons from the model on how smallholder farmer's potential can be harnessed to foster a transition towards the green economy.

Empirical studies on which this data is based were carried out in eastern Zambia, in Chikwa and Kambombo Chiefdoms of the Chama District. The study area is part of the Luangwa Valley Ecosystem where COMACO has been operating. COMACO has been promoting conservation farming as a sustainable farming practice. The farmers are recruited on a voluntary basis and have to pledge to undertake conservation measures on their farm plots as the only condition for joining the COMACO initiative. It is important to note, here, that although there are other organizations promoting conservation farming, the COMACO model was selected for this study because of its unique approach to multifunctional agriculture. Perhaps, the most distinctive features of the model, especially in the context of a green economy, is the linking of conservation farmers to community markets, its focus on value addition, and the use of premium prices to compensate farmers who undertake environmental measures on their farms and avoid deforestation and encroachment into wildlife areas. The products from CF are sold under a label called 'It's Wild'. These features represent an important departure from initiatives solely focused on sustainable land management practices.

The Work of COMACO is concentrated along the Luangwa Valley Ecosystem, home to the South Luangwa and North Luangwa National Parks, two of Zambia's top wildlife parks. The valley is rich in both aquatic and terrestrial biodiversity. In addition to its rich flora, the ecosystem is home to a variety of large fauna such as lions, wild dogs, elephants, water bucks, and Thornicroft's Giraffes which are endemic to the area. The areas where COMACO is operating are presented in Figure 1.

Although the Luangwa Valley area is an ecosystem of outstanding biodiversity, it is also home to among the poorest people in the country. According to Seshaman [27], Chama, the study district, was in 2000 among the poorest 16 (16) in Zambia with incidences of poverty averaging 74.5%. At that time, the county had 72 districts. Mainly, the people in the area depend on rain-fed agriculture for their livelihood. Figure 2 presents Chama as the study area.

In Chama, data for local level studies was collected using a questionnaire survey. In total, 206 questionnaires were administered to households in the study area. Half (50%) of these households are farmers participating in the COMACO model. Theses households were selected from COMACO records kept at COMACO depots and identified in the community with the help of COMACO lead farmers. From the records, there are nearly 20,000 farmers that are participating in the COMACO program in the study area. However, for this study, participants were randomly selected from a population of 8000 farming households that were within an estimated 20 Km radius of COMACO depots in the area. This study was delimited to this area because of accessibility constraints beyond the 20 Km radius. For each, of the respondents involved in the COMACO program, the nearest non-COMACO neighbor was also targeted for the survey. This was done to allow for comparison of findings between households participating in the COMACO model and those not participating.

The questionnaire was used to collect data on farm size, type of crops cultivated, on-farm practices, and benefits or demerits of conservation farming as viewed by community members. This data also allowed the study to test (using Chi-square test with Yates correction) whether there were any significant statistical differences in farming practices between COMACO farmers and non-COMACO farmers. Besides, the questionnaire survey, 20 in-depth interviews were also conducted with key informants associated with the COMACO model and agriculture in general. Interviews were conducted with agricultural officers, wildlife officers, COMACO representatives, lead farmers, and other local community members.

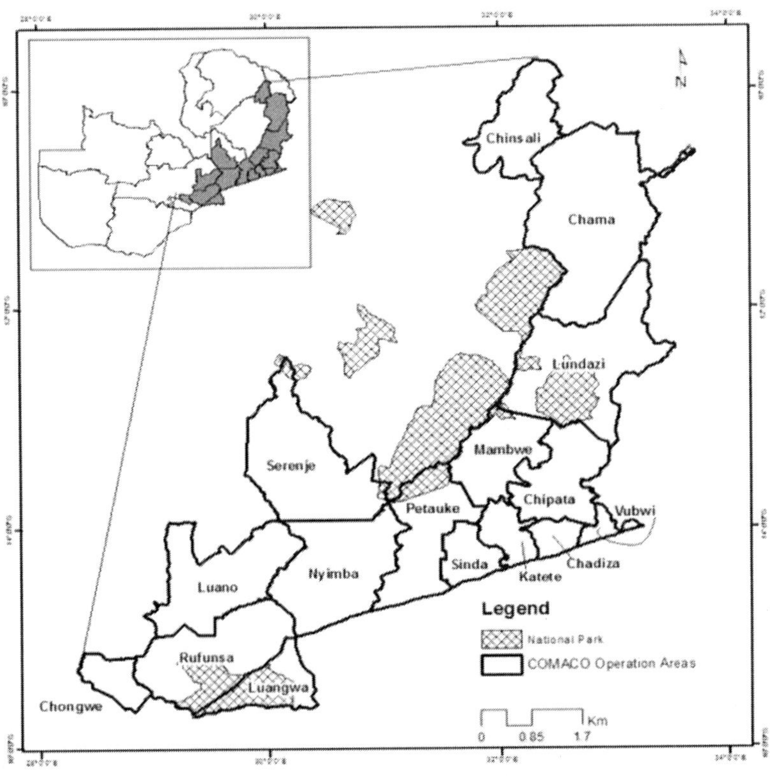

Figure 1. COMACO areas of operation.

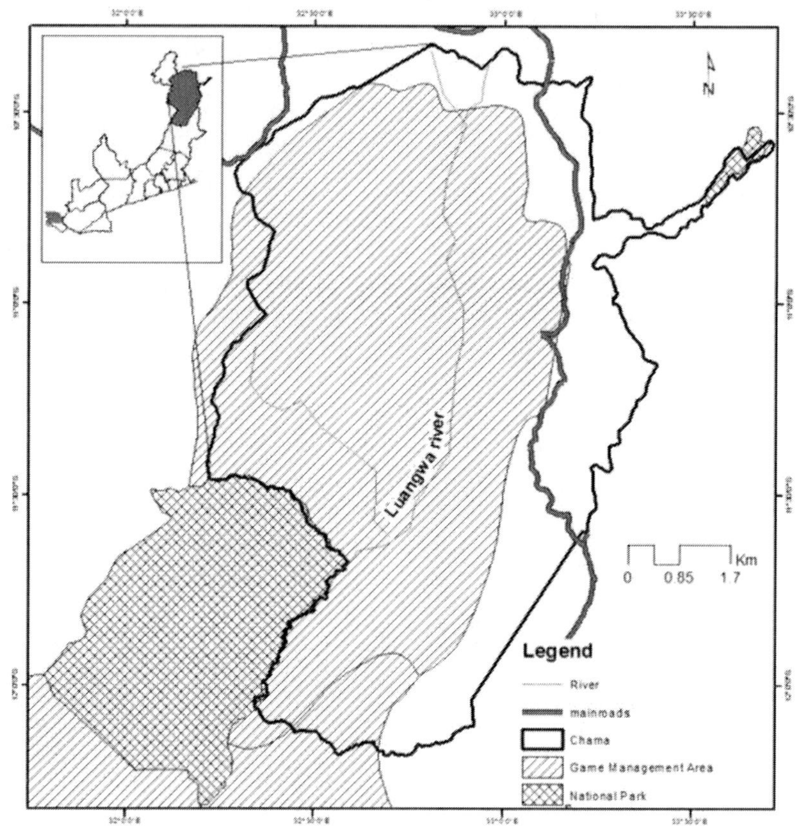

Figure 2. Chama district.

The questionnaire survey and interviews conducted in Chama were complimented by a consumer survey. As the processed products emanating from the programs are sold in the major supermarkets across the country, a consumer survey was conducted with shoppers outside the premises of three major supermarkets. These are Shoprite, PickNPay, and Spar. These supermarkets were selected on the basis that they are the three main supermarkets which also stock COMACO products. In total, 40 consumers participated in the survey. The shoppers who participated in the study constituted those who were willing to respond to questions or fill in the survey form. In the consumer survey, respondents were asked questions relating to their awareness of COMACO products, eco-labelling, and willingness to pay premium prices for environmentally friendly products.

3. RESULTS AND DISCUSSION

3.1. Agricultural Livelihoods in Chama District

Chama District, where this study was carried out, is largely rural with a very small urban center. The majority of the population is highly reliant on crop-based agriculture. In Kambombo and Chikwa Chiefdoms, for example, crop farming represented the most important economic activity with 98% of the research participants identifying it as their main source of livelihood. Other important sources of livelihoods include livestock rearing (mainly small ruminants and poultry) and small scale trading. Crop farming is mainly dominated by maize which is grown both for subsistence and for sale. Nearly all farmers that participated in the survey indicated that they grew maize in addition to other crops such as cotton, rice, ground-nuts, millet, and sorghum. From these results, it is evident that agriculture occupies a very important place in the economy of the district, with maize being the top crop. This mostly stems from the fact that maize is the staple crop in the area.

The agricultural sector is mainly dominated by smallholder farmers. Among the study respondents, the study found that the mean size of land cultivated by farmers is two hectares. Only a few farmers (less than 6%) cultivate land sizes in excess of four hectares. Clearly, the agricultural sector in the area is characterized by an absence of large-scale commercial farmers. In Zambia, this is not unique to Chama. Large-scale farming mostly occurs along the line of rail regions that have easy access to transportation lines, the market, and other services. Chama on the other, like many other rural districts is not connected by rail, has a poor road connection and is far from major market centers. Indeed, among some of the factors that farmers who participated in this study identified as major agricultural hurdles include difficulties of accessing markets of agricultural produce.

3.2. Features of Conventional Farming in Chama

Like other areas of Zambia, conventional farming involving conventional tillage and application of external or inorganic inputs dominates the crop-based agricultural sector in the study area. In this approach, land preparation involves clearing the farm plot of wild vegetation, crop residues (often burnt), and creation of seedbeds. The crops grown, mainly maize, are sustained by application of inorganic fertilizers. In the study, 60% of the farmers who practice conventional farming and participated in this study, indicated that they used fertilizer (mainly nitrogen and phosphorous fertilizer) to support growth of crops. Considering the difficulties that smallholder farmer's face in accessing inorganic fertilizers, this represents a slightly higher level of fertilizer use when compared to the national level. At the national level, statistics show that 55% of smallholder farmers use inorganic fertilizer [28]. Unfortunately, this use of fertilizer does not often translate into higher crop yields among smallholder farmers because of delays in fertilizer application (sometimes due to late access to it), misapplication, and insufficient use [28].

Besides use of inorganic fertilizer, the farming approach is characterized by agricultural extensification in which farmers periodically expand their agricultural activities into virgin lands mostly to replace fertilizer starved lands. As will be seen in the subsequent sections, over 70% of the farmers who participated in the study indicated that they extended their farm plots in the past five years, mostly into virgin territory. For conservationists that were interviewed in this study, this level of agricultural extension is of particular concern as virgin lands often include protected forest areas or wildlife areas.

This trend where farmers have been extending their fields into virgin territory is not new and should be viewed in its historical context. Prior to the 1990s, agriculture in the country was largely sustained by fertilizer subsidies through the socialist policies of the then government. In the 1990s, the socialist policies in Zambia gave way to a free market economy. This saw the withdrawal of many agriculture related subsidies. It is argued that the withdrawal of state intervention in agriculture significantly affected crop and livestock production and created food deficits in rural areas. In turn, this increased the pressure on natural resources and biodiversity as households turned to non-farm products for survival, while others begun to expand into virgin land into protected areas in a bid to acquire fertile virgin lands to replace fertilizer starved soils [29,30,31]. In addition, the loss of Jobs through privatization and public sector reforms triggered an urban-to rural migration which added further pressure on forested and wildlife lands as some of these migrants began to settle on protected lands. Indeed, some studies report an increase in encroachment of protected areas during this period as shrinking urban employment opportunities forced many to take on farming [30,31,32,33]. According to Mbindo [31], forest cover in Zambia, which was at 39,755,000 hectares in 1999, had reduced to 31,346,000 hectares by 2000 while in national parks, poaching was rife as people sought to complement their agricultural livelihood.

It is this type of farming that has mainly been linked to an array of environmental impacts which includes biodiversity loss, poor soil organic matter, disturbance of soil structure through tillage, pollution of aquatic systems from runoff of nitrates, and land-use based GHG emissions. The features of this agriculture are shown in Figure 3.

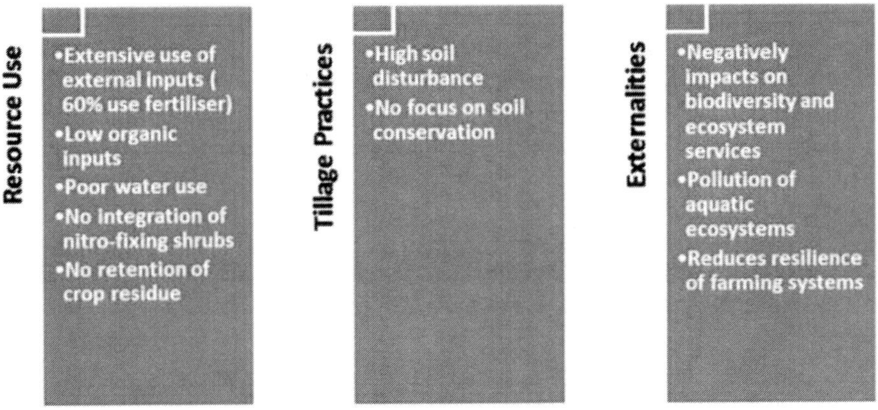

Figure 3. Features of conventional farming in the study area.

From the results of the study and the discussion above, it is evident that this type of agriculture is primarily aimed at production of food and fibers which does not include environmental goals. In this model, there is little space for green economy aspirations such as green products, jobs, or businesses. From a sustainability point of view, this is an example of agriculture that can be considered as unsustainable in the long run. To change this picture and move towards a green economy trajectory requires decoupling crop production from environmental damage and excessive resource use. This, it appears, is what the COMACO model seeks to achieve.

3.3. COMACO's Multifunctional Agricultural Model

As already noted in the preceding section of this work, at the heart of the COMACO model is the use of conservation farming as a sustainable land management strategy. Farmers who embrace conservation farming are expected to adopt a range of sustainable farming practices that include minimum tillage, crop residue retention, and use of permanent planting basins. Ideally, if farmers religiously followed these practices, it is expected that this would lead to high yields, less or no use of external inputs, increased on-farm water conservation, improved soil nutrients, less erosion and have positive effects on on-farm carbon sequestration and storage. From the results of the study, the level of adoption of sustainable farming practices is much higher among COMACO farmers than non-COMACO farmers. For example, all COMACO (100%) respondents were found to practice minimum tillage while only 2.9% of non-COMACO farmers had adopted the practice. Further, COMACO farmers showed higher levels of adoption of other practices such as retention of crop residues on farm plot after harvest, use of animal manure, crop rotation, and intercropping. For these, the Chi-square test results showed that the differences between COMACO farmers and non-COMACO farmers were significant. Table 1 indicates the level of adoption of sustainable farming practices between COMACO farmers and non-COMACO farmers.

Table 1. Differences in Level of Adoption of Sustainable Farming Practices.

Farm Practices	COMACO Farmers (%) (n = 103)	Non-COMACO Farmers (%) (n = 103)
Minimum Tillage ***	100.0	2.9
Mixed Cropping	65.0	53.4
Crop residue Retention ***	60.2	17.5
Manure Use **	38.8	21.4
Integration of Shrubs ***	31.1	1.9
Fertilizer Use **	36.9	60.2
Crop Rotation **	82.5	61.2
Intercropping *	36.9	19.4
Mulching	29.1	22.3

Chi-square test, 1 df: *** Significant at $p < 0.0001$, ** Significant at $p < 0.005$, * Significant at $p < 0.01$.

While the level of adoption of sustainable farming practices is much higher among COMACO farmers, it is important to note that not all practices promoted by COMACO were adopted by COMACO farmers. As can be seen from Table 1, apart from minimum tillage, there is no single practice that was adopted by 100% of COMACO farmers who are practicing conservation farming. Further, it is evident that there are some practices that are less adopted than others. For example, while less than 50% of the respondents' practice intercropping, mulching, and integrate nitro-fixing shrubs in their crop fields, some elements such as crop rotation, mixed cropping, and crop-residue retention are practiced by the majority of the conservation farmers. Moreover, the study found that some practices such as mixed farming were already widespread in the study area before the introduction of conservation farming. In this regard, the study found no significant differences (X^2= 3.372, n = 206, p = 0.663) in levels of adoption of mixed cropping between COMACO and non-COMACO farmers. It appears, here, that COMACO is building on some of the existing indigenous farming practices in the area.

Of critical importance here is the fact that among COMACO farmers, fertilizer use is significantly lower than non-COMACO farmers who are involved in conventional farming. This difference was also found to be significant ($X^2 = 10.28, n = 206, p = 0.003$). Only 36.9% of the conservation farmers indicated that they used fertilizer on their farm plot (cf 60.2% for conventional farming). This reflects a considerable reduction in use of inorganic fertilizers among conservation farmers.

Perhaps, another major observation in this study is that, apart from selective adoption of some of the conservation farming practices, very few farmers place their entire cultivated land under conservation farming. Figure 4 shows the proportion of land that farmers reserve for conservation farming.

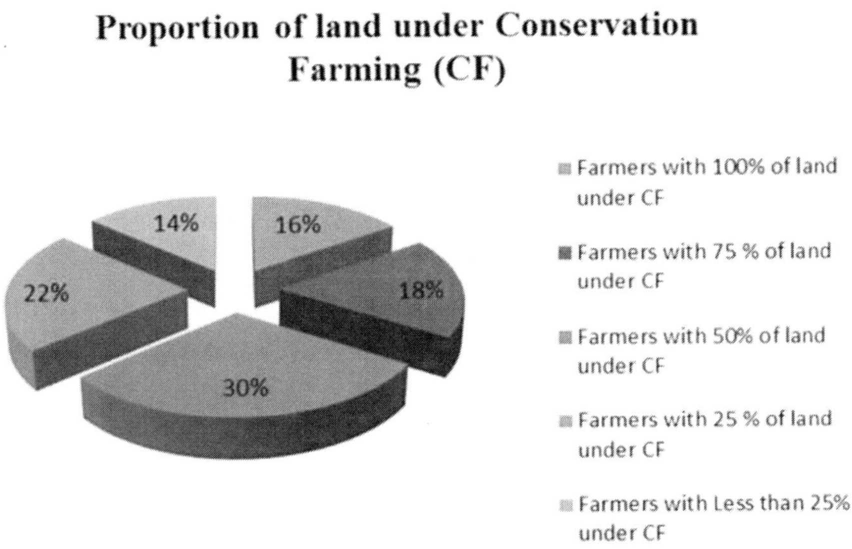

Figure 4. Proportion of land that is placed under conservation farming.

As can be seen from Figure 4, only a small proportion of farmers who participated in the study (16%) have placed their entire cultivated land under conservation farming (CF). For the majority farmers only part of their cultivated land is under CF. This, in reality, means that nearly 85% of COMACO farmers also practice conventional farming. While farmers do point to important benefits associated with CF, including improved crop yields, it is clear that there are important bottlenecks to complete adoption of CF. Among the bottlenecks to full adoption identified by the farmers include too much labor involved in the creation of planting basins, weeds, and lack of advanced technology such as rippers to ease the pressure on family labor. During interviews with some of the key informants, it was found that these factors are among the most important constraints to adoption of conservation farming. Indeed, some of the respondents that still practice conventional farming noted that they were not involved in conservation because it was too labor intensive.

Still, despite the selective adoption of conservation farming practices, the results also show that farmers are highly appreciative of some of the elements of conservation farming. Some of the benefits that farmers identified are presented in Figure 5.

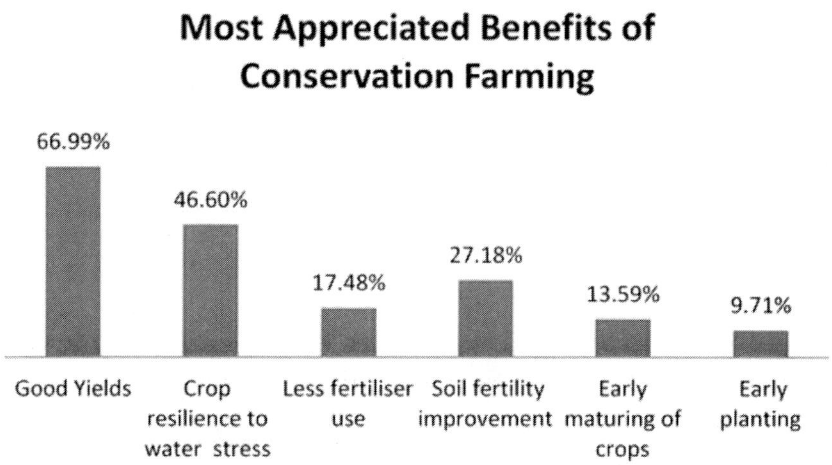

Figure 5. Most Appreciated Benefits of Conservation Farming by COMACO farmers.

As can been seen, over half (67%) of the respondents that participated in the household survey single out 'good yields' as one of the most important benefits they have gained from conservation farming practices. Indeed, in a more extended study than this one, Nshimbi and Vinya [34] claim that they found that more than 80% of COMACO farmers directly attributed increased crop production on their farm plots to conservation farming promoted by COMACO. Besides good yields, Figure 5 also shows that another most important benefit that farmers identified was that their farming systems were becoming more resilient to water stresses most often associated with climatic changes. This is largely the result of use of permanent planting basins which allow for on-farm water conservation and therefore buffer against water stresses induced by prolonged dry spells. In this regard, conservation farming emerges as a climate adaptation technology.

3.4. Conservation Farming Practice and the Environment

As earlier noted, CF as promoted by COMACO seeks to not only improve agricultural productivity but also deliver environmental dividends beyond the farm-site, including halting of deforestation. While representatives of COMACO who participated in the study claim that so far, COMACO has been successful in doing so, evidence on the ground is rather mixed. Certainly as seen in Table 1, there are a number of elements where COMACO's claims can be substantiated. At a micro-level (i.e., farm site level), COMACO has been successful in fostering a reduction in fertilizer use and increase in uptake of a number of sustainable farming practices such as crop rotation, intercropping, and on-farm crop residue retention. All these have potentially beneficial effects on the environment.

There is also a huge claim that because of COMACO efforts, wildlife in the Luangwa Valley Area has increased considerably. However, in this study, we were unable to empirically verify this claim because of two things. First, because we lack objectively collected baseline data and secondly, we were unable to do so without isolating the initiatives of other actors involved in wildlife protection in the area. In this regard, a better picture of how the model has performed in halting wildlife depletion is best provided by Lewis [35] who makes explicit links between the COMACO initiative and wildlife conservation. Further, other studies elsewhere also note that such type of sustainable agricultural approaches can be achieved in harmony with wildlife. On the other hand, our study found that the COMACO approach has been, to some extent, unable to halt agricultural extensification. Our micro-level data show that farmers, including those under CF, continue to expand their farm plots into virgin lands (Table 2). Indeed, the study did not find any significant differences between COMACO and Non-COMACO farmers ($X^2 = 1.58$, $p = 0.207$) in terms of expansion of farm plots. In both groups, more than half of the respondents (68% of COMACO farmers and 80% of non-COMACO farmers) have expanded their fields over the past five years (Table 2). Of all who extended their farm plots, more than 70% did so by opening new plots on 'virgin' (or new) land (Figure 6).

Partly, the inability of the COMACO model to halt expansion of farm plots to virgin land arises from the fact that CF is only partially adopted by farmers in the area.

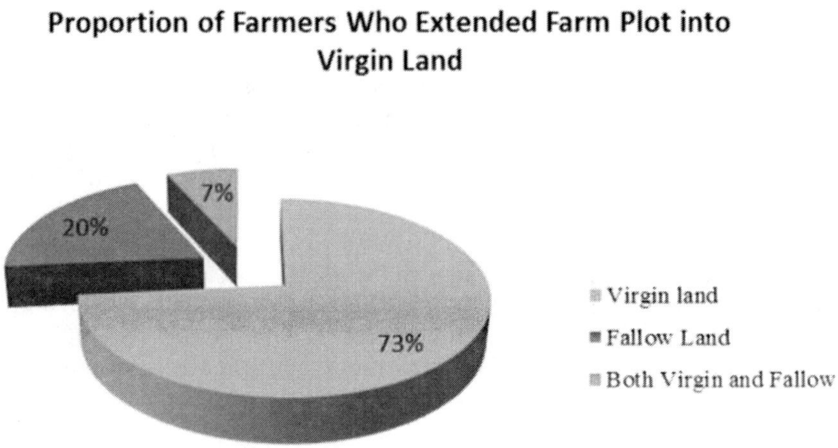

Figure 6. Proportion of Farmers Who Extended Farm Plots to Virgin Land.

Table 2. Differences in Agricultural Extensification.

	COMACO Farmers		Non-COMACO Farmers	
		%		%
Extended Farm Over the Past Five Years	71	68.9	80	77.7
Did Not Extend Farm Over the Past Five Years	32	31.1	23	22.3
Total	103	100	103	100

Chi-square test, 1 degree of freedom (df): not significant at $p = 0.207$. Source: Authors, 2016.

3.5. Innovative Features of the COMACO Model

Notwithstanding the limitations and constraints to conservation farming emerging from the results of the study, the COMACO model still represents an important attempt at greening the agricultural sector through sustainable land management practices. Here, conservation agriculture is used as a multifunctional agricultural strategy. As earlier noted, while COMACO is not the only actor promoting conservation agriculture in Zambia; the COMACO approach is quite unique from other similar attempts. This is because in promoting conservation agriculture, most actors are mainly concerned with increasing crop yields, efficiency in agro-resource use and buffering crop based agriculture against climate related risks. In this regard, conservation agriculture still remains single-sector centered (i.e., focusing on the agricultural sector alone). Further, the promotion of CF mostly focuses on the production side and is delinked from other components of the agricultural value chain such as value addition and marketing.

As a departure from other approaches to conservation farming, the COMACO model assumes a truly multifunctional character with potential to deliver social, economic, and ecological benefits. In the model, conservation farming is conceptualized as an approach for not only improving agricultural productivity and resource use efficiency, but also for fostering biodiversity conservation and halting the expansion of agriculture into forests and wildlife habitats. In this regard, in the model, on-farm practices are accompanied by conservation outreach programs. In this way, the COMACO approach deliberately attempts to bridge the gap between agriculture and conservation. This is important, as traditionally, agriculture and biodiversity conservation have been viewed as two separate domains of policy and practice [9,20]. Thus, we envision here, that if a rural green economy is ever to be a reality, agricultural production, and biodiversity conservation need not be considered as incompatible goals. Many studies dealing with agri-environmental management now prove that agriculture spaces can be important arenas of biodiversity conservation and ecosystem services management [14,16]. These studies render support for shift in conservation, to go beyond protected areas to encompass various types of socio-ecological systems, including agricultural environments, where lands can be used for multiple purposes (see also [20,36]. It is here, where agricultural businesses can make the greatest impacts by integrating biodiversity and ecosystem services protection in their activities. Figure 7 shows the multi-functionality of conservation farming as promoted by the COMACO model.

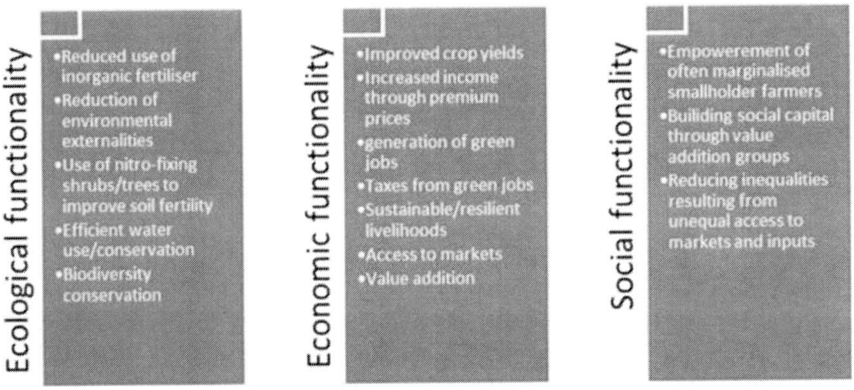

Figure 7. Multifunctionality of the COMACO agricultural model.

Beyond the farm site, COMACO is involved in provision of markets for farmers' produce. Through what are termed as community markets, COMACO provide ready markets for farmers' produce, cutting distances to markets and allowing for quick payments for their produce. Indeed, in this study, the results show that quick payments for produce, and easy access to and reduced distance to the market are among the elements of the COMACO model most appreciated by farmers. Figure 8 presents some of the benefits of the COMACO model that farmers in the study identified beyond simply sustainable land management practices.

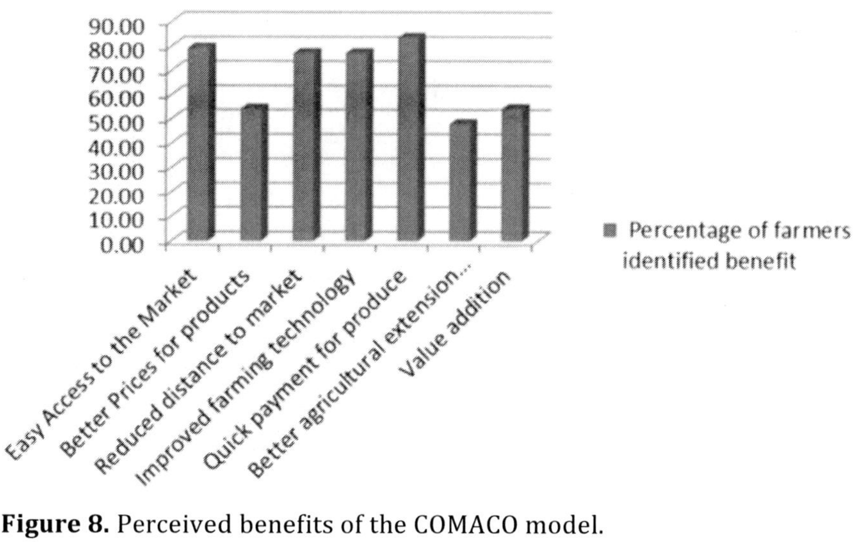

Figure 8. Perceived benefits of the COMACO model.

Further, there is also an attempt to draw on new environmental voluntary instruments such as premium pricing and eco-labelling in order to incentivize conservation actions by farmers. In this regard, to compensate for undertaking environmental measures on their farm plots, products from conservation farming attract what has been termed as 'conservation dividends'. According to a COMACO officer that participated in this study, a premium price is offered to these farmers which are 20% more than the standard crop prices offered on conventional markets. During interviews, it was noted by some of the respondents in the study that the use of premium prices has made it more profitable for farmers to engage in sustainable agricultural practices than poaching or charcoal production.

Within the program, farmers are organized into groups which engage into value addition. The processed products are then packaged and marketed in Zambia's urban markets (as eco-friendly products) under the label *'It's Wild'*. *'It's Wild'* is a label that draws attention to the fact that the products are organically produced and aimed at supporting conservation efforts. In a way, the brand mirrors an eco-label which conveys environmental information to the consumer. From the results of this study, there is no doubt that COMACO has scored some level of success in marketing the label 'It's Wild'. Today, *'It's Wild'* products can be found in major supermarkets in Zambian cities including Lusaka. For example, Shoprite and PickNPay, two of the leading supermarkets all stock these products. Similarly, our consumer survey shows a high level awareness of the existence of these products. In particular, the study shows that consumers are mostly aware of '*It's Wild* Chama Rice', peanut butter and honey products. Thirty six (36), representing 90% of the 40 consumers who participated in the survey had indicated that they had purchased at least one of these products before.

However, an important concern here is that most consumers did not associate the *'It's Wild'* products with environmental concerns. In fact, more than half of the respondents noted that they purchased the products because they liked the taste of the products and only 12% indicated that they were motivated by environmental concerns. This situation, however, shows a lack of awareness among consumers of the importance of environmental information on products they purchase. Indeed, when consumers were asked if they had any knowledge of eco-labels, more than 90% of them expressed ignorance of what they were all about. However, when asked if they were willing to pay a little bit more for a product that was produced in an environmentally friendly manner than a similar product that was not, more than 72% responded 'yes'. Thus, it seems here, that some consumers do have environmental concerns in regard to how a product is produced but are largely unaware that labelling conveys such information.

The lack of awareness of environmental information tools such as eco-labelling and premium pricing also stems from the fact that, at the national level, there have been no deliberate policy and institutional efforts to popularize these instruments, be it in agriculture or other sectors. Moreover, the environmental sector is also characterized by an absence of independent certification schemes. These factors, in addition to constraints affecting the adoption of conservation farming such as the pressure on family labor and lack of appropriate ripping technology are among the hurdles that need to be overcome in order to allow multifunctional agriculture support a transition towards a green economy. It must be recognized that an important pursuit such as the green economy cannot be achieved on the basis of mere rhetoric. Going forward, the country will need to develop green economy policy frameworks that take these factors into cognition. As yet, other than policies supportive of sustainable land management strategies, there are none explicitly focusing on the green economy.

3.6. Lessons From the COMACO model

From the discussion of the COMACO model in the preceding sections of this paper, it is evident that the COMACO model holds great potential for integrating agri-business with biodiversity and ecosystem services concerns. As a rural-based agricultural enterprise operating in some of the poorest (yet biologically rich areas of Zambia), the model has potential to revolutionize the agriculture industry in the country because of its unique focus on sustainable land practices, value addition, and provision of missing infrastructure such as missing markets and hence lowering farmers transaction costs

There are a number of policy lessons that can be drawn from the COMACO model, and are of critical importance to harnessing the potential of smallholder farmers to support the transition towards a green economy. First, as already noted, among the factors that condition success in African agriculture is the issue of access to and use of fertilizers among smallholder farmers ([10]). It is noteworthy, here, that the COMACO model demonstrates the need for minimization of costly inputs in smallholder farming systems in order to reduce farmers' costs, improve resource use, and competitiveness of their produce. Moreover, the model draws our attention to the need to give priority to fostering use of organic fertilizers and other biological means of improving smallholder farming systems. This may have the duo effect of reducing externalities associated with inorganic fertilizer use while also reducing farmer's costs of production. It is important to note here, that besides conservation farming, organic agriculture, which overly emphasizes use of biological means of enhancing soil fertility, is another strategy that can easily play an important role in supporting the greening of agriculture in Zambia.

The potential for organic agriculture in the country is great. Zambia has thousands of farmers already engaged in organic agriculture and was in 2008, ranked as fourth on countries with the highest number of organic farms in Africa [37]. In this paper's view, if supported by a favorable policy environment, both conservation farming and organic agriculture would play an important role in wide-scale development of green businesses and jobs through certification schemes, marketing and sales of products. Already, the COMACO model demonstrates that green businesses and jobs can be created within the rural economy. Value addition, agriculture extension services and development of new markets such as community markets all make it possible for new businesses and jobs to emerge. It is important in this regard, that policy makers give explicit attention to these elements in their quest to promote the greening of the agricultural sector.

Second, the COMACO model also raises the need to buttress smallholder agricultural systems against climate change related stresses and shocks if they are to play a role in the transition towards a green economy. The results demonstrate that undertaking climate change risk reduction strategies need not be very costly as simple measures such as use of permanent planting basins (which promote on-farm water conservation) can equally be effective. It is crucial, however, that in promoting such measures, COMACO, government departments, and organizations promoting sustainable agricultural practices pay particular attention to the demand that such practices place on family labor, a concern raised by most respondents in this study.

The concern with labor brings us to the third point: It is our contention here that a green economy cannot be achieved on the back of technological backwardness of smallholder farmers. As already noted, the results of the study show that while farmers in the COMACO model appreciate a number of elements of multifunctional agriculture, lack of appropriate and less labor intensive technology remains a hindrance to advancing sustainable land use practices among smallholder farmers. Here the COMACO model can take a leaf from the three step model to include a focus on design and fabrication of technology both for tillage, processing, and value addition. In Chama, farmers lack ripping technology and rely on handheld tools to create planting basins and clear weeds on conservation farms. Consequently, farmers only put a fraction of their fields under conservation farming. In fact, this finding is not unique to Chama. In another study on conservation farming practices in Central Zambia, we found that farmers rarely place their entire fields under conservation farming, largely due to family labor constraints and lack of ripping technology [15]. In this regard, it seems that to harness the potential of smallholder farmers towards a green economy, the role of appropriate agricultural technology that meets farmers' expectations is crucial. Thus, green agricultural policies and strategies must also seek to increase farmers' access to appropriate ripping technology.

Far beyond what happens at the level of the farm plot, the COMACO model also demonstrates that to harness the potential of smallholders, access to markets, and premium pricing are of crucial importance to fostering their productivity and incentivizing their participation in multifunctional agriculture. Often, as the case of Chama shows, smallholders located in rural areas have to contend with long distances to markets and unfavorable prices for their produce. Creating appropriate farmer-centered markets such as community markets allows agriculture to deal with these elements and contribute to resolving the perennial problem of unequal access to high paying markets.

4. CONCLUSIONS

With the emergence of the concept of the green economy, the need for agro-initiatives to integrate biodiversity, ecosystem services, and socio-economic goals in agricultural practices has been brought to the fore. However, as business actors, environmentalists, and academics are only getting to grasp the essential elements of a green economy; few practical and successful examples of initiatives that are pursuing these strategies in African countries such as Zambia exist. The COMACO model, in this regard, represents one of the few rural based agri-business models in Zambia that seem to have embraced some of the elements of a green economy. Elsewhere in Africa, similar models that hold great promise include the three step model implemented by the World Agroforestry Centre in Cameroon. Many of the elements in the COMACO model, including a focus on sustainable land management practices, emphasis on conservation, value addition, and local markets for farmer products, represent an important departure from conventional farming that is often solely focused on production of food and fibers. However, COMACO can also benefit from the three step model especially with regard to domestication of indigenous trees and shrubs and skills development in processing and design of value addition equipment and technology.

While it is clear that smallholder agriculture in Zambia is constrained by a variety of factors that include poor productivity, lack of access to inputs, missing markets, and a variety of environmental ills (such as climate change and poor soils), this paper shows that COMACO seem to have turned these constraints into a green agri-business opportunity. By looking at some of the innovative features of the COMACO model, the paper shows that opportunities for harnessing the potential of smallholder farmers towards the green economy not only lie in sustainable land management practices, but also in developing new, accessible markets for farmer products, innovative incentives (such as conservation dividends and premium prices) and value addition. However, lessons from the same model also show that a number of constraints need to be overcome for multifunctional agriculture initiatives to thrive. These include lack of a supportive policy and institutional framework, technological backwardness of the smallholder farmers, and lack of consumer awareness of environmental information instruments.

ACKNOWLEDGMENTS

We thank the United Nations University-Institute for Natural Resources in Africa (UNU-INRA) for their technical and financial support towards the research project that led to the production of this paper. In particular, we thank Calvin Atewamba and anonymous reviewers whose comments were invaluable in improving the quality of the paper. We also thank our Research Assistants in the Chama District who travelled long distances to administer the survey instruments.

AUTHOR CONTRIBUTIONS

Orleans Mfune, Moses Ngongo Chisola, and Ignitius Ziba conceived and designed the experiments; Orleans Mfune, Moses Ngongo Chisola, and Ignitius Ziba performed the experiments; Orleans Mfune, Moses Ngongo Chisola, and Ignitius Ziba analyzed the data; Orleans Mfune contributed materials/analysis tools; Orleans Mfune, Moses Ngongo Chisola, and Ignitius Ziba wrote the paper.

CONFLICTS OF INTEREST

This work has been submitted as a working paper to UNU-INRA who also the funded the research project that led to the production of the paper. In this regard, permission to publish the working paper as a research article has been obtained from UNU-INRA.

REFERENCES

1. United Nations Environment Programme (UNEP). Towards a Green Economy: Pathways to Sustainable Development: A synthesis for Policy Makers; UNEP: Nairobi, Kenya, 2011.
2. Musvoto, C.; Nahman, A.; Nortje, C.; de Wet, B.; Mahumani, B. Agriculture and the Green Economy in South Africa; Council for Scientific and Industrial Research (CSIR): Pretoria, South Africa, 2014.
3. Connolly, K.; Allan, G.J.; Mclintyre, S.G. The Evolution of Green Jobs in Scotland: A Hybrid Approach. Energy Policy2016, 88, 355–360.
4. Yi, H. Green Businesses in a Clean Energy Economy: Analysing Drivers of Business Growth in US States. Energy2014, 68, 922–929.
5. Cai, W.; Wang, C.; Chen, J.; Wang, S. Green Economy and Green Jobs: Myth or Reality? The Case of China's Power Generation Sector. Energy 2011, 36, 5995–6003.

6. Furchtgott-Roth, D. The Elusive and Expensive Green Growth. Energy Econ. 2012, 34, 543–552.
7. Mattison, H.A.; Norris, K. Bridging the Gap between Agricultural Policy, Land Use and Biodiversity. Trends Ecol. Evol.2005, 20, 610–616.
8. Milestad, R.; Ahstrom, J.; Bjorklund, J. Essential Multiple Functions of Farms in Rural Communities and Landscapes. Renew. Agric. Food Systems. 2011, 26, 137–148.
9. Batary, P.; Bald, A.; Kleijn, D.; Tscharntke, T. Landscape-moderated Biodiversity Effects of Agri-environmental Management. Proc. Royal Soc. 2011, 278, 1894–1902.
10. United Nations Environment Programme (UNEP). Agriculture and the Green Economy. 2015. Available online: http://www.unep.ch/etb/ publications /Agriculture/flyer%20UNEP%20FOAM.pdf (accessed on 29 December 2015).
11. Dumanski, J.; Peiretti, R.; Benets, J.; McGarry, D.; Pieri, C. The Paradigm of Conservation Agriculture. Proceedings of the World Association of Soil and Water Conservation, August, 2006; P1:58-64. Available online: http://www.unapcaem.org/publication/ConservationAgri/ParaOfCA.pdf (accessed on 31 January 2016).
12. Cunder, T. The Role of Rural Development Policy in Environmental and Land Management in Slovenia. J. Cent. Eur. Agric. 2007, 8, 237–242.
13. Pearce, D.; Markandya, A.; Barbier, E. Blue Print for a Green Economy; Earthscan Publications Ltd.: London, UK, 1989.
14. Leakey, R.R.R. Living with the Trees of Life: Towards the Transformation of Tropical Agriculture; Centre for Agriculture and Bioscience International (CABI): Wallingford, UK, 2012; p. 224.
15. Gorman, M.; Mannion, J.; Kinsella, J.; Bogue, P. Connecting Environmental Management and Farm Households Livelihoods: The Rural Environment Protection Scheme in Ireland. J. Environ. Policy Plan. 2001, 3, 137–147.
16. Mfune, O. Extending Conservation to Farmlands in Zambia. Prescribed Practices and Reality. J. Sustain. Dev. 2014, 7, 46–59.
17. Jean-Vasile, A. Multi-functional Agriculture and the Green Economy. In Sustainable Practices: Concepts, Methodologies, Tools, and Applications; Information Science Reference: Hershey, PA, USA, 2013; pp. 1701–1722.
18. United Nations Conference on Environment and Development (UNCED). Agenda 21—An Action Plan for the Next Century; United Nations Conference on Environment and Development: New York, NY, USA, 1992.
19. McIntyre, B.D.; Herren, H.; Wakhungu, J.; Watson, R. (Eds.) International Assessment of Agricultural Knowledge Science and Technology for Development: Global Report; Island Press: New York, NY, USA, 2008.
20. Scherr, S.J.; McNeely, J.A. Biodiversity Conservation and agricultural sustainability: Towards a new conservation paradigm of eco-agriculture landscapes. Philos. Trans. Royal Soc. 2008, 363, 477–494.
21. Kareiva, P.; Watts, S.; McDonald, R.; Boucher, T. Domesticated Nature: Shaping Landscapes and Ecosystems for Human Welfare. Science 2007, 316, 1866–1869.
22. Leakey, R.R.B. Agroforestry: A delivery mechanism for multi-functional agriculture. In Handbook of Agroforestry: Management Practices and Environmental Impact; Kellimore, R.L., Ed.; Nova: New York, NY, USA, 2010; pp. 461–471.
23. Asaah, E.K.; Tchoundjeu, Z.; Leakey, R.R.B.; Takousting, B.; Njong, J.; Edang, I. Trees, agroforestry and multifunctional agriculture in Cameroon. Int. J. Agric. Sustain. 2011, 9, 110–119.
24. Leakey, R.R.B. Agroforestry—Participatory domestication of trees. In Encyclopedia of Agriculture and Food Systems; van Alfen, N.K., Ed.; Elsevier Publishers: San Diego, CA, USA, 2014; Volume 1, pp. 253–269.
25. Leaky, R.R.B. The Role of Trees in Agroecology and Sustainable Agriculture in the Tropics. Annu. Rev. Phytopathol.2014, 52, 113–133.
26. Baudron, F.; Mwanza, H.; Triomphe, B.; Bwalya, M. Conservation Agriculture in Zambia: A Case Study of Southern Province; FAO/ACT: Nairobi, Kenya, 2007.

27. Seshaman, V. Deprivation in Zambia: District-Level Profiles and Rankings Based on an Index of Depravation; United Nations International Children's Emergency Fund (UNICEF): Lusaka, Zambia, 2000.
28. Tembo, S.; Sitko, N. Technical Compendium: Descriptive Agricultural Statistics and Analysis for Zambia; IAPRA Working Paper 76; Indaba Agricultural Policy Research Institute: Lusaka, Zambia, 2013.
29. Kajoba, G.M.; Chidumayo, E.N. Degradation of Forest Reserves in Zambia: A Case Study of Muyama in Central Zambia. UNEP Desertification Control Bull. 1999, 35, 59–66.
30. Kajoba, M.G. Squatting in Protected Forest Areas: A Dilemma for Sustainable Natural Resources Management; Research Report; Provincial Forest Action Programme and Forest Department: Ndola, Zambia, 1999.
31. Mbindo, K. Forest Cover Crisis in the Tropics: A Case Study of Zambia. In Proceedings of the XII World Forestry Congress, Quebec, QC, Canada, 21–28 September 2003.
32. Chidumayo, E. Land-cover Transformation in Central Zambia: Role of Agriculture, Biomass Energy and Rural Livelihoods. In Proceedings of the International Symposium on 'Area Studies: Past Experiences and Future Visions', Kyoto, Japan, 19–22 January 2001.
33. GRZ-MTNR. Zambia Forest Action Plan. Challenges and Opportunities; GRZ: Lusaka, Zambia, 1988.
34. Nshimbi, M.; Vinya, R. Impacts of Public-Private Partnerships on Local Livelihood and Natural Re-sources Dynamics: Perceptions from Eastern Zambia. Resources 2014, 3, 471–485.
35. Lewis, D. Community Markets for Conservation (COMACO) links Biodiversity Conservation with Sustainable Improvements in Livelihoods and Food Production. PNAS 2011, 108, 13957–13962.
36. Mery, G.; Alfro, R.; Kanninen, M.; Lobovikov, M. (Eds.) Forests in the Global Balance-Changing Conservation Paradigms; International Union of Forest Research Organisations: Vienna, Austria, 2005.
37. International Federation of Organic Agriculture Movements (IFOAM). One Earth, Many Hands: IFOAM 2008 Annual Report; IFOAM: Bonn, Germany, 2008.

CHAPTER 20

Ecology of Fungus Gnats (*Bradysia* spp.) in Greenhouse Production Systems Associated with Disease-Interactions and Alternative Management Strategies

Raymond A. Cloyd

Department of Entomology, Kansas State University, Manhattan, KS 66506, US

ABSTRACT

Fungus gnats (*Bradysia* spp.) are major insect pests of greenhouse-grown horticultural crops mainly due to the direct feeding damage caused by the larvae, and the ability of larvae to transmit certain soil-borne plant pathogens. Currently, insecticides and biological control agents are being used successively to deal with fungus gnat populations in greenhouse production systems. However, these strategies may only be effective as long as greenhouse producers also implement alternative management strategies such as cultural, physical, and sanitation. This includes elimination of algae, and plant and growing medium debris; placing physical barriers onto the growing medium surface; and using materials that repel fungus gnat adults. This article describes the disease-interactions associated with fungus gnats and foliar and soil-borne diseases, and the alternative management strategies that should be considered by greenhouse producers in order to alleviate problems with fungus gnats in greenhouse production systems.

Keywords: fungus gnats; pest management; plant protection; diseases; greenhouse; cultural control; sanitation; physical barriers; repellents

1. INTRODUCTION

Fungus gnats in the genus *Bradysia* spp. (Diptera: Sciaridae) are major insect pests of greenhouse production systems feeding on a wide-range of horticultural crops [1,2,3] with the commonly encountered species being *Bradysia coprophila*Comstock and *B. impatiens* Johannsen [4,5,6]. Fungus gnats, in general, are ubiquitous in greenhouses [7] but are primarily a problem under conditions of excessive moisture, which commonly occurs during propagation when cuttings and plugs are developing root systems [8,9]. The larval stage, which resides within the growing medium, feeds mainly on decaying plant material and fungi [10]. However, they will also feed on healthy plant roots and tunnel into stems of young cuttings and seedlings [4], and the crowns of mature plants [11]. Therefore, the larvae are primarily responsible for causing direct plant damage by feeding on the roots, thus interfering with the ability of plants to uptake water and nutrients, which results in wilting and stunted growth [3,4,12,13]. Fungus gnat adults are mainly a nuisance causing minimal direct plant damage [9].

A number of insecticides from different chemical classes and various biological control agents including a soil-dwelling predatory mite (*Stratiolaelaps scimitus* (Womersley)), a rove beetle (*Dalotia coriaria* Kraatz), and entomopathogenic nematode (*Steinernema feltiae* Filipjev) have been shown to be effective in suppressing or regulating fungus gnat populations [1,8,14,15,16,17,18,19]. However, these strategies may only be effective if alternative plant protection strategies are implemented simultaneously. Furthermore, sole reliance on insecticides may be an issue as there is the possibility of fungus gnats developing resistance to insecticides [20,21]. This article will discuss the disease-interactions associated with fungus gnats, and the alternative management strategies that should be considered so as to alleviate problems with fungus gnats in greenhouse production systems.

2. Disease-Interactions

Fungus gnat larvae not only cause direct damage during feeding but their feeding may predispose plants to attack from soil-borne plant pathogens by creating wounds, which allow entry of certain soil-borne fungi [12,22,23,24]. Furthermore, larvae are capable of directly transmitting certain fungal diseases including *Pythium* spp., *Fusarium* spp., and *Verticillium*spp., from diseased to non-infected plants [24,25,26,27]. It has been shown that fungus gnat larvae can transmit *Thielaviopsis basicola* Berk & Broome, *Fusarium oxysporum* f. sp. *radices-lycopersici* Jarvis & Shoemaker, *Verticillium albo-atrum* Reinke & Berthold, and *Fusarium avenaceum* (Fr:Fr) [24,27,28,29,30]. It has also been reported that the oospores of certain *Pythium* spp. may survive passage through the digestive tract of *B. impatiens* larvae, which might be intact/viable after being excreted [24]. Furthermore, larvae of the fungus gnat, *B. impatiens* have been shown to ingest the propagules of *Pythium aphanidermatum* (Edson) and macroconidia of *F. avenaceum* that can then be introduced into young healthy plants during the feeding process [13,30]. Under laboratory conditions, fungus gnat larvae are capable of transmitting *Pythium* spp.; however, the extent of larval transmission may be less of a factor under greenhouse conditions, and may vary depending on the specific *Pythium* species [31]. In addition, it has been suggested, based on laboratory experiments, that fungus gnat larval feeding may decrease geranium seedling (*Pelargonium × hortorum* L.) susceptibility to *Pythium* infection or at least there is no enhancement in infection [32]. However, this may not represent what would occur under greenhouse conditions.

Fungus gnat adults are capable of carrying, on their bodies, the aerial conidia of certain foliar and soil-borne plant pathogenic fungi that produce aerial dispersal stages such as *Botrytis cinerea* Pers.:Fr., *F. avenaceum*, *F. acuminatum* (Ellis & Everhart), *T. basicola*, *V. dahliae* Kleb., and *V. albo-atrum*, which can then be transmitted to healthy plants [25,27,28,29,33]. Adult fungus gnats may not be able to vector *Pythium* spp., because any sexual and/or asexual reproductive structures are produced belowground (there are no aerial dispersal stages), which are not accessible to fungus gnat adults for acquisition because adults live aboveground [34,35].

Although adults do not fly very well and tend to reside near the growing medium, they may still be able to disperse spores of foliar plant pathogens in localized areas of the greenhouse [33]. The potential for both life stages of fungus gnats to transmit diseases means that the tolerance level for the presence of this pest may be very low, which is similar to other insect pests that vector diseases such as the western flower thrips, *Frankliniella occidentalis* Pergande [36]. Moreover, this means that intensive plant protection practices need to be implemented including alternative management strategies.

3. ALTERNATIVE MANAGEMENT STRATEGIES

Moisture management and sanitation are important in reducing problems with fungus gnats in greenhouse production systems. The use of well-drained growing media and not over-watering plants may avoid issues with soil-borne plant pathogenic fungi, thus diminishing the possibility of disease transmission by the larvae and adult. The accumulation of water and presence of algae may lead to abundant populations of fungus gnats and thus damage to greenhouse-grown crops [37]. Therefore, the installation of cement flooring may eliminate problems with algae, and consequently fungus gnats. However, this may not be economically feasible. A more cost-effective strategy may be to use landscape or weed fabric barriers, which are geotextile, non-biodegradable materials that when installed underneath benches may reduce algae growth and thus mitigate problems with fungus gnats [11]. It is also important to remove plant material and growing medium debris immediately from the greenhouse or place into containers, located within greenhouses, with tight-sealing lids, as fungus gnat adults may emerge from disposed growing medium and migrate onto the main crop [38].

An alternative strategy that should be considered in order to alleviate problems with fungus gnats is the use of physical barriers. Physical barriers are designed to alter the environment in order to make it inaccessible to insect pests [39]. Physical barriers do not reduce insect pest populations immediately, which is typical of most insecticides [40]; however, placing a physical barrier on the surface of the growing medium may inhibit fungus gnat adult emergence and also possibly reduce female egg-laying or egg survival [41].

It has been hypothesized that physical barriers such as sand or diatomaceous earth placed on the surface of the growing medium would negatively impact fungus gnat adults when they emerge and/or prevent females from laying eggs [42]. Diatomaceous earth is composed of siliceous skeletonized diatoms [43], which remove the insect cuticle waxes, absorb oils and waxes on the outer cuticle, or disrupt the integrity of the insect cuticle resulting in extensive loss of water from the insect body [44]. However, it has been shown that placing diatomaceous earth or sand on the growing medium surface did not have any effect on fungus gnat adult emergence or inhibit females from laying eggs because these physical barriers contain small openings that allow larvae to pupate, and adult females to lay eggs [41,45].

Growstones™ (Growstone, Inc., Albuquerque, NM, USA) aggregates, which are processed from 100% recycled glass and are used as a substrate in hydroponic systems [46] have been used as physical barriers and shown to reduce fungus gnat adult emergence over time and may also reduce egg-laying or egg survivability although this depends on the thickness of the Growstones™ layer [47]. It is important to note, that a physical barrier is not a stand-alone practice and therefore must be used as part of a holistic plant protection strategy [39]. This strategy may be more effective with plants growing in large containers whereas physical barriers may not be feasible under propagation due to the excessively moist conditions and small size of the containers.

Another potential alternative strategy to minimize plant damage due to fungus gnats is by repelling fungus gnat adults away from the growing medium thus reducing egg-laying. Repellent compounds have been used against other insects such as mosquitoes [48,49]. It was discovered, under laboratory conditions, that Bounce® original brand fabric softener dryer sheets (Outdoor Fresh Scent; Procter and Gamble, Cincinnati, OH) actually repel fungus gnat adults [50]. The dryer sheets contain several volatile constituents including linalool (3,7-dimethyl-1,6-octadien-3-ol), a colorless monoterpene alcohol that is present naturally in plants such as lavender (*Lavandula angustifolia* Mill.), marjoram (*Origanum vulgare* L.), and basil (*Ocimum basilicum* L.) [51,52,53]. In fact, greenhouse producers are actually inserting dryer sheets into the growing medium to repel fungus gnat adults [11]. However, one problem with this strategy is that fungus gnat adults are not killed directly and may migrate into other areas of the greenhouse. Therefore, dryer sheets must be placed throughout the greenhouse in order to protect plants from fungus gnat adults.

Finally, the use of jasmonate has been investigated to determine if stimulating wound signaling pathways may protect plants against fungus gnat larvae. This is based on the premise that signaling pathways, when activated by insects with chewing mouthparts, may allow plants to defend themselves by producing secondary metabolites [54]. However, it was found that applications of methyl jasmonate to *Arabidopsis thaliana* plants resulted in no mortality or reduced feeding activity of fungus gnat larvae [55].

4. CONCLUSIONS

It is important to understand that fungus gnats, both larvae and adults, may transmit certain soil-borne plant pathogens (e.g., fungi) to greenhouse-grown horticultural crops, either directly or indirectly thus potentially resulting in an economic loss. Although insecticides and biological control agents are effective against fungus gnat larval populations, alternative plant protection strategies should also be implemented. This includes cultural, sanitation, physical barriers, and use of repellent materials. These strategies may contribute to enhancing the efficacy of insecticides and biological control agents in suppressing or regulating fungus gnat populations in greenhouse production systems.

ACKNOWLEDGMENTS

The author wishes to acknowledge Amy L. Raudenbush, Amy Dickinson, Erik R. Echegarary, Karen A. Marley, and Richard A. Larson from the University of Illinois (Champaign, IL, USA) and Kansas State University (Manhattan, KS, USA) for all of their collective assistance in conducting research with fungus gnats. The author also wants to thank Mary Beth Kirkham, Professor from the Department of Agronomy at Kansas State University for reviewing an initial draft of the manuscript.

CONFLICTS OF INTEREST

The author declares no conflict of interest.

REFERENCES

1. Hamlen, R.A.; Mead, F.W. Fungus gnat larval control in greenhouse plant production. J. Econ. Entomol. 1979, 72, 269–271.
2. Jagdale, G.B.; Casey, M.L.; Grewal, P.S.; Lindquist, R.K. Application rate and timing, potting medium, and host plant effects on the efficacy of Steinernema feltiae against the fungus gnat, Bradysia coprophila, in floriculture. Biol. Control2004, 29, 296–305.
3. Vaughan, M.M.; Tholl, D.; Tokuhisa, J.G. An aeroponic culture system for the study of root herbivory on Arabidopsis thaliana. Plant Methods 2011, 7, 5–14.
4. Wilkinson, J.D.; Daugherty, D.M. The biology and immature stages of Bradysia impatiens (Diptera: Sciaridae). Ann. Entomol. Soc. Am. 1970, 63, 656–660.
5. Alberts, S.A.; Kennedy, M.K.; Carde, R.T. Pheromone-mediated anemotactic flight and mating behavior of the sciarid fly Bradysia impatiens. Environ. Entomol. 1981, 10, 10–15.

6. Lindquist, R.K.; Faber, W.R.; Casey, M.L. Effect of various soilless root media and insecticides on fungus gnats. HortScience 1985, 20, 358–360.
7. Harris, M.A.; Gardner, W.A.; Oetting, R.D. A review of the scientific literature on fungus gnats (Diptera: Sciaridae) in the genus Bradysia. J. Entomol. Sci. 1996, 31, 252–271.
8. Chambers, R.J.; Wright, E.M.; Lind, R.J. Biological control of glasshouse sciarid flies (Bradysia spp.) with the predatory mite, Hypoaspis miles, on cyclamen and poinsettia. Biocontrol Sci. Technol. 1993, 3, 285–293.
9. Cloyd, R.A. Fungus gnat and shore fly management strategies: Panel discussion. In Proceedings of the 16th Conferences Insect and Disease Management Ornamentals, Society of American Florists, Alexandria, VA, USA, 19–21 February 2000; King, A.I., Greene, I.D., Eds.; Society of American Florists: Alexandria, VA, USA, 2000; pp. 57–59.
10. Kennedy, M.K. Survival and development of Bradysia impatiens (Diptera: Sciaridae) on fungal and non-fungal food sources. Ann. Entomol. Soc. Am. 1974, 67, 745–749.
11. Cloyd, R.A.; Horticultural Entomology/Plant Protection, Department of Entomology, Kansas State University, Manhattan, KS, USA. Personal observation. 2014.
12. Fawzi, T.H.; Kelly, W.C. Cavity spot of carrots caused by feeding of fungus gnat larvae. J. Am. Soc. Hort. Sci. 1982, 107, 1177–1181.
13. Jarvis, W.R.; Shipp, J.L.; Gardiner, R.B. Transmission of Pythium aphanidermatum to greenhouse cucumber by the fungus gnat Bradysia impatiens (Diptera: Sciaridae). Ann. Appl. Biol. 1993, 122, 23–29.
14. Harris, M.A.; Oetting, R.D.; Gardner, W.A. Use of entomopathogenic nematodes and a new monitoring technique for control of fungus gnats, Bradysia coprophila (Diptera: Sciaridae), in floriculture. Biol. Control 1995, 5, 412–418.
15. Gouge, D.H.; Hague, N.G.M. Glasshouse control of fungus gnats, Bradysia paupera, on fuchsias by Steinernema feltiae. Fund. Appl. Nematol. 1995, 18, 77–80.
16. Ludwig, S.W.; Oetting, R.D. Evaluation of medium treatments from management of Frankliniella occidentalis(Thripidae: Thysanoptera) and Bradysia coprophila (Diptera: Sciaridae). Pest Manag. Sci. 2001, 57, 1114–1118.
17. Carney, V.A.; Diamond, J.C.; Murphy, G.D.; Marshall, D. The potential of Atheta coriaria Kraatz (Coleoptera: Staphylinidae), as a biological control agent for use in greenhouse crops. IOBC/WPRS Bull. 2002, 25, 37–40.
18. Cloyd, R.A.; Dickinson, A. Effect of Bacillus thuringiensis subsp. israelensis and neonicotinoid insecticides on the fungus gnat Bradysia sp. nr. coprophila (Lintner) (Diptera: Sciaridae). Pest Manag. Sci. 2006, 62, 171–177.
19. Cloyd, R.A.; Chiasson, H. Activity of an essential oil derived from Chenopodium ambrosioides on greenhouse insect pests. J. Econ. Entomol. 2007, 100, 459–466.
20. Brewer, K.K.; Keil, C.B. A mixed function oxidase factor contributing to permethrin and dichlorvos resistance in Lycoriella mali (Fitch) (Diptera: Sciaridae). Pestic. Sci. 1989, 26, 29–39.
21. Bartlett, G.R.; Keil, C.B.O. Identification and characterization of a permethrin resistance mechanism in populations of the fungus gnat Lycoriella mali (Fitch) (Diptera: Sciaridae). Pestic. Biochem. Physiol. 1997, 58, 173–181.
22. Leath, K.T.; Newton, R.C. Interaction of a fungus gnat, Bradysia sp. with Fusarium spp. on alfalfa and red clover. Phytopathology 1969, 59, 257–258.
23. Graham, C.L.; McNeill, M.J. Soybean crown and root damage by Bradysia coprophila. J. Econ. Entomol. 1972, 65, 597–599.
24. Gardiner, R.B.; Jarvis, W.R.; Shipp, J.L. Ingestion of Pythium spp. by larvae of fungus gnat Bradysia impatiens (Diptera: Sciaridae). Ann. Appl. Biol. 1990, 116, 205–212.
25. Kalb, D.W.; Millar, R.L. Dispersal of Verticillium albo atrum by the fungus gnat (Bradysia impatiens). Plant Dis. 1986, 70, 752–753.
26. Favrin, R.J.; Rahe, J.E.; Mauza, B. Pythium spp. associated with crown rot and cucumbers in British Columbia greenhouses. Plant Dis. 1988, 72, 683–687.
27. Gillespie, D.R.; Menzies, J.G. Fungus gnats vector Fusarium oxysporum f. sp. Radicis-lycopersici. Ann. Appl. Biol. 1993, 123, 539–544.

28. Harris, M.A. Dissemination of the Phytopathogen Thielaviopsis basicola by the Fungus gnat Bradysia coprophila and Biological Control of these Pests by Fusarium proliferatum and steinernematid Nematodes. Ph.D. Thesis, University of Georgia, Athens, GA, USA, 1995.
29. James, R.L.; Dumroese, R.K.; Wenny, D.L. Botrytis cinerea carried by adult fungus gnats (Diptera: Sciaridae) in container nurseries. Tree Planters Notes 1995, 46, 48–53.
30. El-Hamalawi, Z.A.; Stangellini, M.E. Disease development on lisianthus following aerial transmission of Fusarium avenaceum by adult shore flies, fungus gnats, and moth flies. Plant Dis. 2005, 89, 619–623.
31. Braun, S.E.; Sanderson, J.P.; Wraight, S.P. Larval Bradysia impatiens (Diptera: Sciaridae) potential for vectoring Pythiumroot rot pathogens. Phytopathology 2012, 102, 283–289.
32. Braun, S.E.; Sanderson, J.P.; Nelson, E.B.; Daughtrey, M.L.; Wraight, S.P. Fungus gnat feeding and mechanical wounding inhibit Pythium aphanidermatum infection of geranium seedlings. Phytopathology 2009, 99, 1421–1428.
33. El-Hamalawi, Z.A. Acquisition, retention and dispersal of soilborn plant pathogenic fungi by fungus gnats and moth flies. Ann. Appl. Biol. 2008, 153, 195–203.
34. Hyder, N.; Coffey, M.D.; Stanghellini, M.E. Viability of oomycete propagules following ingestion and excretion by fungus gnats, shore flies, and snails. Plant Dis. 2009, 93, 720–726.
35. Braun, S.E.; Castrillo, L.A.; Sanderson, J.P.; Daughtrey, M.L.; Wraight, S.P. Transstadial transmission of Pythium in Bradysia impatiens and lack of adult vectoring capacity. Phytopathology 2010, 100, 1307–1314.
36. Parrella, M.P.; Jones, V.P. Development of integrated pest management strategies in floricultural crops. Bull. Entomol. Soc. Am. 1987, 33, 28–34.
37. Keates, S.E.; Sturrock, R.N.; Sutherland, J.R. Populations of adult fungus gnats and shore flies in British Columbia container nurseries as related to nursery environment, and incidence of fungi on the insects. New For. 1989, 3, 1–9.
38. Hogendorp, B.K.; Cloyd, R.A. Insect management in floriculture: How important is sanitation is avoiding insect problems? HortTechnology 2006, 16, 633–636.
39. Banks, H.J. Physical control of insects—Recent developments. J. Aust. Entomol. Soc. 1976, 15, 89–100.
40. Boiteau, G. Physical Barriers for the Control of Insect Pests. In Encyclopedia of Pest Management; Pimentel, D., Ed.; Marcel Dekker, Inc.: New York, NY, USA, 2002; pp. 616–618.
41. Cloyd, R.A.; Dickinson, A.; Kemp, K.E. Effect of diatomaceous earth and Trichoderma harzianum T-22 (Rifai Strain KRL-AG2) on the fungus gnat Bradysia sp. nr. coprophila (Diptera: Sciaridae). J. Econ. Entomol. 2007, 100, 1353–1359.
42. Quarles, W. Diatomaceous earth for pest control. IPM Practitioner 1992, 14, 1–11.
43. Ebeling, W. Sorptive dusts for pest control. Ann. Rev. Entomol. 1971, 16, 123–158.
44. Korunic, Z. Diatomaceous earths, a group of natural insecticides. J. Stored Prod. Res. 1998, 34, 87–97.
45. Cloyd, R.A.; Dickinson, A. Effects of growing media containing diatomaceous earth on the fungus gnat Bradysia sp. nr. coprophila (Lintner) (Diptera: Sciaridae). HortScience 2005, 40, 1806–1809.
46. Evans, M.R. Physical properties of and plant growth in peat-based root substrates containing glass-based aggregate, perlite, and parboiled fresh rice hulls. HortTechnology 2011, 21, 30–34.
47. Raudenbush, A.L.; Cloyd, R.A.; Echegaray, E.R. Effect of physical barrier on adult emergence and egg survival associated with the fungus gnat, Bradysia sp. nr. coprophila (Diptera: Sciaridae), under laboratory conditions. HortScience 2014, 49, 905–910.
48. Rutledge, L.C.; Collister, D.M.; Meixsell, V.E.; Eisenberg, G.H.G. Comparative sensitivity of representative mosquitoes (Diptera: Culicidae) to repellents. J. Med. Entomol. 1983, 20, 506–510.

49. Fradin, M.S.; Day, J.F. Comparative efficacy of insect repellents against mosquito bites. N. Engl. J. Med. 2002, 347, 13–18.
50. Cloyd, R.A.; Marley, K.A.; Larson, R.A.; Arieli, B. Bounce® fabric softener dryer sheets repel fungus gnat, Bradysia sp. nr. coprophila (Diptera: Sciaridae), adults. HortScience 2010, 45, 1830–1833.
51. Mansour, F.; Ravid, U.; Putievsky, E. Studies of the effects of essential oils isolated from 14 species of Labiatae on the carmine spider mite, Tetranychus cinnabarinus. Phytoparasitica 1986, 14, 137–142.
52. Morales, M.R.; Simon, J.E.; Charles, D.J. Comparison of essential oil content and composition between field and greenhouse grown genotypes of methyl cinnamate basil (Ocimum basilicum L.). J. Herbs. Spices Med. Plants 1993, 1, 25–30.
53. Narusuye, K.; Kawai, F.; Matsuzaki, K.; Miyachi, E. Linalool suppresses voltage-gated currents in sensory neurons and cerebellar purkinije cells. J. Neural Transm. 2005, 112, 193–203.
54. Ryan, C.A. Protease inhibitors in plants: Genes for improving defenses against insects and pathogens. Annu. Rev. Phytopathol. 1990, 28, 425–429.
55. McConn, M.; Creelman, R.A.; Bell, E.; Mullet, J.E.; Browse, J. Jasmonate is essential for insect defense in Arabidopsis. Proc. Natl. Acad. Sci. USA 1997, 94, 5473–5477.

CHAPTER 21

An Innovative Adaptive Control System to Regulate Microclimatic Conditions in a Greenhouse

Giuseppina Nicolosi, Roberto Volpe and Antonio Messineo

Faculty of Engineering and Architecture, Kore University of Enna, Cittadella Universitaria, Enna 94100, Italy

ABSTRACT

In the recent past home automation has been expanding its objectives towards new solutions both inside the smart home and in its outdoor spaces, where several new technologies are available. This work has developed an approach to integrate intelligent microclimatic greenhouse control into integrated home automation. Microclimatic control of greenhouses is a critical issue in agricultural practices, due to often common sudden daily variation of climatic conditions, and to its potentially detrimental effect on plant growth. A greenhouse is a complex thermodynamic system where indoor temperature and relative humidity have to be closely monitored to facilitate plant growth and production. This work shows an adaptive control system tailored to regulate microclimatic conditions in a greenhouse, by using an innovative combination of soft computing applications. In particular, a neural network solution has been proposed in order to forecast the climatic behavior of greenhouse, while a parallel fuzzy scheme approach is carried out in order to adjust the air speed of fan-coil and its temperature. The proposed combined approach provides a better control of greenhouse climatic conditions due to the system's capability to base instantaneous solutions both on real measured variables and on forecasted climatic change. Several simulation campaigns were carried out to perform accurate neural network and fuzzy schemes, aimed at obtaining respectively a minimum forecasted error value and a more appropriate fuzzification and de-fuzzification process. A Matlab/Simulink solution implemented with a combined approach and its relevant obtained performance is also shown in present study, demonstrating that through controlled parameters it will be possible to maintain a better level of indoor climatic conditions. In the present work we prove how with a forecast of outside temperature at the next time-instant and rule-based controller monitoring of cooling or heating air temperatures and air velocities of devices that regulate the indoor micro-climate inside, a better adjustment of the conditions of comfort for crops is achievable.

Keywords: energy management; forecast climatic variable; greenhouse automation; artificial neural network; fuzzy logic control

1. INTRODUCTION

A home control system (HCS) is a tool to monitor and manage the main features of a residential home to improve its general efficiency and reduce its environmental impact. HCS serves three main functions: security and safety monitoring; efficiency of space heating/cooling, water heating and lighting; and monitoring external environment and weather. An important aspect is represented by the control of outdoor environment in terms of home security, garden or sunroom management, external lighting, etc. A significant amount of literature has been produced in the field of architectural design of gardens. In particular, in cases of limited outdoor space, vertical gardens and home-grown for-food cultures have recently attracted significant attention. Meanwhile, the use of greenhouses for agricultural production (both in large or small gardens) has increased worldwide, especially in areas that are climatically favorable, such as in the Mediterranean basin.

Greenhouse cultivation is widely used in agriculture because it provides plants protection from unfavorable weather, and, in recent years, it has become a means to achieve a controlled agricultural production. Greenhouses, in fact, directly influence the microclimate and the biological cycle of plants with direct effects on their production cycles and productivity. All plants' growing phases may be managed through control of air temperature, relative humidity, CO_2 concentration, irradiation and irrigation [1]. The daily variation of these parameters is not always favorable to plant growth, therefore, maintaining favorable climatic conditions in the greenhouse across the seasons becomes necessary. Meanwhile, technologies such as automatic irrigation are also useful in order to improve production. In general terms, experience has demonstrated how effective automatic microclimate control procedures can be on plant production.

Several strategies have been proposed to monitor and control the parameters that influence crop growth and production [2,3,4,5]. Best available practices aim at creating the ideal conditions to optimize the yield of cultivated product and reduce energy consumption [6,7,8,9]. For example, technologies able to control the ambient temperature via cooling devices based on water evaporation typically named fogging [10], fan-pad [11], and package [12] have been applied. However, air conditioning technologies for temperature and humidity control are attracting more significant interest among scholars and practitioners. Heat pump-type refrigerating cycle machines undoubtedly represent the most advanced technological choice, if compared to conventional evaporative type cooling systems. The latter, in fact, can prove economically inviable due the high electricity consumption in larger greenhouse systems. In general, several research efforts focusing on different approaches to greenhouse internal climate control have been produced. The problem presents a two-fold objective: (1) prediction of microclimatic parameters aimed at improving energy saving and/or crop production; and (2) greenhouse system management, which in turn influences those microclimatic parameters. Both issues need to be tackled concurrently in order to attain a good compromise and obtain good overall performance results.

Conventional control techniques, such as those using simulation approaches [13], used to predict and to manage microclimatic requirements have been developed over the years [1,2]. However, the microclimate in agricultural greenhouses is a complex and nonlinear system and optimal parameters greatly depend on crop "comfort". Existing models are ill-suited to predicting greenhouse energy requirements for instance. Moreover, no examples of integration of crop production to integrated greenhouse control systems, such as those related to energy management and energy savings, are currently available. Recently Chen and co-authors [14] investigated potential differences between a greenhouse energy model and a buildings energy model. Meanwhile, different computational approaches have been applied for various aspects of temperature prediction [15,16].

Temperature prediction methods can be primarily divided into two groups: (1) physical methods based on mathematical theory; and (2) black box methods based on modern computational technology. Over the last decades, numerous physical greenhouse temperature methods have been proposed [16]. Generally, the physical methods present a high degree of complexity due to the high number of parameters to be determined [17] as well as to the difficulty in measuring those parameters. Other approaches based on the particle swarm optimization (PSO) [18] algorithm, (a heuristic global optimization method) are adopted to create temperature predictions based on a least squares support vector machine (LSSVM) model with parameters optimized by PSO. That work shows the importance of an accurate model that can predict temperature values several hours before they occur and therefore optimize operational costs. However, the basic PSO does not always ensure convergence to an optimal solution and easily suffers from partial optimization [19]. Moreover this approach does not allow to adapt the indoor thermal condition dynamically with respect to the variation of external climatic change, therefore a PSO optimization process [20] is needed, and that increases system overheads.

As opposed to the described methods, the "black box" method, typical of soft computing technology, uses a prediction method [21,22]. Therefore, it does not need to determine every parameter value as is needed in physical-type approaches. In addition, in a context of micro-climatic control where the unplanned conditions could cause rapid variations, an artificial neural network (ANN) used to predict climatic parameters, combined with a fuzzy logic controller (FLC) used to adjust the fan coil speed and its temperature, could be the better solution in order to meet the typical greenhouse requirements.

The ANN can be a valuable tool for the prediction of microclimate parameters, while the use of rule-based controllers (FLC) enables the implementation of multi-criteria control strategies. In fact, the use of smart setting and tuning techniques for these controllers can improve the crop production as demonstrated by authors in [23]. For this reason, the FLCs, based on linguistic rules instead of inflexible reasoning, can be the right choice to describe greenhouse systems and to increase crop production.

In any case, whatever the method used, weather and climate parameters need to be forecasted, and this can be done by a neural network that is well suited for this purpose. In these terms, it would be good to include information on the fuzzy logic control rules. Moreover, a single FLC [22] for fan coil regulation would be not sufficient to meet the greenhouse requirements in terms of crops production; a better control of inlet air speed and its temperature could improve production.

The main scope of our work is the implementation of an innovative intelligent microclimate controller for a greenhouse (small or large) by using a parallel fuzzy scheme for the calculation of the inlet air speed and temperature. The main aim is to facilitate the human greenhouse management and to obtain good results in terms of production. This work aims to prove that control of the "micro-climate" inside the greenhouse can be improved by using a pro-active and dynamic control—a control that takes into account instant forecasts of environmental parameters, in a dynamic and flexible way. A series of sensors measure the physical greenhouse environmental parameters at time *t*, and ANN estimates prediction also on instant *t*; this allows the FLC to control (at time *t*) the indoor micro-climate depending on the forecasted development of the system (immediate $t + 1$). This approach is applicable for all weather-climatic parameters that influence the behavior of the micro-climate inside the greenhouse. The present work is organized as follows. Section 2 provides an analysis of microclimatic behavior and typical control approaches found in literature are described. Subsequently, a methodology able to determine the parameters involved in the proposed approach is explained in Section 3. In Section 4 the combined approach based on a neural network model and a parallel control fuzzy scheme (PCFS) are described. Finally, an implementation of the proposed approach and its performance are shown.

2. A PRELIMINARY GREENHOUSE ANALYSIS

The microclimate control in a greenhouse is a complex task since it is affected by several interdependent variables [24]; this makes the application of the conventional techniques for indoor climate regulation (cooling, ventilation, heating) often difficult.

While the indoor thermal conditions of residential building depend on objective factors (type of construction) and subjective factors (individual comfort), the optimal climate conditions for the growth of a crop in a greenhouse may be considered objective because they depend exclusively on the type of crop. However, the greenhouse internal climate conditions are subject to quick variations due to unpredictable and unstable external climate parameters. For this reason, some researchers proposed the use of innovative techniques based on artificial intelligence, such as fuzzy logic and neural networks, which allow the achievement of an efficient climate regulation inside the greenhouse [25].

Fuzzy logic is quite widespread as a technique for the control and monitoring of microclimate parameters within greenhouses [2,26].

Zhang et al. [27] used fuzzy control to automatically adjust climatic parameters, such as temperature and relative humidity, according to the needs of crops growth. In [28] the authors establish a dynamic ventilation control system and study the fuzzy temperature control systems based on programmable logic controller (PLC) and industrial wireless local area network technologies.

The air temperature in protected systems depends on heat exchanges (irradiance, convection, conduction, etc.) occurring with the external environment through the walls of the greenhouse and the ground. Due to their capability to handle both numerical data and linguistic information, fuzzy logic systems are suitable to model greenhouse climate. In fact, fuzzy logic allows for making decisions regarding greenhouse management depending on online measurements of the quality parameters. By using fuzzy logics, dynamic and automatic control become feasible, owing to the system's ability for coping, for instance, with rapid temperature changes, which can occur in any moment of the year.

In the recent years some relevant aspects of greenhouse management have been formalized through mathematical models [29,30,31,32]. These studies have investigated the relationships between greenhouse air temperature and relative humidity and several outdoor parameters such as wind speed, solar radiation [17], and heat exchanges between the indoor air and various elements of the greenhouse (cover plant, soil, etc.) [32,33,34,35].

Control of the internal temperature takes place through natural ventilation, while in the presence of high outside temperatures (30–35 °C) during long periods it is necessary to use artificial cooling in order to maintain a functional microclimate for the cultivated plant species.

Nowadays, most greenhouses use traditional air conditioning systems for regulation of the inner temperature, but they often cause high electricity consumption.

This is one of the reasons why research efforts have been focused on innovative efficient climate regulation, to guarantee optimal conditions for the crop, at the same time minimizing energy consumption as the literature has done in smart-home [36,37].

3. METHODOLOGY

In order to test the proposed approach, a greenhouse scenario need to be formalized. Several conventional designs [38,39] are reported in the literature. They are usually have five steps. The first step deals with understanding the physical system and its control requirements; subsequently it is possible to develop a model, which includes the plant, sensors and actuators. The third step is on the use of a linear control representation in order to determine a detailed version of the controller. The fourth step aims at simplifying the complex previous representation. The control system is divided in several sub-systems. The calculation proceeds with the development of different algorithms, one for each sub-system controlled. The last step deals with the simulation of the designed system including the effects of non-linearity, errors, and parameter variations. If the performance is not satisfactory then the system is remodeled, the controller re-designed, the algorithms rewritten and then tested again. In some cases, it is possible to verify each sub-system and finally to test the entire controlled system [35].

Using fuzzy logic, it is possible to rely on a more simplified procedure. It starts with understanding and characterizing the system behavior by using the knowledge and the experience of a real scenario. Subsequently it is possible to directly design the control algorithm using fuzzy rules, which describe the principles that guide the controller´s regulation in terms of the relationship between inputs and outputs. The last steps are the simulation and debugging. In case of un-satisfactory performance, some fuzzy rules can be adjusted and a new simulation can be run.

The fuzzy-logic-based design methodology simplifies the design loop which results in the following significant advantages:

- It reduces the design development cycle, simplifying design complexity;
- It improves time to market, provides a better alternative solution to non-linear control and improves control performance;
- It simplifies implementation and reduces hardware costs.

An important aspect of using fuzzy control is related to the creation of membership functions and the relevant optimization techniques [40]. In [41] a new application of adaptive neuro-fuzzy inference system (ANFIS) for control of greenhouses is introduced. This approach is well-suited to the great deal of uncertainty that exists in such a system and improves its general performance using a genetic algorithm (GA). However, this approach is useful for adapting control parameters, such as the number and shape of the membership functions used.

The exclusive use of a fuzzy controller may be useful for accuracy purposes, but it has been shown in [42,43] that such control can be improved if predicted values at time '$t + 1$' are used as time "t" input parameters. However such an approach foresees a preprocessing of input data using an ANN assuming that the parameters affecting the micro-climate inside a greenhouse can be studied and can determine an ANN, useful for the prediction of the short-term behavior following the date of measurement where an FLC indicates their choice of control.

In most cases the greenhouse crops have a good climate condition if the temperature is between 17 °C and 27 °C. However several indoor temperature values are sometimes detected within the minimum and maximum values of 10 °C and 35 °C. An important role in these particular systems is also played by air humidity, which affects various biological and physiological processes in the plants. The causes of variation of the relative humidity in a confined environment are mainly due to evapo-transpiration of plants, the internal temperature of the greenhouse, and ventilation, both natural and artificial.

The relationship between the temperature, relative humidity and absolute humidity is defined by the Mollier diagram, as explained in [39] and depends on the air temperature.

In order to validate the system proposed in this paper, we shall analyze a scenario where the FLC receives the temperature value as input. In fact our approach is applicable both to the prediction of a single parameter and to that of n parameters. In addition, the correlation between those variables has already been demonstrated in some previous studies [44]. Moreover, whilst some adaptive neuro-fuzzy methods [45,46] produce target temperature as an output, the present approach produces two parameters that allow a rapid and more effective adjustment of microclimate control: air inlet speed (hot or cold) and its temperature, defined during design phase. This approach allows an easier control of the temperature target value, since both the airflow and its temperature are controlled. The proposed approach is bound to generate energy savings compared to more traditional methods. In fact, the target temperature is maintained via a number of "soft" approaches towards the target temperature, attained also via using the value expected at time "$t + 1$" as an input value and therefore reducing the numbers of temperature "corrections" and overshoots typical of more traditional methods.

The data used to determine the control parameters (air temperature of cooling/heating and speed of the air fan coil) are the following:

- $\Delta T(t)$ = Difference between inside greenhouse temperature and optimum temperature that must be maintained in the greenhouse (usually varying from −10 °C to 10 °C);

- $S(t)$ = Through a range from 0 to 1 values, $S(t)$ shows incident solar radiation on the greenhouse roof (usually varying from 0 W/m^2 to 1000 W/m^2);
- $C(t)$ = Cloudiness, which reduces the solar radiation (usually it varies from 0% to 100%);
- $W(t)$ = Wind speed (it varies in the range from 0 m/s to about 45 m/s), here shown in a range between 0% and 100%;
- E = % of sensing system error due to sensor measurement; a percentage parameter that varies randomly in the interval [−4%, +4%]. In the simulation scenario it is introduced in order to simulate the effect of measurement errors of the sensors system.

The approach can be achieved using sensors available on the market such as the Grove Sensor of temperatures and humidity or Moisture Sensor (SKU: SEN0114) for temperature only. The latter is best suited to high humidity scenarios. Both sensors are connected to an Arduino board that allows integration to ANN and FLC. The approach in present work was implemented and validated in Matlab/Simulink. Temperature measurement inside the greenhouse, was ensured by sensors placed at a point at one-third the length of the greenhouse from the front end, at the center of the greenhouse, and a point at one-third the length of the greenhouse from the rear end. The size of the greenhouse was simulated to be 6.5-m wide, 15-m long, and 3.8-m high. The temperature values were computed and recorded at 10-min intervals; a mean value was extrapolated each hour in order to allow the FLC to change the state of the cooling system or heating every six sampling intervals. The proposed approach allows to act at each measurement.

4. SYSTEM MODEL

The operating principle of the proposed system is shown in Figure 1. It is based on a parallel control fuzzy scheme (PCFS).

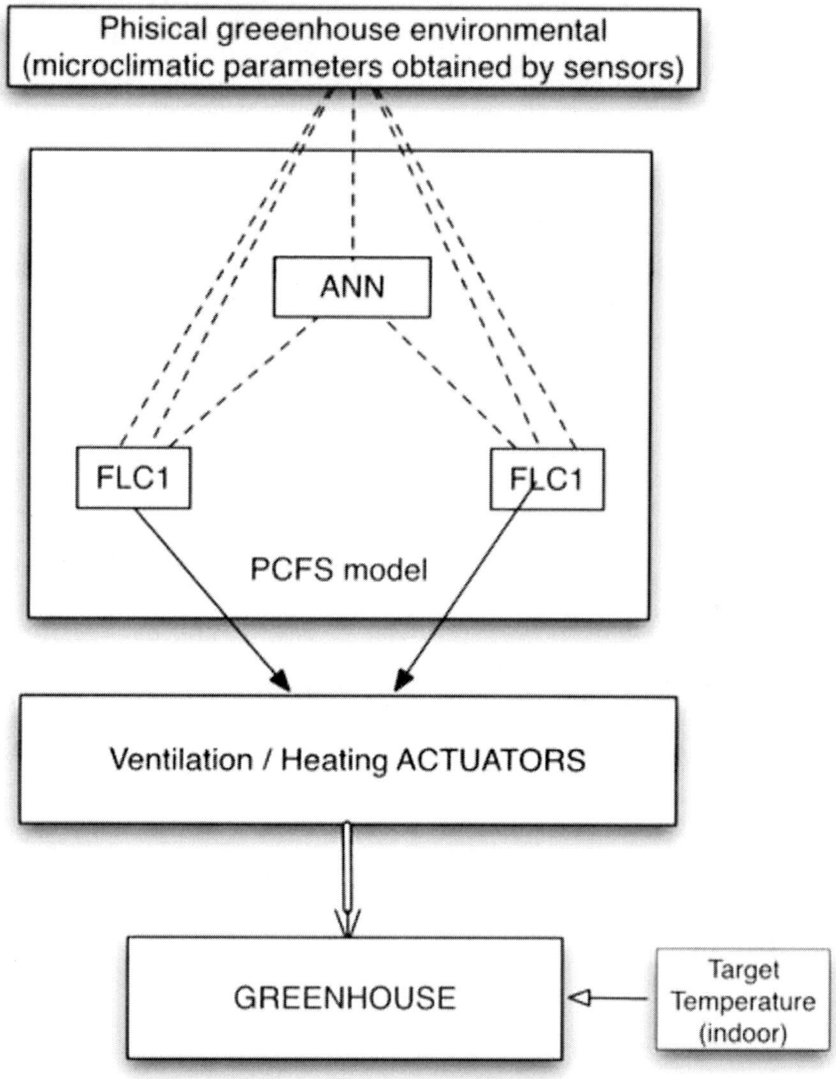

Figure 1. PFCS System Model. ANN: artificial neural network; and FLC: fuzzy logic controller.

The advantage of this model is that the fuzzy logic is characterized by linguistic rules, rather than complex mathematical formulations. Climate parameters are recorded in time steps preceding the time of forecast ($t + 1$); they are used in order to infer the trend of indoor temperature within the greenhouse. The outputs of the ANN are therefore the forecasted indoor temperatures, calculated by model at each time step. The coupling of ANN and two FLCs allows a more accurate design of the greenhouse, which was based on the control strategy not only on the past trend of the temperatures, but also on the forecasted ones. In this way it will operate similarly to a typical retroactive feed fuzzy system.

This work aims at proving that "micro-climate" control inside a greenhouse can be improved if a pro-active and dynamic control is in place, one that takes forecasting of environmental parameters into a dynamically and flexible control. The sensors measure physical greenhouse environmental parameters at time *t*, and ANN (Figure 1) estimates a prediction at time *t*; this allows FLC to establish (at time *t*) the indoor micro-climate depending on what is forecasted for an immediate *t*+ 1 time. This approach is applicable to all weather-climatic parameters that influence the behavior of the micro-climate inside the greenhouse. Figure 2 shows an example on which the tests described are based and the relative results which validate the proposed method.

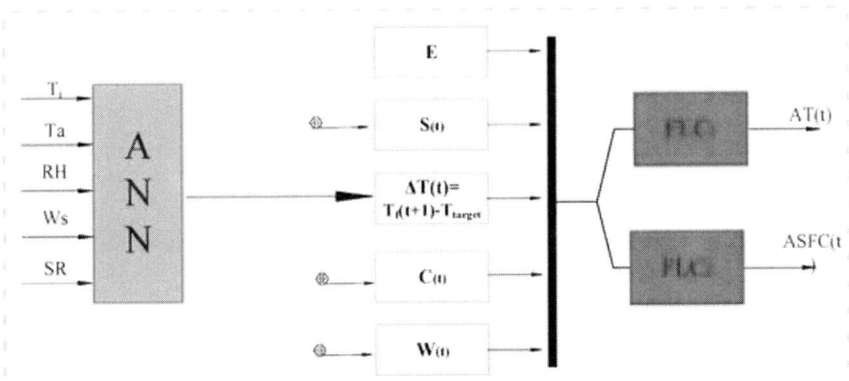

Figure 2. Parallel control fuzzy scheme (PCFS).

The approach described proves that just by forecasting outside temperature at the time *t* + 1 (created by ANN) and by controlling internal temperature through FLC, a better control of the optimal conditions can be achieved. The approach could be easily extended to the prediction of variation and control of other typical parameters of a greenhouse, i.e., humidity.

The parameters were measured, every hour, by using a weather station installed in a farm located in the countryside close to the city of Enna, in the south of Italy.

4.1. PCFS Model

The PCFS handles the switching on or off of the intelligent temperature regulation system, by using two FLC that are involved in the regulation of the inlet air speed of fan coil and its air temperature. The other climatic parameters depend mainly of temperature: by varying the temperature inside the greenhouse it is possible to determine the best conditions for the crops.

The PCFS (Figure 2) receives as inputs some physical parameters, such as indoor air temperature (Ti) and the outdoor air temperature (Ta), the relative humidity (RH), the wind speed (Ws) and the incident solar radiation (S) on a horizontal layer placed at the height of the greenhouse's roof and it adjusts the output speed of the system for cooling/heating, based on two fuzzy logic controllers. Both FLCs (1 and 2) receive an input at a given time (t) of the following parameters: $S(t)$, $C(t)$, $W(t)$, E, $\Delta T(t)$. The last will be calculated by using the difference of temperature values, between the forecasted ($t+1$) indoor temperature of the greenhouse $T(t+1)$ and Topt, the latter being the optimal temperature value that must be maintained in the greenhouse. Topt is constant and depends on the crop. The temperature $T(t+1)$ will be forecasted by an ANN.

The output parameters are the following: air speed of fan-coil ($ASFC$) at time t, named $ASFC(t)$, and its temperature (here called air temperature, AT) at time t, named $AT(t)$.

The controller takes as input crisp values of $S(t)$, $C(t)$, $W(t)$, E, $\Delta T(t)$, which are converted into linguistic values by using a chosen set of membership functions. The use of linguistic rules and not complex mathematical formulations can be a valuable tool for adjusting thermo-hygrometric parameters in indoor environments like a greenhouse and overcome issues of nonlinear and multivariable systems [5,47].

The membership functions used for the input parameters are shown in Table 1, while the graphical representations are shown in Figure 3 and Figure 4.

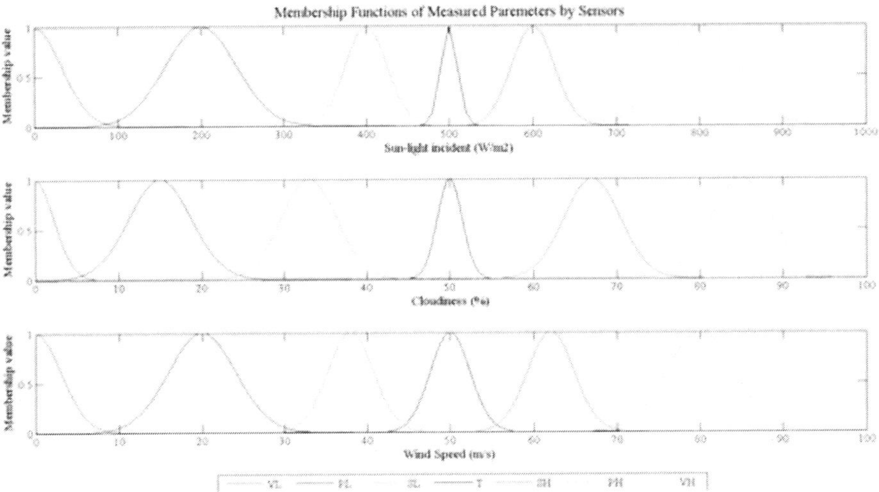

Figure 3. Membership functions of $S(t)$, $C(t)$ and $W(t)$.

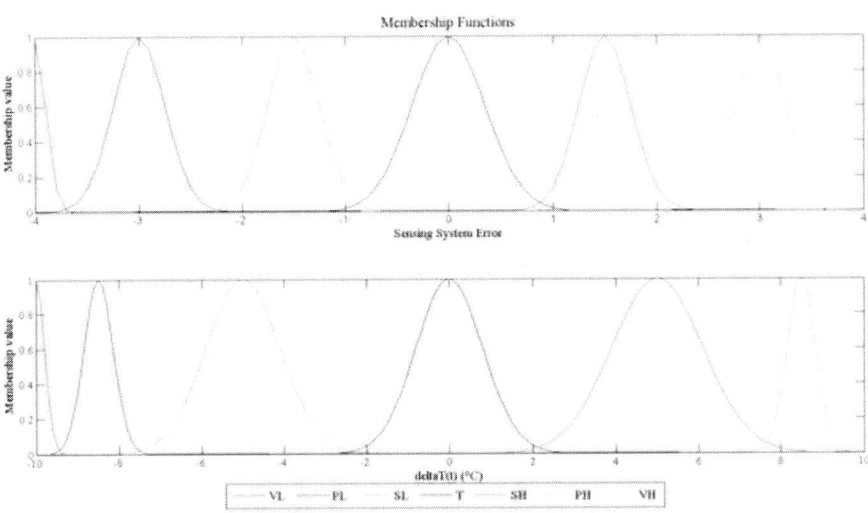

Figure 4. Membership functions of E and $\Delta T(t)$.

Table 1. Linguistic values of the FLC.

Linguistic Value	Incident Solar Radiation ($S(t)$)	Cloudiness ($C(t)$)	Wind Speed ($W(t)$)	Error (E)	Difference of Temperature ($\Delta T(t)$)
Very low (VL)	$0 \div 10$	$0 \div 10$	$0 \div 20$	$-4 \div -3$	$-10 \div -8.6$
Pretty low (PL)	$10 \div 30$	$11 \div 20$	$21 \div 32$	$-2.99 \div -1.51$	$-8.59 \div -5$
Slightly low (SL)	$31 \div -49$	$21 \div 49$	$33 \div 49$	$-1.5 \div -0.1$	$-0.1 \div -4.99$
Target (T)	50	50	50	0	0
Slightly high (SH)	$51 \div 70$	$51 \div 70$	$51 \div 77$	$0.1 \div 1.5$	$0.1 \div 4.99$
Pretty high (PH)	$71 \div 89$	$71 \div 80$	$78 \div 90$	$1.51 \div 2.99$	$5 \div 8.59$
Very high (VH)	$90 \div 100$	$81 \div 100$	$91 \div 100$	$3.0 \div 4$	$8.6 \div 10$

The aim of the FLCs is to elaborate on these linguistic values by using an inference mechanism based on a set of if-then rules. These rules are combined in the FLCs, which return membership functions, represented, in this work, by Gaussian functional shapes [39], following Equation (1):

$$f(x, w, f_c) = e^{\frac{-(x-f_c)^2}{2\,w^2}} \quad (1)$$

where f_c is the center of the membership function and w determines its width. It allows determining the linguistic output, which can assume one of the above parameter values. The Gaussian model [48], used in this approach, takes inputs and produces outputs based on four steps:

- mapping each of the inputs into a fuzzy variable (*fuzzification*);
- determination of the output of each rule given its fuzzy antecedents;
- determination of the aggregate output(s) of all of the fuzzy rules;
- mapping of the fuzzy output(s) to crisp output(s) (*defuzzification*).

The fuzzy output variables associated to the state of the actuators of management control parameters (Figure 3): $ASFC(t)$, and its temperature AT(t) that is associated to the following rules:

- Zero: not activated;
- Low: activated at low speed;
- Medium: activated at medium speed;
- High: activated at maximum speed.

While the air speed $ASFC(t)$ can be fixed in four activating states (by using output de-fuzzification rules of FLC_1), the fan coil temperature ($AT(t)$ parameter) is the best temperature value that could be used in order to increase or decrease the internal greenhouse temperature. The fan coil temperature can be determined in a range (by using de-fuzzification rules of FLC_2), with an established design time, and depends on specific crop. In a simulated scenario (shown in the next section) three fixed values are established, equal to 5 °C, 10 °C and 15 °C. While this is only an example, the approach can be run by using different temperature values and/or ranges. The target temperature (see Figure 1) is typical of a specific crop. The approach addresses the problem showing a generalized solution, and could be valuable for any kind of crop. The following performance section shows an example by using 25 °C as a target temperature.

Through the inference mechanism it is possible to determine the correct output according to the fuzzy inference rules presented. The fuzzy logic rules, implemented by using Matlab/Simulink, allow PCFS to set the better control decision to obtain the target temperature.

4.2. Neural Network Model

The most important feature of ANNs is their ability to learn models through experience, expressed via sets of recorded data [43], without needing an explicit mathematical representation of the studied phenomenon [49,50].

The ANN used in the application described in this paper is a multilayer perceptron (MLP) with three layers: one input, one hidden layer and one output. The number of neurons in the hidden layer (13 neurons) was chosen through sensitivity analysis as the number of neurons that minimizes the prediction error (E) and maximizes the values of the coefficient of determination (R_0^2).

A sigmoidal function was chosen in order to represent the transfer function (TF) for the hidden layer and a linear function was chosen for the output layer.

The considered data for learning phase were measured and recorded in a weather station located in Enna (near the greenhouse used by tests), while the indoor climate parameters were measured directly by using sensors located inside of the greenhouse. Data input are (Figure 2):

- Indoor air temperature (Ti (°C)), measured inside the greenhouse;
- Outdoor air temperature (Ta (°C));
- Relative humidity (RH (%));
- Wind speed (Ws (m/s));
- Solar radiation (S (W/m^2)).

The network output is the air Temperature forecasted: $Tf(t+1)$, measured inside the greenhouse, and this output will be the input of the intelligent controller described above.

Before training the ANN, the data are normalized between 0 and 1 by using Equation (2):

$$\frac{1}{Nt}\sum_{i=1}^{Nt}\frac{\left|Yi-\ddot{Y}i\right|}{max_{i=1}^{Nt}(Yi)}100 \quad (2)$$

Finally, the ANN was trained using the early stopping criterion. In order to test the performance of each phase of ANN, the data were divided into three subsets: the training set (70% of data set), validation set (15% of data set) and test set (15% of data set). The accuracy of the networks predictions is evaluated by using the control indices shown in Table 2. Several tests have been carried out in order to determine the better combination in terms of hidden and input neural networks.

Table 2. Error parameters.

Error Index	Equation
Mean Squared Error (MSE)	$MSE = \text{mean}\left(e_1^2\right)$
Mean Absolute Percentage Error ($MAPE$)	$MAPE = \text{mean}\left(\lvert pi \rvert\right)$
Mean Absolute Error (MAE)	$MAE = \frac{1}{N}\sum_{i=1}^{N}\lvert C_1 - O_i\rvert$
Standard Error (SE)	$SE = \frac{\sqrt{\sum_{i=1}^{N}(C_i-O_i)^2}}{N-1}$

5. PERFORMANCE EVALUATION OF THE PCFS

In order to validate the proposed approach, a simulated scenario was implemented by using a coupling model (ANN–PCFS) in Matlab/Simulink.

By using a fuzzy controller logic (implemented in PCFS scheme) it was possible to obtain an automatic control of heating/cooling and ventilation in a greenhouse, in order to have an accurate control taking into account the forecasted climatic parameters (i.e., the PCFS regulates the microclimatic conditions at time t taking into account the forecasted parameters at time $t+1$). Thus, the PCFS regulated automatically both the air speed and the temperature of a fan coil. The implemented scenario considers real climate parameters of a greenhouse located in the center of Sicily (Italy) in a typical summer season, where the oscillations between $T_{\min}$ (during the night) and $T_{\max}$ (during the day) can be 10–12 °C.

Results show that PCFS dynamically manages the system during 24 h (a typical summer day, with medium values), in order to guarantee an optimal indoor temperature, which is set to 25 °C.

In order to validate the proposed approach, two different performance evaluations have been carried out. The first step consists of implementing an auto-regressive neural network with external inputs (NNARX). The aim of this step is the selection of the network with the lowest forecast error. Several NNARX models have been tried, by using the same input parameters, as described in Section 4.1, using the summer data set. The NNARX models have been created by varying both the number of input parameters and the number of hidden layer neurons, through a sensitivity analysis it was chosen the network with the best performance.

In Table 3 the simulation campaign is summarized. Each row of Table 3 shows the best performance obtained with the specified number of input neurons and number of neurons in hidden layer. The grey row shows the best performances obtained, referring to the network with 13 neuron input neurons and 35 neurons in the hidden layer.

Table 3. ANN performance indices-the chosen number of neurons is the one in the row with a grey background. NNARX: auto-regressive neural network with external inputs.

# Input Neurons NNARX	# Neurons in Hidden Layer	*MSE*	*MAPE*	*MAE*	*SE*
5	45	3.280×10^{-2}	4.749	4.102×10^{-2}	0.327
8	15	6.097×10^{-5}	5.456	0.456×10^{-1}	0.248
10	55	0.097×10^{-1}	6.789	3.127×10^{-2}	0.525
12	25	5.801×10^{-3}	5.329	09879×10^{-1}	0.236
13	35	1.097×10^{-5}	4.321	0.021×10^{-2}	0.223
14	35	2.096×10^{-2}	6.453	0.004	0.670
15	75	0.419×10^{-2}	7.234	6.001×10^{-1}	0.228
16	45	4.001×10^{-5}	8.345	0.003×10^{-2}	0.471
17	10	4.915×10^{-3}	7.123	0.087×10^{-1}	0.512
18	15	6.001×10^{-5}	6.198	0.025	0.318
19	20	2.168×10^{-3}	7.002	0.295×10^{-1}	0.258
20	30	1.005×10^{-4}	8.201	0.025×10^{-1}	0.310

The better solutions of the proposed ANN model are tested by using the values of a summer period, in order to compare with a complex literature approach [22], that used several multilayer feed-forward neural networks that are trained to model each subsystem in order to achieve a multiple neural control of the greenhouse. It introduces a multiple neural control for the greenhouse micro-climate which consists of the division of the greenhouse control phase in periods where a suitable controller is selected to drive the internal climate.

A good performance was achieved by using a classifier in order to extract local models from the greenhouse database and obtain a multiple neural control in order to improve the greenhouse performances by using an Elman neural network.

Figure 5 shows how the proposed ANN achieves comparable results, in terms of prediction error during learning phase, by using a very simple model that could be implemented in embedded devices used in controlled greenhouses.

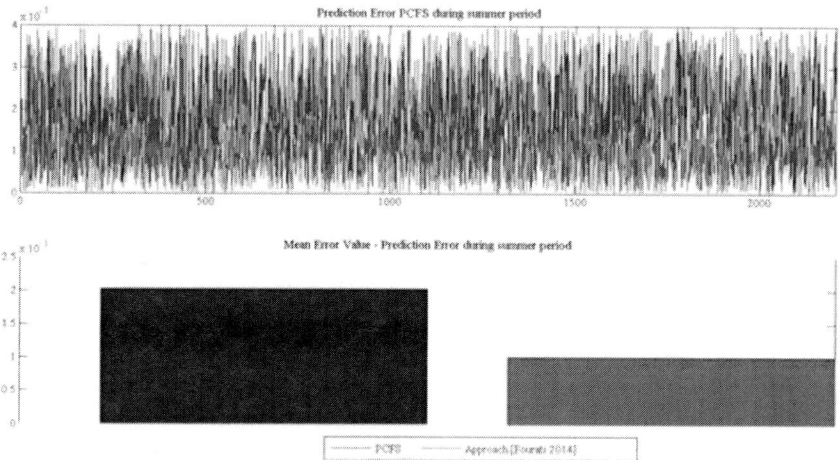

Figure 5. Error comparison during learning test.

The approach of Fourati [22] is more complex, indeed the possible implementation is not shown in the paper, instead, usually several literature papers show the possibilities to implement simple ANN, such as the proposed model, by using (Commercial) Off-the-Shelf (COTS) devices [51,52]. The difference between the proposed approach and Fourati approach is equal to approximately 1×10^{-1}. This appears to be a very low error in greenhouse scenarios.

Figure 6 shows the error obtained calculating the parameter as the difference between forecasted indoor greenhouse temperature and target temperature (specific for the crop, and in this evaluated scenario fixed to 25 °C), called in the graph "dTEMPERATURE" and referred to henceforth as $\boldsymbol{\Delta T}$.

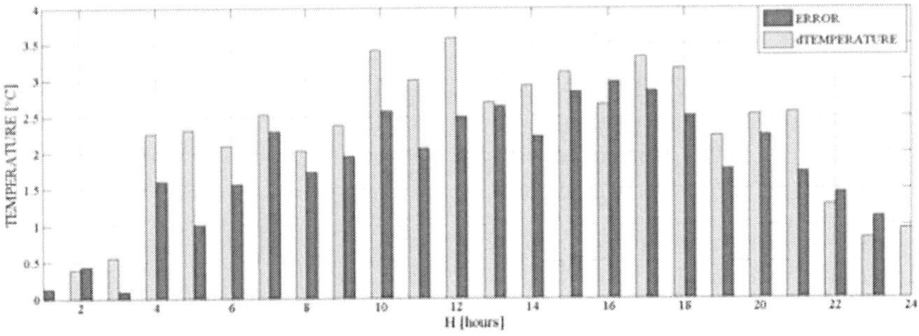

Figure 6. NNARX performance.

After the validation of the NNARX model, through the implemented scenario it was possible to show the evolution of the inlet air speed injected by the fan coils and the inside air temperature of greenhouse forecasted by the NNARX model. Figure 6 and Figure 7 show the performance of PCFS during the management decisions in a typical summer day, exactly during every struck hour.

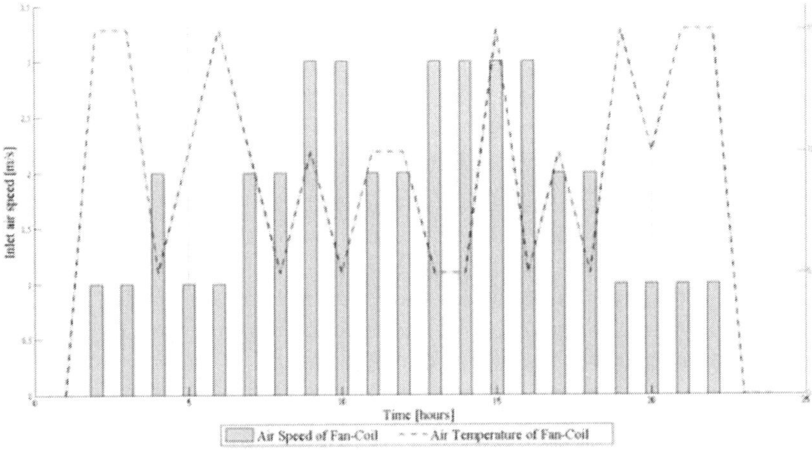

Figure 7. PCFS performance.

The shown values do not refer to the progressive behavior of 24 h, in other words the situation at given time t does not depend on the value shown in the graph at $t - 1$ and moreover it does not influence the value at $t + 1$.

The figures show what happens when a certain ΔT occurs at a given time t, calculated by using forecasted temperature at time $t + 1$. For example, at 1:00 p.m., if the forecasted inside temperature is 25 °C, the inlet air speed and its temperature have to be set.

Figure 7 shows air speed and temperature as a function of the time during an evaluated summer day. At 1:00 p.m., the shown control values (inlet air speed and its temperature) do not appear to depend on the action taken at 12:00 a.m. In fact, the values appear as if the system turns on for the first time at a given hour (at 12:00 a.m.). It is necessary to underline that the inlet air speed has been set to the default value of 2 m/s, while, as shown, when ΔT (calculated by using neural network forecasts) is different to 25 °C, the PCFS (in order to obtain an adjustment of the inside climatic condition) drives the system: both air temperature (5–10–15 °C) and air velocity (1, 2, 3 m/s) of the fan coils.

Figure 8 shows a simulation scenario of the greenhouse-controlled temperature during a one-week simulation time. The temperature is kept around its target value, that is chosen equal to 25 °C as a requirement for an optimal plant growth environment. Notably, when internal temperature exceeds its target value, around midday, the system starts operating to remove hot air and push in cooler outside air. When the controller parameters exceed the target value by too much, the system runs at maximum air speed (in the simulated scenario the maximum value is set) and its temperature is very low (in the simulated scenario 5 °C) in order to keep the internal temperature to the target value.

Figure 8 shows also a comparison with results on [23]. Although [23] is probably the most influential literature work that addresses smart climatic control in greenhouses, that work is focused on energy management through photovoltaic energy in order to minimize the use of conventional electrical grid and to lower costs of agriculture production. However, a comparison based on climatic control is shown exclusively. The diagram shown on Figure 8 demonstrates that the proposed approach achieves better performances with respect to both method and results shown on [23] and to the "traditional" control approach, (such as a basic thermostat). The combination of ANN and parallel fuzzy scheme, allows us to dynamically respond to outdoor and indoor climatic variations.

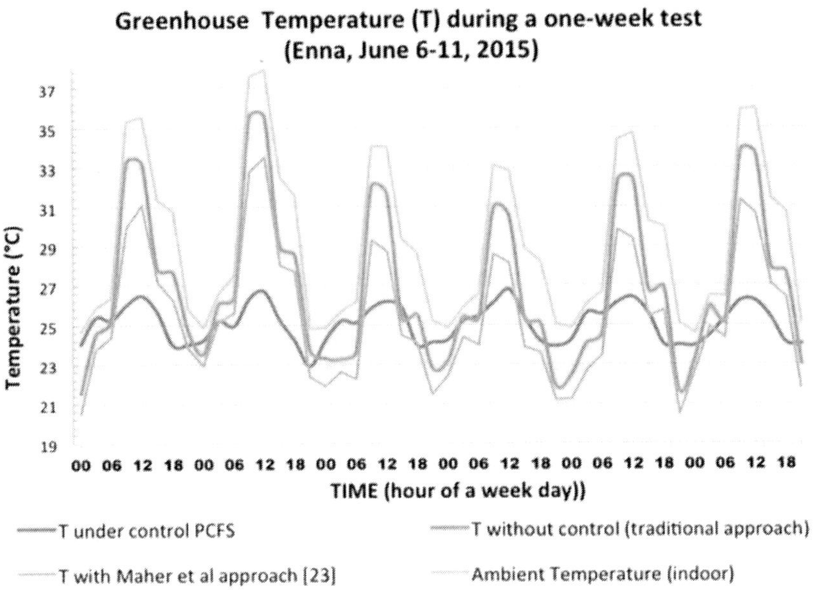

Figure 8. Greenhouse temperature (summer week).

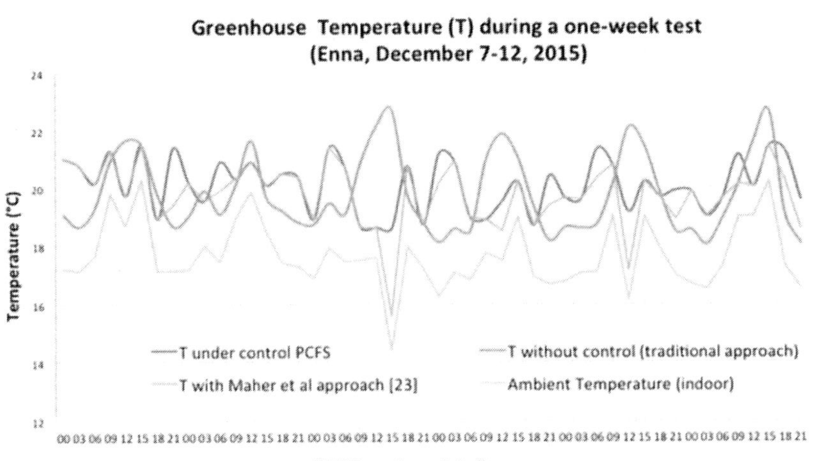

Figure 9. Greenhouse temperature (winter week).

In order to better assess the behavior of PCFS approach, Figure 9 shows the trend of temperatures during a typical winter week. In this case, the temperature inside the greenhouse is kept at around 20 °C, considered as the optimal target for plant growth.

In particular, if the temperature falls below this value, the system starts working by self-setting with one of three speeds and one of three hot air input temperatures, as was the case with the scenario in the summer case.

Finally the results show that the neural network model performs off-line learning in order to set up PCFS. Without the proposed approach, a traditional ON/OFF system can be used at the same air speed velocity and with a fixed air temperature of fan coils. The PCFS takes as input the values of the trained NNARX model and, through the membership functions, improves the control, managing the switching on or off of the system and adjusting the inlet air speed and its temperature.

6. CONCLUSIONS

In this paper, an application of a combined neuro-fuzzy model for dynamic and automatic control of greenhouse climatic parameters has been proposed.

The approach shows a couple of FLCs fed by a climatic temperature predictor. The temperature forecast is carried out by an NNARX model, which attained good forecasting performances. In fact, results show a low error related to data approximation. Moreover, the results of proposed approach allow us to strike a good compromise between the fan coil speed and its temperature in order to increase the crop productivity. This solution could be used for different crops as it is feasible to adapt it in different production greenhouses and allows for shaping the control mechanism of PCFS in several scenarios. The results show that the efficient and dynamical regulation of the on/off times of the air speed of fan-coil and its temperature allows us to achieve a more efficient use of energy. Some experiments in a real test-bed over several days could be realized in a follow up of this work.

Moreover, a further application of an integration in a smart home system of an improved controller, based on other characteristics which affect the crops of a garden, such as humidity, cloudiness, and different types of solar radiation, is a goal of our future research works.

AUTHOR CONTRIBUTIONS

All of the authors have contributed toward developing the ideas. All of the authors have been involved in performance evaluation and in the definition of reference scenarios. All of the authors have been involved in preparing the manuscript. All of authors are responsible for the integrity of the work as a whole.

CONFLICTS OF INTEREST

The authors declare no conflict of interest.

REFERENCES

1. Van Straten, G.; Van Willigenburg, G.; Van Henten, E.; Van Ooteghem, E. Optimal Control of Greenhouse Cultivation; CRC Press: London, UK, 2010; p. 326. ISBN 978-1-42-005961-8.
2. Lafont, F.; Balmat, J.F. Optimized fuzzy control of a greenhouse. Fuzzy Sets Syst. 2002, 128, 47–59.
3. Li, X.; Strezov, V. energy and greenhouse gas emission assessment of conventional and solar assisted air conditioning systems. Sustainability 2015, 7, 14710–14728.
4. Hahn, F. Fuzzy controller decreases tomato cracking in greenhouses. Comput. Electron. Agric. 2011, 77, 21–27.
5. Bennis, N.; Duplaix, J.; Enéa, G.; Haloua, M.; Youlal, H. Greenhouse climate modelling and robust control. Comput. Electron. Agric. 2008, 61, 96–107.
6. Heidari, M.D.; Omid, M. Energy use patterns and econometric models of major greenhouse vegetable productions in Iran. Energy 2011, 36, 220–225.
7. Gupta, M.J.; Chandra, P. Effect of greenhouse design parameters on conservation of energy for greenhouse environmental control. Energy 2002, 27, 777–794.
8. Cossu, M.; Murgia, L.; Ledda, L.; Deligios, P.A.; Sirigu, A.; Chessa, F. Solar radiation distribution inside a greenhouse with south-oriented photovoltaic roofs and effects on crop productivity. Appl. Energy 2014, 133, 89–100.
9. Mohammadi, A.; Omid, M. Economical analysis and relation between energy inputs and yield of greenhouse cucumber production in Iran. Appl. Energy 2010, 87, 191–196.
10. Leyva, R.; Constán-Aguilar, C.; Sánchez-Rodríguez, E.; Romero-Gámez, M.; Soriano, T. Cooling systems in screenhouses: Effect on microclimate, productivity and plant response in a tomato crop. Biosyst. Eng. 2015, 129, 100–111.
11. Banik, P.; Ganguly, A. Thermal modeling and economical analysis of a solar desiccant assisted distributed fan-pad ventilated greenhouse. Lect. Notes Eng. Comput. Sci. 2014, 2, 1274–1279.
12. Castilla, N.; Hernandez, J. Greenhouse technological packages for high-quality crop production. Int. Soc. Hortic. Sci.2006, 761, 285–297.
13. West, S.R.; Ward, J.K.; Wall, J. Trial results from a model predictive control and optimisation system for commercial building HVAC. Energy Build. 2014, 72, 271–279.

14. Chen, J.; Yang, J.; Zhao, J.; Xu, F.; Shen, Z.; Zhang, L. Energy demand forecasting of the greenhouses using nonlinear models based on model optimized prediction method. Neurocomputing 2016, 174, 1087–1100.
15. Paniagua-Tineo, A.; Salcedo-Sanz, S.; Casanova-Mateo, C.; Ortiz-García, E.G.; Cony, M.A.; Hernández-Martín, E. Prediction of daily maximum temperature using a support vector regression algorithm. Renew. Energy 2011, 36, 3054–3060.
16. Van Beveren, P.J.M.; Bontsema, J.; Van Straten, G.; Van Henten, E.J. Minimal heating and cooling in a modern rose greenhouse. Appl. Energy 2015, 137, 97–109.
17. He, F.; Ma, C. Modeling greenhouse air humidity by means of artificial neural network and principal component analysis. Comput. Electron. Agric. 2010, 71, S19–S23.
18. Yu, H.; Chen, Y.; Hassan, S.G.; Li, D. Prediction of the temperature in a Chinese solar greenhouse based on LSSVM optimized by improved PSO. Elsevier Comput. Electron. Agric. 2016, 122, 94–102.
19. Hasni, A.; Taibi, R.; Draoui, B.; Boulard, T. Optimization of greenhouse climate model parameters using particle swarm optimization and genetic algorithms. Energy Proc. 2011, 6, 371–380.
20. Chang, W. A modified particle swarm optimization with multiple subpopulations for multimodal function optimization problems. Appl. Soft Comput. 2015, 33, 170–182.
21. Lpez, J.; Baille, A.; Bonachela, S.; Real, M.G.; Prez-Parra, J. Predicting the energy consumption of heated plastic greenhouses in south-eastern Spain. Span. J. Agric. Res. 2006, 4, 289–296.
22. Fourati, F. Multiple neural control of a greenhouse. Neurocomputing 2014, 139, 138–144. Available online: http://dx.doi.org/10.1016/j.neucom.2014.02.052 (accessed on 15 March 2017).
23. Maher, A.; Kamel, E.; Enrico, F.; Atif, I.; Abdelkader, M. An intelligent system for the climate control and energy savings in agricultural greenhouses. Energy Effic. 2016, 9, 1241–1255.
24. Kolokotsa, D.; Saridakis, G.; Dalamagkidis, K.; Dolianitis, S.; Kaliakatsos, I. Development of an intelligent indoor environment and energy management system for greenhouses. Energy Convers. Manag. 2010, 51, 155–168.
25. Castañeda-Miranda, R.; Ventura-Ramos, E., Jr.; del Rocío Peniche-Vera, R.; Herrera-Ruiz, G. Fuzzy Greenhouse climate control system based on a field programmable gate array. Biosyst. Eng. 2006, 94, 165–177.
26. Nachidi, M. Stabilization of T-S Fuzzy Systems with Application on a Greenhouse: Static Output-Feedback Control Design Problem; Lap Lambert Academic Publishing GmbH KG: Saarbrucken, Germany, 2012; Available online: https://books.google.lu/books?id=-VWHMgEACAAJ (accessed on 15 March 2017).
27. Zhang, D.; Wu, X.; Zhang, C. The application of fuzzy control in greenhouse environment control. Appl. Mech. Mater.2014, 543–547, 1432–1435.
28. Bayindir, R.; Cetinceviz, Y. A water pumping control system with a programmable logic controller (PLC) and industrial wireless modules for industrial plants—An experimental setup. ISA Trans. 2011, 50, 321–328.
29. Fitz-Rodríguez, E.; Kubota, C.; Giacomelli, G.A.; Tignor, M.E.; Wilson, S.B.; McMahon, M. Dynamic modeling and simulation of greenhouse environments under several scenarios: A web-based application. Comput. Electron. Agric.2010, 70, 105–116.
30. Singh, G.; Singh, P.P.; Lubana, P.P.S.; Singh, K.G. Formulation and validation of a mathematical model of the microclimate of a greenhouse. Renew. Energy 2006, 31, 1541–1560.
31. Boulard, T.; Baille, A. A simple greenhouse climate control model incorporating effects of ventilation and evaporative cooling. Agric. For. Meteorol. 1993, 65, 145–157.
32. Song, Y.; Huang, X.; Feng, Y. A Kind of Temperature and Humidity Adaptive Predictive Decoupling Method in Wireless Greenhouse Environmental Test Simulation System. Adv. J. Food Sci. Technol. 2013, 5, 1395–1403.

33. Dariouchy, A.; Aassif, E.; Lekouch, K.; Bouirden, L.; Maze, G. Prediction of the intern parameters tomato greenhouse in a semi-arid area using a time-series model of artificial neural networks. Measurement 2009, 2, 456–463.
34. Haykin, S. Neural Networks and Learning Machines, 3rd ed.; Prentice Hall: Upper Saddle River, NJ, USA, 2009.
35. Franco, A.; Valera, D.L.; Pea, A. Energy efficiency in greenhouse evaporative cooling techniques: Cooling boxes versus cellulose pads. Energies 2014, 7, 1427–1447.
36. Collotta, M.; Pau, G. An Innovative Approach for Forecasting of Energy Requirements to Improve a Smart Home Management System Based on BLE. IEEE Trans. Green Commun. Netw. 2017, 1, 112–120.
37. Marvuglia, A.; Messineo, A.; Nicolosi, G. Coupling a neural network temperature predictor and a fuzzy logic controller to improve thermal comfort regulation in an office building. Build. Environ. 2014, 72, 287–299.
38. Uchida Frausto, H.; Pieters, J.G. Modelling greenhouse temperature using system identification by means of neural networks. Neurocomputing 2004, 5, 423–428.
39. Pasgianos, G.D.; Arvanitis, K.G.; Polycarpou, P.; Sigrimis, N. A nonlinear feedback technique for greenhouse environmental control. Comput. Electron. Agric. 2003, 40, 153–177.
40. Collotta, M.; Pau, G.; Maniscalco, V. A fuzzy logic approach by using particle swarm optimization for effective energy management in IWSNs. IEEE Trans. Ind. Electron. 2017, in press.
41. Mohamed, S.; Hameed, I.A. A GA-based adaptive neuro-fuzzy controller for greenhouse climate control system. Alex. Eng. J. 2015.
42. Salazar, R.; López, I.; Rojano, A. A Neural Network Model to Predict Temperature and Relative Humidity in a Greenhouse. Proceedings of International Symposium on High Technology for Greenhouse System Management (Greensys 2007), Naples, Italy, 4–6 October 2007.
43. Hyndman, R.J.; Koehler, A.B. Another look at measures of forecast accuracy. Int. J. Forecast. 2006, 22, 679–688.
44. Atia, D.M.; El-madany, H.T. Analysis and design of greenhouse temperature control using adaptive neuro-fuzzy inference system. J. Electr. Syst. Inf. Technol. 2016, in press.
45. López-Cruz, I.L.; Hernández-Larragoiti, L. Neuro-fuzzy models for air temperature and humidity of a greenhouse. ISHS Acta Hortic. 2010, 927, 611–617.
46. Lauchouri, C.; Lafifi, M.; Mausouri, K.; Belmeguenai, A. Adaptive neuro-fuzzy inference systems for modeling greenhouse climate. Int. J. Adv. Comput. Sci. Appl. 2016, 7, 96–100.
47. Stiles, M.R. A design model of transient temperature performance for a green greenhouse. Distrib. Gener. Altern. Energy J. 2012, 27, 56–76.
48. Iliev, O.L.; Sazdov, P.; Zakeri, A. A fuzzy logic based approach for integrated control of protected cultivation. Facta Univ. 2012, 11, 119–128.
49. Olunloyo, V.O.S.; Ajofoyinbo, A.M.; Ibidapo-Obe, O. On development of fuzzy controller: The case of gaussian and triangular membership functions. Sci. Res. 2011, 2, 257–265.
50. Marvuglia, A.; Messineo, A. Using recurrent artificial neural networks to forecast household electricity consumption. Energy Procedia 2012, 14, 45–55.
51. Forssell, M. Hardware Implementation of Artificial Neural Networks. Available online: https://users.ece.cmu.edu/~pgrover/teaching/files/NeuromorphicComputing.pdf (accessed on 15 March 2017).
52. Hasan, R.; Taha, T.M.; Yakopcic, C.; Mountain, D.J. High throughput neural network based embedded streaming multicore processors. In Proceedings of the 2016 IEEE International Conference on Rebooting Computing (ICRC), San Diego, CA, USA, 17–19 October 2016.

Index